Advances in Polymer Science

Recently Published and Forthcoming Volumes

Polysaccharides II
Volume Editor: Klemm, D.
Vol. 205, 2006

Neodymium Based Ziegler Catalysts – Fundamental Chemistry
Volume Editor: Nuyken, O.
Vol. 204, 2006

Polymers for Regenerative Medicine
Volume Editor: Werner, C.
Vol. 203, 2006

Peptide Hybrid Polymers
Volume Editors: Klok, H.-A., Schlaad, H.
Vol. 202, 2006

**Supramolecular Polymers ·
Polymeric Betains · Oligomers**
Vol. 201, 2006

Ordered Polymeric Nanostructures at Surfaces
Volume Editor: Vancso, G. J., Reiter, G.
Vol. 200, 2006

Emissive Materials · Nanomaterials
Vol. 199, 2006

Surface-Initiated Polymerization II
Volume Editor: Jordan, R.
Vol. 198, 2006

Surface-Initiated Polymerization I
Volume Editor: Jordan, R.
Vol. 197, 2006

Conformation-Dependent Design of Sequences in Copolymers II
Volume Editor: Khokhlov, A. R.
Vol. 196, 2006

Conformation-Dependent Design of Sequences in Copolymers I
Volume Editor: Khokhlov, A. R.
Vol. 195, 2006

Enzyme-Catalyzed Synthesis of Polymers
Volume Editors: Kobayashi, S., Ritter, H., Kaplan, D.
Vol. 194, 2006

Polymer Therapeutics II
Polymers as Drugs, Conjugates and Gene Delivery Systems
Volume Editors: Satchi-Fainaro, R., Duncan, R.
Vol. 193, 2006

Polymer Therapeutics I
Polymers as Drugs, Conjugates and Gene Delivery Systems
Volume Editors: Satchi-Fainaro, R., Duncan, R.
Vol. 192, 2006

Interphases and Mesophases in Polymer Crystallization III
Volume Editor: Allegra, G.
Vol. 191, 2005

Block Copolymers II
Volume Editor: Abetz, V.
Vol. 190, 2005

Block Copolymers I
Volume Editor: Abetz, V.
Vol. 189, 2005

Intrinsic Molecular Mobility and Toughness of Polymers II
Volume Editor: Kausch, H.-H.
Vol. 188, 2005

Intrinsic Molecular Mobility and Toughness of Polymers I
Volume Editor: Kausch, H.-H.
Vol. 187, 2005

Neodymium Based Ziegler Catalysts – Fundamental Chemistry

Volume Editor: Oskar Nuyken

With contributions by
R. Anwander · A. Fischbach · L. Friebe · O. Nuyken · W. Obrecht

The series *Advances in Polymer Science* presents critical reviews of the present and future trends in polymer and biopolymer science including chemistry, physical chemistry, physics and material science. It is adressed to all scientists at universities and in industry who wish to keep abreast of advances in the topics covered.

As a rule, contributions are specially commissioned. The editors and publishers will, however, always be pleased to receive suggestions and supplementary information. Papers are accepted for *Advances in Polymer Science* in English.

In references *Advances in Polymer Science* is abbreviated *Adv Polym Sci* and is cited as a journal.

Springer WWW home page: springer.com
Visit the APS content at springerlink.com

Library of Congress Control Number: 2006929798

ISSN 0065-3195
ISBN-10 3-540-34809-3 Springer Berlin Heidelberg New York
ISBN-13 978-3-540-34809-2 Springer Berlin Heidelberg New York
DOI 10.1007/11761013

This work is subject to copyright. All rights are reserved, whether the whole or part of the material is concerned, specifically the rights of translation, reprinting, reuse of illustrations, recitation, broadcasting, reproduction on microfilm or in any other way, and storage in data banks. Duplication of this publication or parts thereof is permitted only under the provisions of the German Copyright Law of September 9, 1965, in its current version, and permission for use must always be obtained from Springer. Violations are liable for prosecution under the German Copyright Law.

Springer is a part of Springer Science+Business Media

springer.com

© Springer-Verlag Berlin Heidelberg 2006

The use of registered names, trademarks, etc. in this publication does not imply, even in the absence of a specific statement, that such names are exempt from the relevant protective laws and regulations and therefore free for general use.

Cover design: WMXDesign GmbH, Heidelberg
Typesetting and Production: LE-TEX Jelonek, Schmidt & Vöckler GbR, Leipzig

Printed on acid-free paper   02/3100 YL – 5 4 3 2 1 0

Volume Editor

Prof. Dr. Oskar Nuyken
Lehrstuhl für Makromolekulare Stoffe
TU München
Lichtenbergstraße 4
85747 Garching, Germany
oskar.nuyken@ch.tum.de

Editorial Board

Prof. Akihiro Abe
Department of Industrial Chemistry
Tokyo Institute of Polytechnics
1583 Iiyama, Atsugi-shi 243-02, Japan
aabe@chem.t-kougei.ac.jp

Prof. A.-C. Albertsson
Department of Polymer Technology
The Royal Institute of Technology
10044 Stockholm, Sweden
aila@polymer.kth.se

Prof. Ruth Duncan
Welsh School of Pharmacy
Cardiff University
Redwood Building
King Edward VII Avenue
Cardiff CF 10 3XF, UK
DuncanR@cf.ac.uk

Prof. Karel Dušek
Institute of Macromolecular Chemistry, Czech
Academy of Sciences of the Czech Republic
Heyrovský Sq. 2
16206 Prague 6, Czech Republic
dusek@imc.cas.cz

Prof. W. H. de Jeu
FOM-Institute AMOLF
Kruislaan 407
1098 SJ Amsterdam, The Netherlands
dejeu@amolf.nl
and Dutch Polymer Institute
Eindhoven University of Technology
PO Box 513
5600 MB Eindhoven, The Netherlands

Prof. Jean-François Joanny
Physicochimie Curie
Institut Curie section recherche
26 rue d'Ulm
75248 Paris cedex 05, France
jean-francois.joanny@curie.fr

Prof. Hans-Henning Kausch
Ecole Polytechnique Fédérale de Lausanne
Science de Base
Station 6
1015 Lausanne, Switzerland
kausch.cully@bluewin.ch

Prof. Shiro Kobayashi
R & D Center for Bio-based Materials
Kyoto Institute of Technology
Matsugasaki, Sakyo-ku
Kyoto 606-8585, Japan
kobayash@kit.ac.jp

Prof. Kwang-Sup Lee
Department of Polymer Science &
Engineering
Hannam University
133 Ojung-Dong
Daejeon 306-791, Korea
kslee@hannam.ac.kr

Prof. L. Leibler
Matière Molle et Chimie
Ecole Supérieure de Physique
et Chimie Industrielles (ESPCI)
10 rue Vauquelin
75231 Paris Cedex 05, France
ludwik.leibler@espci.fr

Prof. Timothy E. Long
Department of Chemistry
and Research Institute
Virginia Tech
2110 Hahn Hall (0344)
Blacksburg, VA 24061, USA
telong@vt.edu

Prof. Ian Manners
School of Chemistry
University of Bristol
Cantock's Close
BS8 1TS Bristol, UK
ian.manners@bristol.ac.uk

Prof. Martin Möller
Deutsches Wollforschungsinstitut
an der RWTH Aachen e.V.
Pauwelsstraße 8
52056 Aachen, Germany
moeller@dwi.rwth-aachen.de

Prof. Oskar Nuyken
Lehrstuhl für Makromolekulare Stoffe
TU München
Lichtenbergstr. 4
85747 Garching, Germany
oskar.nuyken@ch.tum.de

Prof. E. M. Terentjev
Cavendish Laboratory
Madingley Road
Cambridge CB 3 OHE, UK
emt1000@cam.ac.uk

Prof. Brigitte Voit
Institut für Polymerforschung Dresden
Hohe Straße 6
01069 Dresden, Germany
voit@ipfdd.de

Prof. Gerhard Wegner
Max-Planck-Institut
für Polymerforschung
Ackermannweg 10
Postfach 3148
55128 Mainz, Germany
wegner@mpip-mainz.mpg.de

Prof. Ulrich Wiesner
Materials Science & Engineering
Cornell University
329 Bard Hall
Ithaca, NY 14853, USA
ubw1@cornell.edu

Advances in Polymer Science
Also Available Electronically

For all customers who have a standing order to Advances in Polymer Science, we offer the electronic version via SpringerLink free of charge. Please contact your librarian who can receive a password or free access to the full articles by registering at:

springerlink.com

If you do not have a subscription, you can still view the tables of contents of the volumes and the abstract of each article by going to the SpringerLink Homepage, clicking on "Browse by Online Libraries", then "Chemical Sciences", and finally choose Advances in Polymer Science.

You will find information about the

- Editorial Board
- Aims and Scope
- Instructions for Authors
- Sample Contribution

at springer.com using the search function.

Preface

Even though Ziegler catalysts have been known for almost half a century, rare earth metals (Ln), particularly neodymium (Nd)-based Ziegler catalyst systems, only came into the focus of industrial and academic research well after the large scale application of titanium, cobalt and nickel catalyst systems. As a direct consequence of the late recognition of the technological potential of rare earth metal Ziegler catalysts, these systems have attracted much attention.

Considerable progress has been made in this field as a result of intensive work performed during the last few years. Worth mentioning is the structural identification of a variety of Ln/Al heterobimetallic complexes and the role of alkyl aluminum cocatalysts in molar mass control. Furthermore, a deeper understanding of the polymerization mechanism, such as the living character of neodymium-catalyzed diene polymerization associated with the reversible transfer of living polymer chains between Nd and Al, was revealed quite recently. In spite of the vast number of patents and publications mainly issued during the last decade, a comprehensive review that covers the scientific as well as the patent literature has been missing until now.

In this volume we try to review the available literature by two independent approaches to Nd and Ln-catalyzed diene polymerizations. In the first part of the volume, which is entitled "Neodymium-Based Ziegler/Natta Catalysts and their Application in Diene Polymerization", a polymer chemist's view is given with strong emphasis on Nd-based catalyst systems. Also technological and industrial aspects of Nd-catalyzed diene polymerizations are addressed. In the second part of the volume, which is entitled "Rare-Earth Metals and Aluminum Getting Close in Ziegler-type Organometallics", a more organometallic perspective is given and Ln-based catalyst systems are addressed. By the synopsis of these different perspectives, the reader will comprehend the complexity of Ln-based Ziegler catalyst systems and their application to the polymerization of dienes.

This volume also gives a description of the evolution in Nd-catalyzed polymerization of dienes from the early works to the current state of the art. The authors highlight the tremendous variety of investigated catalyst systems and both articles order the catalyst systems according to the type of anions: carboxylates, alcoholates, halides, hydrides, phosphates, phosphonates, allyls, tetraalkylaluminates, cyclopentadienyl complexes, amides, acetylacetonates, and siloxides. In the whole volume special attention is paid to the role of aluminum alkyl cocatalysts. While in the first part the focus is on the dependence of diene polymerization on cocatalyst types and on Nd/Al-ratios, in the second part of the review the catalyst intermediates that could be isolated from the reaction of Ln precursors with organoaluminum compounds are structurally characterized. Furthermore, in the first part of the volume the influence of temperature, solvents, amount of cocatalyst, etc. on polymerization characteristics are reviewed. Also polymerization processes such as polymerization in bulk, slurry and gas phase as well as the diene homopolymerization in the presence of the monomer styrene are addressed. Supported catalyst systems are summarized in both parts of this volume.

This review does not cover the application of Ln-polymerization catalysis to polar monomers. A comprehensive review on this topic is urgently required. Nevertheless, we hope that this volume will become the future key reference in Ln and especially in Nd-based catalyst systems as well as in Nd-catalyzed polymerization of dienes. As a starting point for future work unsolved and open questions are summarized in a separate chapter of the first part of this volume. We really hope that this list of open questions will inspire and stimulate further research in this interesting field of catalysis.

Garching, June 2006   *Oskar Nuyken*

Contents

Neodymium Based Ziegler/Natta Catalysts and their Application in Diene Polymerization
L. Friebe · O. Nuyken · W. Obrecht 1

Rare-Earth Metals and Aluminum Getting Close in Ziegler-type Organometallics
A. Fischbach · R. Anwander 155

Author Index Volumes 201–204 283

Subject Index 285

Neodymium-Based Ziegler/Natta Catalysts and their Application in Diene Polymerization

Lars Friebe[1,2] (✉) · Oskar Nuyken[2] · Werner Obrecht[3]

[1]Department of Chemistry, University of Toronto, 80 St. George Street, Toronto, Ontario M5S 3H6, Canada
lars.friebe@gmail.com

[2]Lehrstuhl für Makromolekulare Stoffe, TU München, Lichtenbergstraße 4, 85747 Garching, Germany
lars.friebe@gmail.com

[3]Lanxess Deutschland GmbH, Business Unit TRP, LXS-TRP-APD-PD, Building F 41, 41538 Dormagen, Germany

1	Introduction	5
1.1	Ziegler/Natta Catalysts in Diene Polymerization	5
1.2	Butadiene Rubber (BR) and Neodymium Butadiene Rubber (Nd-BR)	7
2	**Polymerization in Solution**	12
2.1	Catalyst Systems and their Components	12
2.1.1	Neodymium Components and Respective Catalyst Systems	13
2.1.2	Cocatalysts/Activators	32
2.1.3	Halide Donors	35
2.1.4	Molar Ratio $n_{Cocatalyst}/n_{Nd}$	39
2.1.5	Molar Ratio n_{Halide}/n_{Nd}	42
2.1.6	Addition Order of Catalyst Components, Catalyst Preformation and Catalyst Aging	47
2.1.7	Supported Catalysts	54
2.1.8	Other Additives	55
2.2	Technological Aspects of the Polymerization in Solution	58
2.2.1	Solvents	59
2.2.2	Monomer Concentration	63
2.2.3	Moisture and Impurities	64
2.2.4	Monomer Conversion, Shortstop and Stabilization of Polymers	64
2.2.5	Formation of Dimers	65
2.2.6	Post-Polymerization Modifications	66
2.2.7	Polymerization Temperature	68
2.2.8	Control of Molar Mass	74
2.2.9	Miscellaneous	81
2.3	Homo- and Copolymerization in Solution	81
2.3.1	Homopolymerization of Isoprene	82
2.3.2	Copolymerization of Butadiene and Isoprene	84
2.3.3	Homopolymerization and Copolymerization of Substituted Butadienes (other than Isoprene)	85
2.3.4	Copolymerization of Butadiene and Styrene	88
2.3.5	Copolymerization of Butadiene with Ethylene or 1-Alkenes	91

3	Other Polymerization Technologies	93
3.1	Polymerization in Bulk/Mass and in Suspension	93
3.2	Polymerization in the Gas Phase	94
3.3	Homopolymerization of Dienes in the Presence of Other Monomers	98
4	Kinetic and Mechanistic Aspects of Neodymium-Catalyzed Butadiene Polymerization	99
4.1	Kinetic Aspects	99
4.2	Active Species and its Formation	101
4.3	Polyinsertion Reaction and Control of Microstructure	111
4.4	Living Polymerization	115
4.5	Molar Mass Regulation	124
5	Open Questions	127
6	Evaluation of Nd-BR-Technology	131
7	Remarks on Present Developments in Nd-Technology and Speculations about Future Trends	134
	References	137

Abstract This article reviews the polymerization of dienes by neodymium (Nd) based Ziegler/Natta-catalyst systems. Special attention is paid to the monomer 1,3-butadiene (BD). The review covers scientific as well as patent literature which was published during the last decade to 2005. For a better understanding of the recent developments the early work on lanthanide-catalyzed diene polymerization is also addressed. The most important product obtained by Nd catalysis, butadiene rubber (Nd-BR) is introduced from an industrial as well as from a material scientist's point of view. Strong attention is paid to the great variety of Ziegler/Natta type Nd-catalyst systems which are often referred to as binary, ternary and quaternary systems. Different Nd-precursors, cocatalysts, halide donors and other additives are reviewed in detail. Technological aspects such as solvents, catalyst addition order, catalyst preformation, polymerization temperature, molar mass control, post-polymerization modifications etc., are presented. A considerable part of this review discusses variations of the molar ratios of the catalyst components and their influence on the polymerization characteristics. Non-established polymerization technologies such as polymerization in bulk, slurry and gas phase as well as the homopolymerization in the presence of other monomers are addressed. Also the copolymerizations of butadiene with isoprene, styrene and alkenes are reviewed. Mechanistic aspects such as formation of the active catalyst species, the living character of the polymerization, mode of monomer insertion, and molar mass control reactions are also explained. In the summary Nd technology is evaluated in comparison with other established technologies for the production of high *cis*-1,4-BR. Unsolved and open questions about Nd-catalyzed diene polymerization are also presented.

Keywords Diene polymerization · Mechanism · Neodymium catalysis · Rubber · Ziegler/Natta catalysts

Abbreviations
6PPD $\quad$ N-1,3-dimethylbutyl-N'-phenyl-p-phenylendiamine
7PPD $\quad$ N-1,4-dimethylpentyl-N'-phenyl-p-phenylendiamine

77PD	N,N'-bis-1,4-(1,4-dimethylpentyl)-p-phenylendiamine
ABS	acrylonitrile butadiene styrene terpolymer
AFM	atomic force microscopy
at	atactic
BD	1,3-butadiene
BHT	2,6-di-*tert*-butyl-*p*-kresol
BIT	black incorporation time
BPH	2,2-methylene-bis-(4-methyl-6-*tert*-butylphenol)
BR	butadiene rubber
BuLi	butyl lithium
BzCl	benzyl chloride
c_i	concentration of compound i
CL	ε-caprolactone
Co-BR	butadiene rubber obtained by cobalt catalysis
Cp	cyclopentadienyl
Cp*	pentamethylcyclopentadienyl
Cp*'	$C_5Me_4{}^n$propyl ligand
d	day
D	electron donor
DEAC	diethylaluminum chloride
DEAH	diethylaluminum hydride
Di	didymium
DIBAC	diisobutylaluminum chloride
DIBAH	diisobutylaluminum hydride
DMDPS	dimethyl-di-2,4-pentadienyl-(E,E)-silane
DMF	N,N-dimethylformamide
DSC	differential scanning calorimetry
DSV	dilute solution viscosity
E_a	activation energy
EADC	ethylaluminum dichloride
EASC	ethylaluminum sesquichloride
E-BR	butadiene rubber produced by emulsion polymerization
EPDM	ethylene propylene diene copolymer-based rubber
EPM	ethylene propylene copolymer-based rubber
eq.	equivalents
GPC	gel permeation chromatography
h	hour
Hex	n-hexane or hexyl
HIBAO	hexaisobutyl alumoxane
HIPS	high-impact polystyrene
HMPTA	hexamethylphosphoric acid triamide
HV	versatic acid
IISRP	International Institute of Synthetic Rubber Producers
Ind	indenyl
IP	isoprene (2-methyl-1,3-butadiene)
IPPD	N-isopropyl-N'-phenyl-p-phenylendiamine
IR	isoprene rubber
it	isotactic
JSR	Japan Synthetic Rubber
k_a	apparent rate constant

k_p	polymerization rate constant
Li-BR	butadiene rubber obtained by alkyl lithium initiation
Ln	lanthanide
MAO	methylalumoxane
MCH	methyl cyclohexane
MDI	diphenylmethanediisocyanate
min	minute(s)
MMAO	modified methylalumoxane
MMD	molar mass distribution
M_n	number average molar mass
M_v	viscosity average molar mass
MU	Mooney units
M_w	weight average molar mass
NdA	neodymium(III) neopentanolate
Nd-BR	butadiene rubber obtained by neodymium catalysis
Nd^iO	neodymium(III) isooctanoate
NdO	neodymium(III) octanoate
NdN	neodymium(III) naphthenate
NdP	neodymium(III) bis(2-ethylhexyl)phosphate ($Nd(P_{204})_3$)
NdV	neodymium(III) versatate
n_i	molar amount of compound i
n_i/n_j	molar ratio of compound i and j
Ni-BR	butadiene rubber obtained by nickel catalysis
NR	natural rubber
PDI	polydispersity index M_w/M_n
$p_{exp.}$	formal polymer chain number per Nd atom (determined experimentally)
ppm	parts per million
PSD	particle size distribution
r_i	copolymerization parameter for monomer i
r_p	polymerization rate
SBR	styrene butadiene rubber
SSC	single site catalyst
st	syndiotactic
St	styrene
T	temperature
TBB	*tert*-butyl benzene
TBP	tributyl phosphate
TEA	triethylaluminum
THF	tetrahydrofuran
TIBA	triisobutyl aluminum
TIBAO	tetraisobutylalumoxane
Ti-BR	butadiene rubber obtained by titanium catalysis
T_g	glass transition temperature
TMA	trimethylaluminum
TMEDA	tetramethyl ethylene diamine
TOF	turnover frequency
UCC	Union Carbide Corporation
VCH	vinyl cyclohexene
wt.%	weight percent
X	halide

1
Introduction

1.1
Ziegler/Natta Catalysts in Diene Polymerization

The discovery of coordinative polymerization [1] is one of the best examples in science which demonstrates that fundamental research can result in a new and highly successful technology that is applied in large scale and has an enormous impact on modern life [2, 3]. The decisive experiment which initiated this development was carried out in Mülheim (Ruhr)/Germany in October 26, 1953. The respective patent application which claims a "Process for the Synthesis of High Molecular Poly(ethylene)s" was filed on November, 18 of the same year [4]. This patent caused a revolution in the chemical industry as "Ziegler catalysts" quite unexpectedly allowed for the polymerization of alkenes in mild conditions compared to former techniques. The subsequent discovery of diastereomeric poly(propylene)s in March 1954 by Natta [5–7] allowed access to stereoregular polymers which until then were considered a monopoly of nature. In 1963, ten years after the first of these two discoveries Karl Ziegler and Giulio Natta were awarded the Nobel prize for their basic invention and the benefits of the "Ziegler/Natta polymerization" [8].

The industrial potential of their inventions was fully recognized soon after Ziegler's and Natta's achievements. Besides the polymerization of alkenes, Ziegler/Natta type catalysts were also applied to the polymerization of conjugated dienes. Goodrich-Gulf Chemicals found that the coordinative polymerization of isoprene (IP) results in either *cis*-1,4-poly(isoprene) (IR = isoprene rubber) [9–12] or *trans*-1,4-poly(isoprene) [13, 14]. The synthesis of *cis*-1,4-poly(butadiene) (BR = butadiene rubber) was also claimed in a series of patents [15–20] as well as the preparation of *trans*-1,4-poly(butadiene) [21–24] and 1,2-poly(butadiene) [25–30]. After these first patents on the use of Ziegler/Natta-catalysts for the polymerization of conjugated dienes had been filed, the large-scale industrial application of Co- and Ti-based catalysts for the production of high *cis*-1,4-BR began in the early 1960s.

From the early 1960s onwards, the use of lanthanide (Ln) based catalysts for the polymerization of conjugated dienes came to be the focus of fundamental studies [31]. The first patent on the use of lanthanides for diene polymerization originates from 1964 and was submitted by Union Carbide Corporation (UCC) [32, 33]. In this patent the use of binary lanthanum and cerium catalysts is claimed. Soon after this discovery by UCC, Throckmorton (Goodyear) revealed the superiority of ternary lanthanide catalyst systems over binary catalyst systems. The ternary systems introduced by Throckmorton comprise a lanthanide compound, an aluminum alkyl cocatalyst and a halide donor [34]. Out of the whole series of lanthanides Throckmorton accidentally selected Ce-catalysts, the residues of which have a negative influence on the aging performance of the respective BR vulcanizates [35]. Contrary to neodymium residues cerium residues catalyze oxidation of raw BR and BR-based vulcanizates [36]. As a consequence of the poor aging performance of Ce-based BR vulcanizates Goodyear abandoned further developments in this area for many years.

In the late 1970s and early 1980s work on Ln catalysis was resumed, first by Anic (later: Enoxy, Enimont, Enichem, Polimeri) and soon after by Bayer (now: Lanxess). Both companies focused on Nd- rather than on Ce-catalysis, due to the superior aging resistance of the obtained vulcanizates. In addition, Nd-precursors are readily available for modest prices and Nd-catalysts exhibit the highest activity within the lanthanide series (Fig. 2 in Sect. 1.2). Nd-catalysts yield poly(diene)s with higher *cis*-1,4-contents than the Ti- and Co-based catalysts which were commercially established at that time. In their catalyst development work, Anic/Eni focused on Nd-alcoholates [37, 38] while Bayer concentrated on Nd-carboxylates [39, 40]. The large-scale industrial application of Nd-catalysis for BR production was established in the early to mid 1980s, first by Anic/Eni and shortly after by Bayer.

It has to be mentioned that shortly before attention returned to Ln-catalysts actinides came into the focus of industrial research when the potential of uranium-based catalysts was recognized by Eni and later by

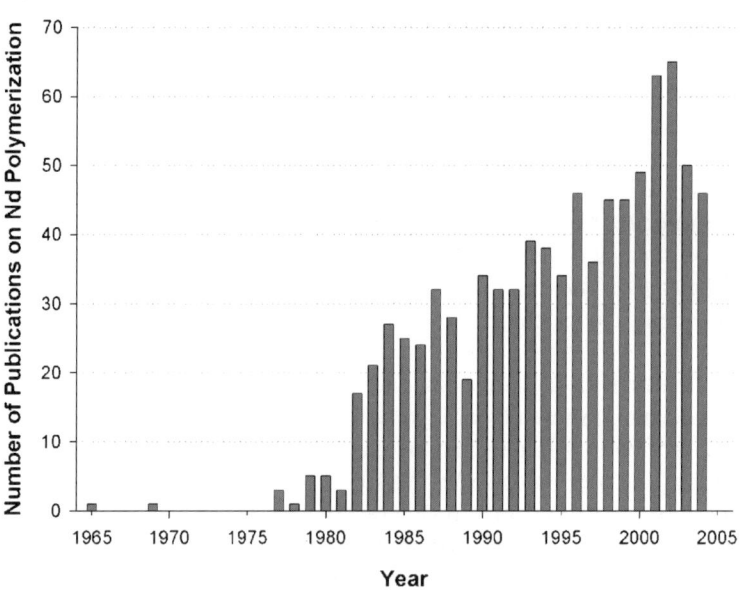

Fig. 1 Number of publications (scientific papers and patents) in the field of neodymium-catalyzed polymerization in the period 1965 to 2004 (*SciFinder® Scholar™* inquiry in December/2005: research topic "neodymium polymerization")

Bayer [41-49]. Uranium-based catalysts yield BR and IR with a significantly higher *cis*-1,4-content than the established Co- and Ti-catalysts. Because of radioactive residues present in the respective polymers, however, the efforts aiming at the large-scale application of uranium catalysts were abandoned soon after by both companies.

A comprehensive review on the whole field of polymerization of conjugated dienes by transition-metal catalysts was compiled by Porri and Giarrusso in 1989 [50].

Industrial exploitation of Nd-catalysis instantly attracted attention to Nd-polymerization-catalysis, as demonstrated by the vast increase of the number of publications starting in the early 1980s (Fig. 1).

A considerable percentage of the publications retrieved by the *SciFinder® Scholar™* inquiry includes patent literature. In this review, patents as well as scientific articles are equally acknowledged.

1.2
Butadiene Rubber (BR) and Neodymium Butadiene Rubber (Nd-BR)

Butadiene Rubber (BR)

Beside styrene butadiene rubber (SBR), BR is the most important synthetic general-purpose rubber. BR accounts for an annual consumption of ca. 2.8 million metric tons. In terms of annual production SBR and BR are only outnumbered by natural rubber (NR) with a production of ca. 6.7 million metric tons a year [51][1].

BR is used in four major areas. By far the largest portion is applied in tires ($\sim$ 70%), especially tire treads and side walls. The second biggest use of BR is thermoplast modification ($\sim$ 25%). Here the two main products are high-impact poly(styrene) (HIPS) and acrylonitrile-butadiene-styrene terpolymer (ABS). BR is used to a much smaller extent in technical rubber goods ($\sim$ 4%), such as conveyor belts, hose roll covers, shoe soles and seals. The smallest application area of BR is golf ball cores ($\sim$ 1%). Although 1% appears to be a small figure, BR consumption for golf ball cores adds up to $\sim$ 30 000 metric tons a year. The importance of BR in golf ball cores is highlighted by an extract of the large number of patents filed in recent years [52-79].

Since the start-up of industrial Ziegler/Natta-BR production in the 1960s, BR has continuously grown, mainly due to the general expansion of tire, HIPS and ABS production. Regarding the different types of tires and various tire parts, there has been some substitution between different tire rubbers (NR, SBR and BR), but BR has kept its overall share [80].

Commercial BR is comprised of a broad range of different BR grades. These grades differ in microstructure (Scheme 1: *cis*-1,4-polymer, *trans*-

---

[1] The International Institute of Synthetic Rubber Producers (IISRP, Houston, Texas, USA) is an international non-profit association of synthetic rubber producers with 41 corporate members domiciled in 20 countries. IISRP represents $\sim$95% of the world supplies of synthetic rubber.

1,4-polymer, 1,2-polymer), molar mass, molar mass distribution, degree of branching and end-group functionalization. Furthermore, special oil-extended[2] grades with exceptionally high molar masses are used in the tire industry. For thermoplast modification, clear grades with extremely low gel content and low solution viscosities are applied.

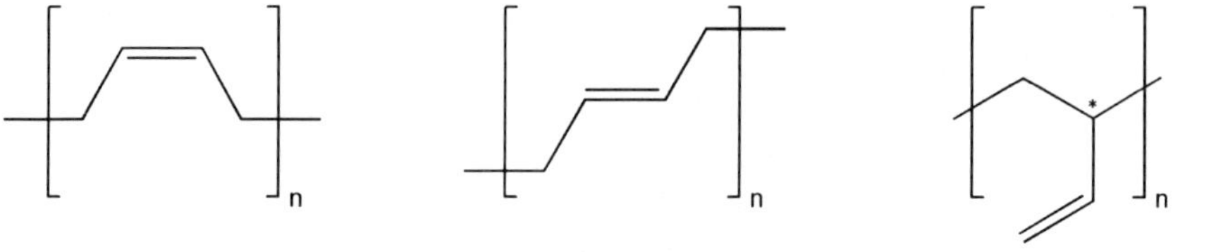

cis-1,4-Poly(butadiene)   *trans*-1,4-Poly(butadiene)   at / it / st - 1,2-Poly(butadiene)

Scheme 1 Poly(butadiene) isomers (at = atactic, it = isotactic, st = syndiotactic)

Today's commercially available BR grades can be classified according to the type of polymerization technology and initiators/catalysts used:

- E-BR: (Emulsion-BR, radical polymerization in aqueous emulsion)
- Li-BR: (Lithium-BR, anionic polymerization in solution)
- Co-BR: (Cobalt-BR, coordinative polymerization in solution)
- Ti-BR: (Titanium-BR, coordinative polymerization in solution)
- Ni-BR: (Nickel-BR, coordinative polymerization in solution)
- Nd-BR: (Neodymium-BR, coordinative polymerization in solution)

As seen in this list, BR is produced by emulsion polymerization (E-BR), by anionic polymerization with lithium alkyls (Li-BR) and by Ziegler/Natta-technology using titanium-, cobalt-, nickel- and neodymium-catalysts (Ti-BR, Co-BR, Ni-BR, Nd-BR). Only by the use of Ziegler/Natta-catalysts BR grades with *cis*-1,4-contents > 93% are obtained. The composition of the different Ziegler/Natta-catalyst systems, the applied catalyst concentrations, the catalyst activities, the BR microstructure, the molar mass distribution (MMD) and the glass transition temperature (T_g) of the respective BR grades are listed in Table 1 [81, 82].

As a consequence of differing *cis*-1,4-content, the various BR-grades exhibit different physical and mechanical properties. Contrary to BR with medium *cis*-1,4-content, (e.g. Li-BR: *cis*-1,4-content = 40%) [83] BR grades with a high *cis*-1,4-content (> 93%) show remarkably low glass transition temperatures (below – 100 °C) and high building tack[3]. The respective vulcanizates which are based on high *cis*-1,4-BR exhibit good low-temperature

---

[2] Rubber grades with high molar mass to which oil is added during the finishing stage of the raw rubber production cycle. By the incorporation of oil the viscosity of the raw rubber is reduced and rubber processing is facilitated.
[3] Stickiness of a rubber to other rubbers. The building tack is important in tire production where layers of different rubber compounds are manually stuck together.

properties, high resilience over a broad temperature range, low heat-build-up on repeated deformation and high abrasion resistance.

Neodymium Butadiene Rubber (Nd-BR)

Nd-BR exhibits the highest *cis*-1,4-content (Table 1) of the four Ziegler/Natta-type BR grades. According to the numerous publications on Nd-BR, the highest *cis*-1,4-contents of this rubber are in the range of 97–99%. However, in this context it has to be mentioned, that the reported *cis*-1,4-content depends somewhat on the analytical method applied [84]. Because of the high *cis*-1,4-content raw (unvulcanized) Nd-BR as well as the respective rubber compounds, which contain fillers, oil, antioxidants, vulcanization aids etc., and the vulcanizates obtained from the rubber compounds by heat treatment, exhibit spontaneous crystallization and strain-induced crystallization. Spontaneous crystallization is temperature dependent. The rate of crystallization of raw (unvulcanized) BR exhibits a maximum at – 50 °C [85]. Strain-induced crystallization results in high building tack of unvulcanized rubber compounds which is important for tire construction [86]. Strain-induced crystallization also gives superior tensile strength, good abrasion resistance and

Table 1 Industrially applied Ziegler/Natta-catalysts in BR production [81, 82], reproduced with permission of Wiley-VCH Verlag GmbH & Co. KGaA

Catalyst system (molar composition)	Concentration of metal M/ (mg·L^{-1})	Yield of BR/ kg (BR)· g^{-1} (M)	*cis*-1,4-content/ %	*trans*-1,4-content/ %	1,2-content/ %	MMD	T_g °C
TiCl$_4$/ I$_2$/ Al^{i}Bu$_3$ (1/1.5/8)	50	4–10	93	3	4	medium	– 103
Co(OCOR)$_2$/ H$_2$O/ AlEt$_2$Cl (1/10/200)	1–2	40–160	96	2	2	medium	– 106
Ni(OCOR)$_2$/ BF$_3$·OEt$_2$/ AlEt$_3$ (1/7.5/8)	5	30–90	97	2	1	broad	– 107
Nd(OCOR)$_3$/ Et$_3$AlCl$_3$/ Al^{i}Bu$_2$H (1/1/8)	10	7–15	98	1	1	very broad	– 109

excellent dynamic performance in vulcanizates [87–91]. Nd-BR is highly linear and unbranched [92]. Molar mass distributions range from PDI = 2.8 [91] to PDI = 3.2–4.0 [89]. This gives Nd-BR a desirable balance of properties, particularly for tire applications. Until recently the major drawback of Nd-BR was high solution viscosities. This limitation was recently overcome by special Nd-BR grades which meet the viscosity requirements for the use in HIPS (grades from Lanxess and Petroflex). The even more demanding viscosity requirements for ABS-applications are not yet met by special Nd-BR grades. Minor drawbacks of Nd-BR tire grades include a high cold flow[4] of raw Nd-BR, a long black incorporation time[5] (BIT) during the preparation of rubber compounds and a poor extrudability of Nd-BR compounds. These drawbacks are counterbalanced by the proper tuning of the MMD, particularly by the presence of a high molar mass fraction. Nd-BR grades from different suppliers vary mainly in this respect. Because the shape of MMD curves is influenced by catalyst composition and catalyst preparation, as well as by post-polymerization reactions such as "molar mass jumping" or "Mooney jumping" (Sect. 2.2.6), in commercial operations great attention is given to these aspects.

The primary use of Nd-BR is in tires. This application of Nd-BR accounts for only $\sim$ 15% of the total amount of BR used in this field. Minor amounts of Nd-BR are used in technical rubber goods and in golf ball cores. To date, only special Nd-BR grades meet the viscosity requirements for rubber modification of HIPS. Mainly due to high solution viscosities Nd-BR is not yet used as the rubber component in ABS.

Neodymium-based catalysts are favored over other Ln metals because they are highly active and the catalyst precursors are readily available for reasonable prices. In addition, Nd catalyst residues do not catalyze aging of the rubber. The use of didymium catalyst systems is also reported in the literature. Didymium consists of a mixture of the three lanthanides: neodymium (72 wt. %), lanthanum (20 wt. %) and praseodymium (8 wt. %).

The high activity of Nd-based catalysts was reported by Shen et al. in 1980 [92]. In this publication, the polymerization activity of the whole lanthanide series was studied. Ln halogen-based binary catalyst systems ($LnCl_3/EtOH/AlEt_3$ or $LnCl_3 \cdot (TBP)_3/Al^iBu_3$), as well as Ln-carboxylate-based ternary catalyst systems ($Ln(naphthenate)_3/Al^iBu_3/EASC$) were used. The activity profile for the entire series of lanthanides is depicted in Fig. 2. Two years later, Monakov et al. confirmed in a similar study that Nd is the most active Ln element [93, 94].

In the lanthanide series samarium (Sm) and europium (Eu) exhibit surprisingly low polymerization activities. Contrary to the other lanthanides, Sm

---

[4] Deformation through gravity of the raw rubber during storage which leads to the deformation of rubber bales.

[5] Time consumed for mixing carbon black into a rubber.

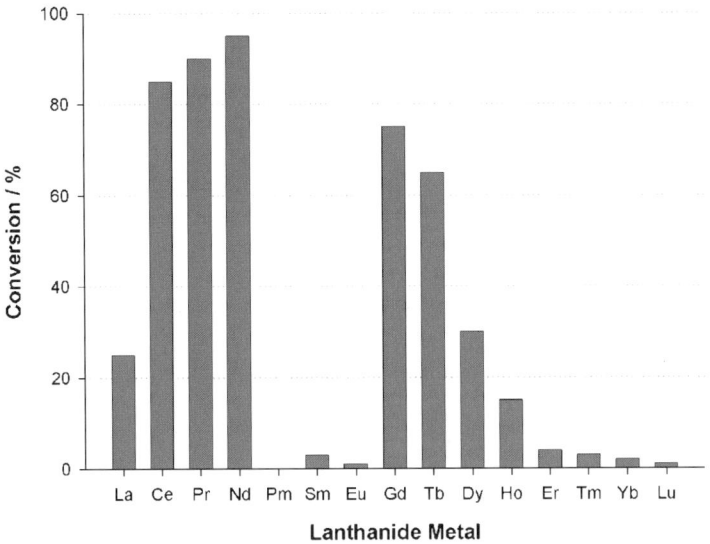

Fig. 2 Activity profile of lanthanide metals in diene polymerization catalysis [92], reprinted with permission of John Wiley & Sons, Inc.

and Eu are reduced from the +III oxidation state to +II. The reduction is accomplished by aluminum alkyls and is only observed for Sm and Eu. It can therefore be concluded that lanthanides must remain in the +III oxidation state in order to maintain high polymerization activity throughout the course of the polymerization. The dependence of the polymerization activities on atomic number and the high level of activities observed for Ce, Pr, Nd, Gd and Tb have not been conclusively explained to the present day. Initial discussions focused on complexation of the metals with diene ligands and on the resulting energy differences [92]. Later, metal-ion radii and charges on the catalytic metal center were considered to be decisive parameters [87].

It is speculated that the use of heterogeneous or partially heterogeneous Nd catalyst systems results in gel formation. Due to this reason, Nd-systems which are soluble in hydrocarbon solvents are preferred today, especially in large-scale operations. The soluble catalysts are usually based on ternary systems which consist of Nd salts with anions bearing long-chain aliphatic groups, an alkyl aluminum cocatalyst and a halide donor.

Today, Nd-BR is industrially produced in Brazil, China, France, Germany, Italy, Japan, Russia, South Africa, South Korea, Taiwan and the USA. The current producers of Nd-BR are listed in alphabetical order: Chi Mei, Japan Synthetic Rubber [95], Jinzhou Petrochemical Co. (part of PetroChina), Karbochem, Korea Kumho, Lanxess (formerly Bayer), Nizhnekamskneftekhim, Petroflex, and Polimeri Europa (formerly Enichem etc.). Amongst these producers Lanxess and Polimeri have been operating at full production since the early to mid 1980s. Chi Mei, Japan Synthetic Rubber, Jinzhou Petrochemical Co., Karbochem, Korea Kumho [96], Nizhnekamskneftekhim [97] and Petroflex started large-scale production quite recently. Lanxess produces Nd-BR in three production sites: Dormagen/Germany, Port Jérôme/France and Orange, Texas/USA. The other producers apply Nd-BR-technology in one plant only.

To date, the original patents on Nd-BR and the respective production processes have expired and a lot of new patent activities can be observed in the field of Nd-catalyzed polymerization of dienes and Nd catalysis in general. As evidenced by their filing of several patents, the following companies have been or are active in this area (given in alphabetic order): Asahi, BASF, Bayer, Bridgestone, Chi Mei, China Petrochemical, Dow, Elf Atochem, Goodyear, Japan Synthetic Rubber (JSR), Kansai, Korea Kumho, Lanxess (formerly Bayer), Michelin, Mitsui, Nippon Zeon, Petroflex, Polimeri Europa (formerly Enichem etc.), Rhodia, Riken, Showa Denko, Spalding Sports Worldwide, Sumitomo, Ube Industries, Union Carbide Chemicals (UCC), and Yokohama Rubber. As indicated by growing patent activity, it can be speculated that even more companies will pursue Nd-BR-technology in the future.

2
Polymerization in Solution

The majority of literature on Nd-mediated diene polymerization is concerned with polymerization in solution. This technology was developed at an early stage of Nd polymerization technology and many basic principles elaborated for solution processes have been adopted in the development of Nd-BR production. Therefore, the "Polymerization in Solution" and various aspects associated with it are reviewed first. Other polymerization technologies such as polymerization in bulk (or mass), suspension (or slurry) and gas phase are addressed in separate Sects. 3.1 and 3.2 at a later stage.

2.1
Catalyst Systems and their Components

Standard Nd-based catalysts comprise binary and ternary systems. Binary systems consist of Nd chloride and an aluminum alkyl or a magnesium alkyl compound. In ternary catalyst systems a halide free Nd-precursor such as a Nd-carboxylate is combined with an Al- or Mg-alkyl plus a halide donor. By the addition of halide donors to halide-free catalyst systems catalyst activities and cis-1,4-contents are significantly increased. In quaternary catalyst systems a solubilizing agent for either the Nd-salt or for the halide donor is used in addition to the components used in ternary systems. There are even more complex catalyst systems which are described in the patent literature. These systems comprise up to eight different catalyst components.

The various components of a catalyst system are either dosed separately to the monomer solution (in simultaneous or consecutive way) or are premixed prior to the addition to the monomer solution (often referred to as "preformed" and/or "aged", see Sect. 2.1.6). Beside the chemical nature of the catalyst components their order of addition (Sect. 2.1.6) plays an important role regarding the heterogeneity of the catalyst system, the polymerization activity and the polydispersity of the resulting polymer.

2.1.1
Neodymium Components and Respective Catalyst Systems

The vast majority of Nd-catalysts are based on Nd in the oxidation state +III. To the best of our knowledge, there is only one recent paper in which a Nd-based polymerization catalyst in the oxidation state +II is mentioned (investigated catalysts: NdI_2 and NdI_2/AlR_3) [98]. In this study it is remarkable that NdI_2 can initiate IP polymerization in hexane without addition of a cocatalyst. In the mid 1980s some investigations on Nd(II)-compounds were carried out. In these experiments the following two Nd(II)-compounds were combined with aluminum alkyl cocatalysts: $NdCl_2/THF$ [99–102] and Ph-NdCl [103, 104]. One study is available on diene polymerization with a Nd(0) compound. The respective Nd(0) species, $(C_6H_6)_3Nd_2$, was obtained from the reaction of Nd metal vapor with benzene [105, 106]. It is not clear from the studies on Nd(II) and Nd(0) catalysts whether Nd remains in the oxidation state +II or 0, respectively, or if disproportionation or oxidation reactions yield Nd(III) species.

In scientific and patent literature on diene polymerization the following Nd(III)-salts are most often cited as catalyst precursors: halides, carboxylates, alcoholates, phosphates, phosphonates, allyl compounds, cyclopentadienyl derivatives, amides, boranes and acetylacetonates. The catalyst systems based on these Nd-sources are reviewed in the following subsections.

2.1.1.1
Neodymium Halides

Nd(III)-halides were the first Nd-compounds applied in diene polymerization [31]. The first systems comprised binary catalyst systems of the type NdX_3/AlR_3 (X = halide, R = alkyl or H). These catalyst systems are heterogeneous and can be very active. In 1985 a neodymium chlorido hydroxide $Nd(OH)_{2.4}Cl_{0.6}$ was reported to exhibit a high activity. However, the heterogeneity of this catalyst leads to the formation of 35% gel during polymerization [107, 108]. As NdX_3-based catalyst systems are often heterogeneous and as the formation of gelled polymer is usually attributed to catalyst heterogeneity, binary catalyst systems do not seem to play a role in the large-scale polymerization of dienes today. Nevertheless, due to their high catalytic activity, Nd-halide systems still attract considerable interest, even after the industrial introduction of the preferred ternary Nd-catalysts, which are based on non-halide Nd precursors. Investigations on Nd-halide-based catalyst systems mainly focus on the increase of the catalysts' performance.

Addition of Electron Donors
The most important progress with NdX_3-based systems is the addition of appropriate ligands or electron donors (D). Work in this area has been continued since 1980, when the first experiments were performed in China [92, 109, 110]. A first review of this work was published by Shen in 1987 [111].

The solubility of NdX_3 catalysts is improved by the addition of electron donors (D). Catalyst activity is remarkably increased without substantial deterioration of the *cis*-1,4-content. Typical donor ligands applied in $NdX_3 \cdot D_n/AlR_3$ type systems are alcohols such as EtOH [92, 112, 113], 2-ethylhexanol [114] or various pentanol isomers [115]. Furthermore, tetrahydrofuran (THF) [35], tributyl phosphate (TBP) [116–119], alkyl sulfoxides [116, 117, 120, 121], propion amide [122, 123], $B(O-CH_2-CH_2-O-CH_2-CH_2-OH)_3/B(O-CH_2-CH_2-O-C_2H_5)_3$ [124, 125], pyridine [126] and N-oxides [127, 128] are applied as donors. The increase in catalyst activity by donor ligands is attributed to the improved solubility of the active species in hydrocarbon solvents [129, 130].

An important aspect is the mode of donor addition. Rao et al. reported on the separate preparation of the NdX_3-donor systems [e.g. $NdCl_3 \cdot D_n$ (D = 2-ethylhexanolate, n = 1.5 and 2.5)] prior to the addition to the monomer solution. This strategy yields a higher *cis*-1,4-content compared to the sequential addition of NdX_3 and donor to the monomer solution [114]. Iovu et al. found increased catalyst activities by the separate preparation of $NdCl_3 \cdot TBP$ prior to the preformation with TIBA. Catalyst activities were further increased by the preformation of $NdCl_3 \cdot TBP/TIBA$ at 20 °C/60 min in the presence of a small amount of BD (n_{BD}/n_{Nd} = 2) [131]. In contrast, Shen et al. did not find any differences when comparing the systems $NdCl_3 \cdot 3EtOH/TEA$ and $NdCl_3/3EtOH/TEA$ [92].

NdCl$_3$ Nanoparticles
A recent effort to increase the activity of $NdCl_3$-based system comprises the use of $NdCl_3$ nanoparticles. Kwag et al. succeeded in the preparation of nanosized $NdCl_3 \cdot 1.5$ THF (particle size $\geq$ 92 nm). The nanosized $NdCl_3$ was prepared in THF medium by dissolution of anhydrous $NdCl_3$. Nanosized $NdCl_3$ particles were obtained in a colloidal formation step during which THF is slowly replaced by cyclohexane upon addition [132]. Catalyst activity is inversely proportional to the size of the $NdCl_3$ nanoparticles. The activation of $NdCl_3$ nanoparticles with DIBAH/TIBA results in catalyst activities which match the activities of standard Nd-carboxylate-based ternary sys-

tems. However, within all $NdCl_3$-based catalyst systems the described $NdCl_3$ nanoparticle-based system exhibits a comparatively high activity.

Cocatalysts in Nd-Halide-Based Systems

Alkyl aluminum cocatalysts are usually applied in amounts of 4–20 equivalents (relative to one eq. of NdX_3) in Nd-halide catalyst systems. According to Marina et al., further increases in the amount of alkyl aluminum do not lead to increased catalyst activities [129, 130]. In contrast, Iovu et al. report on a significant increase when the amount of cocatalyst is increased from $n_{Al}/n_{NdX3} = 20$ to 100 [133]. According to Yang et al., the microstructure of BR is not influenced by variations of the amount of cocatalyst [35]. Hsieh et al. find that the type of cocatalyst has an influence on the *cis*-1,4-content which decreases in the following order of tested cocatalysts: TIBA > DIBAH > DEAH > TEA [134].

In general, aluminum alkyl cocatalysts favor *cis*-1,4-polymerization whereas magnesium alkyl cocatalyst containing systems such as $NdCl_3 \cdot 3TBP/Mg(^nC_4H_9)(^iC_8H_{17})$ lead to the formation of *trans*-1,4-poly(diene)s with a *trans*-1,4-content as high as 95% [135, 136]. In aliphatic solvents the add-

Table 2 Selection of Nd-halide based catalyst systems used for diene polymerization

Neodymium halide	n Donor	Cocatalyst	*cis*-1,4-Content/%	Refs.
$NdCl_3$	2 THF	TEA	97	[35]
$Nd(OH)_{2.4}Cl_{0.6}$	ethylene diamine	TEA	97 (35% gel)	[107, 108]
$NdCl_3$	EtOH	TEA	75–99	[139]
$NdCl_3$	3 TBP	TIBA	≤ 97	[133, 140]
$NdCl_3$	3 TBP	TIBA	98–99	[131]
$NdCl_3$	2 THF	$Mg(C_3H_5)Cl$ MAO	–	[141]
$NdCl_3$	3 propion amide	TIBA DIBAH TEA	> 98	[122, 123]
$NdCl_3$	3 pentane-1-ol 3 pentane-2-ol 3 pentane-3-ol	TEA	> 99	[115]
$NdCl_3$	2 THF	$Sn(C_3H_5)_4$/ BuLi/MAO	96.8	[142, 143]
$NdCl_3$	2 THF	MgR_2/MAO	96–98	[137, 138]
$NdCl_3$	3 TBP	$Mg(^nC_4H_9)(^iC_8H_{17})$	> 5	[135, 136]
NdI_2	THF	TIBA	"high-*cis*"	[98]
$NdCl_3$ (nanoparticles)	THF	DIBAH/TIBA	96	[132]

ition order of the two components $NdCl_3 \cdot 3TBP$ and $Mg(^nC_4H_9)(^iC_8H_{17})$ has no significant effect on the *trans*-1,4-content, but a strong influence of the addition order is observed when aromatic solvents are used. The *trans*-1,4-content can be adjusted in a certain range by the choice of the solvent, by appropriate catalyst preparation and by the molar ratio n_{MgR2}/n_{NdX3} [135, 136].

The *trans*-1,4-stereospecificity of NdX_3/MgR_2 systems is reversed by the addition of methylalumoxane (MAO) and BR with a high *cis*-1,4-content (96–98%) is obtained [137, 138].

Obviously, the addition of halide donors is not an issue in NdX_3-based systems since halides are already present in sufficiently high quantities. A representative selection of Nd-halide-based catalyst systems is given in Table 2.

Preparation of Anhydrous NdCl₃

$NdCl_3$ is commercially available as hexahydrate ($NdCl_3 \cdot 6H_2O$). Complexation with donor ligands and elimination of crystal water are important factors in order to increase the solubility in hydrocarbons and in order to increase polymerization activities. The preparation of anhydrous $NdCl_3$ cannot be accomplished by straightforward drying of $NdCl_3 \cdot 6H_2O$ at elevated temperature because Nd_2O_3 is formed. The classical method for the preparation of anhydrous $NdCl_3$ comprises the sublimation of a mixture of $NdCl_3 \cdot 6H_2O$ and NH_4Cl at 200–300 °C in vacuo [144]. A modern variant of this procedure is the application of azeotropic distillation [145, 146] which is also performed in the presence of appropriate donors [147–149]. The water content of Nd-halides is usually determined by Karl–Fischer titration. However, more reproducible results are obtained by spectroscopy [150, 151].

2.1.1.2
Neodymium Carboxylates

According to the number of citations in patents and in scientific literature neodymium(III) carboxylates have found widespread application. These systems were first reported by Monakov et al. in 1977 [152, 153]. In 1980 the use of the highly soluble neodymium(III) versatate (NdV) was patented by Bayer [154, 155]. At sufficiently low concentrations the most commonly used Nd-carboxylates (Scheme 2) are completely soluble in hydrocarbon solvents. Because of this, Nd-carboxylates are the focus of many studies and are often referred to in the literature.

- Neodymium(III) versatate (NdV), the mixture of isomers of different α,α-disubstituted decanoic acids is also referred to as neodymium neodecanoate.
- Neodymium(III) octanoate (NdO).
- Neodymium(III) isooctanoate (Nd^iO), also designated as neodymium 2-ethylhexanoate.

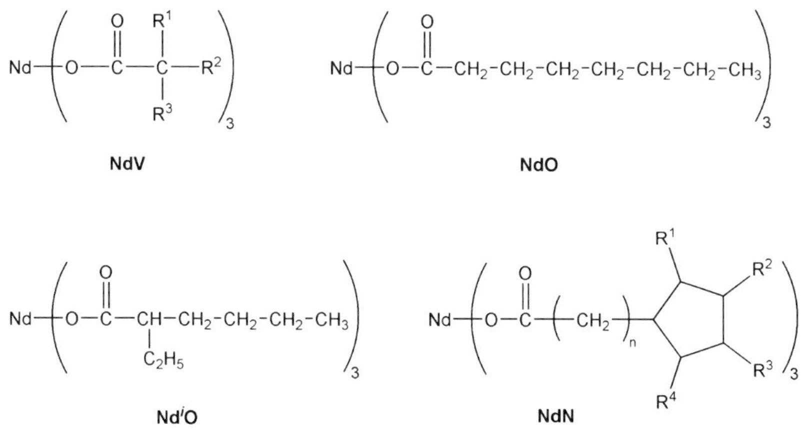

Scheme 2 Most commonly used neodymium(III) carboxylates: Nd versatate (NdV), Nd octanoate (NdO), Nd isooctanoate (Nd^{i}O), Nd naphthenate (NdN)

- Neodymium(III) naphthenate (NdN), an isomeric mixture of substituted cyclopentyl- and cyclohexyl-containing carboxylates.

Both NdO and Nd^{i}O are quite often designated by the term "Nd octanoate" in the literature. In order to avoid confusion one has to read carefully the experimental sections of the respective publications.

Numerous studies on Nd-carboxylate-based catalyst systems address hydrocarbon solubility, catalyst activity, stereospecificity and control of molar mass [49, 89, 111, 141, 156–183][6].

The influence of the chemical structure of different Nd-carboxylates on hydrocarbon solubility and on polymerization activity was investigated in detail by Wilson [183]. This study showed that neocarboxylates ($Nd(OCOCR_3)_3$) have a higher activity than isocarboxylates ($Nd(OCOCH_2R)_3$). Also the length of the aliphatic moieties has an impact on catalyst activity. Within the isocarboxylates, anions with longer hydrocarbon chains exhibit higher polymerization activities. For the respective neocarboxylates this systematic trend could not be confirmed [183]. In another study Wilson compared the performance of NdV and NdN. In ternary catalyst systems of the type Nd(carboxylate)$_3$/DIBAH/^{t}BuCl NdV was more active than NdN [89].

Kobayashi et al. studied catalyst systems in which the chemical composition of the Nd-carboxylates was varied in terms of their electron-withdrawing properties: $Nd(OCOR)_3$ (R = CF_3, CCl_3, $CHCl_2$, CH_2Cl, CH_3) [177]. In this study the highest activity was found for $Nd(OCOCCl_3)_3$-based systems.

---

[6] Scheme 3 in [179] is incorrect; the correct Scheme can be found in [180] (Scheme 2 therein) or in this review (Scheme 33 herein).

Cocatalysts in Nd-Carboxylate-Based Systems

Various cocatalysts are used in Nd-carboxylate-based systems. Most commercially available aluminum alkyls were studied in detail: $AlMe_3$ (TMA) [174, 184–186], $AlEt_3$ (TEA) [159, 187], Al^iBu_3 (TIBA) [175, 179, 188] and $AlOct_3$ [189, 190]. One of the most referenced cocatalysts is Al^iBu_2H (DIBAH), e.g. in [178, 179, 187, 191]. Some of the aluminum alkyl cocatalysts were studied comparatively [49, 174, 179, 189, 190, 192, 193]. Some of these studies report results and trends which seem to be contradictory. Since there are so many factors which have an influence on polymerization characteristics and on polymer properties, the discrepancies between the results of different research groups in many cases can be reconciled on the basis of different experimental conditions.

In Nd-carboxylate systems AlR_3 and AlR_2H cocatalysts are also replaced by alumoxanes, which were described by Sinn and Kaminsky et al. in 1980 [194–196]. Most of these studies focus on methyl alumoxane (MAO) [175, 197–200]. Also tetraisobutyl dialumoxane (TIBAO) was investigated [175]. Differences between aluminum alkyl- and alumoxane-based cocatalysts are also addressed in Sect. 2.1.2.

In addition to aluminum alkyls and alumoxanes, magnesium alkyl compounds and alkyl lithium are applied as cocatalysts in Nd-carboxylate-based systems, e.g. Bu-Mg-^{i}Bu [173], $MgBu_2$ [157] and butyl lithium [201]. The combination of Nd-carboxylates with Mg-alkyls yields catalyst systems with high *trans*-1,4-selectivities. For example, the catalyst system NdV/$MgBu_2$ polymerizes BD to 1,4-poly(butadiene) with 93.8% *trans*-units (66% conversion after 6 h at 50 °C) [157]. By the addition of alkyl aluminum chlorides to NdV/$MgBu_2$ the *trans*-selectivity is reversed to *cis*-selectivity [157, 202, 203]. As opposed to aluminum and magnesium cocatalysts, the use of butyl lithium as a cocatalyst results in heterogeneous catalyst systems. By the application of butyl lithium the polymerization activities are reduced [201].

The amounts of Al-based cocatalysts used in Nd-carboxylate systems range from 8 eq. ($n_{DIBAH}/n_{NdV} = 8$) [81] to 100 eq. ($n_{DIBAH}/n_{NdV} = 100$) [178]. Alumoxanes are even used in excess up to 264 eq. [175].

Table 3 gives a representative selection of studies on Nd(III)-carboxylate-based catalyst systems.

Preparation of Nd-Carboxylates with Reduced Water Content

An early synthesis of water-free Nd carboxylates was reported by Roberts in 1961 [210]. Nd_2O_3 is reacted with an excess of carboxylic acid to yield the Nd carboxylate trihydrate by recrystallization in water. Subsequent dehydration is achieved by storage of the trihydrate over "anhydrone" (i.e. magnesium perchlorate) in vacuo. Dehydration of the trihydrate is also achieved by heating it at 150–180 °C for 2–3 h under a flow of dry nitrogen at reduced pressure [176]. Today, the amide route is the most commonly used laboratory method for the preparation of anhydrous Nd-carboxylates. In this route

Table 3 Selection of investigated Nd(III)-carboxylate based catalyst systems

Neodymium carboxylate	Cocatalyst	Halide donor	cis-1,4-Content/%	Refs.
NdV	DIBAH	$SiCl_4$	97	[187]
NdV	TIBA	EASC	–	[204]
NdV	$MgBu_2$	DEAC	97.8	[157]
NdV	DIBAH	^tBuCl	–	[158]
NdV	DIBAH	EASC	88.2–97.6	[178]
NdV	TIBA	EASC	92.6–96.5	[179, 205]
NdO	TIBA	DEAC	94.9–99.3	[188]
NdO	TIBA	DEAC	< 98	[161]
NdO	TIBA	EASC	99	[163, 169]
Nd^iO	AlR_3 (R = Me, Et, ^iBu)	DEAC	–	[162]
Nd^iO	TIBA	$^iBu_2AlCl$	88–96	[160]
Nd^iO	TEA	$AlBr_3$	83–95	[159]
Nd^iO	TIBA	$EtAlCl_2$	97–98	[206, 207]
Nd^iO	TEA/TIBA	$Et_3Al_2Br_3$	92.1/96.9	[49]
NdN	DIBAH	$^iBu_2AlCl$	–	[208, 209]
NdN	DIBAH	^tBuCl	99	[164]
NdN	AlR_3/AlR_2H	AlR_2X	95	[92]
$Nd(OOCR)_3$ (R = CF_3, CCl_3, $CHCl_2$, CH_2Cl, CH_3)	TIBA	DEAC	69.7–96.1	[177]
Nd(OOC-Ar)	TMA	DEAC	> 99	[16, 184, 185]

$Nd(N(SiMe_3)_2)_3$ is reacted with the respective carboxylic acids and the anhydrous Nd-carboxylate is directly obtained [211, 212].

For the large-scale production of Nd-carboxylates three major routes are described in patent literature. These routes start from Nd_2O_3, $Nd_2(CO_3)_3$ and $Nd(NO_3)_3$. By means of all three routes the Nd carboxylates are obtained by the reaction with the respective carboxylic acids. Through application of the oxide-route, Nd_2O_3 is reacted with the respective carboxylic acid in an organic solvent (hexane or cyclohexane). The reaction is performed in the presence of a large molar excess of acid (n_{acid}/n_{Nd2O3} = 6/1–15/1). In order to start the reaction between Nd_2O_3 and carboxylic acid, diluted hydrochloric acid is added as a catalyst. After the completion of the reaction the mixture is settled and the lower aqueous phase is removed. The upper organic phase contains Nd-carboxylate in a concentration of 38–46 wt. %. The water content of this phase reaches up to 20 000 ppm. The wet Nd-carboxylate containing solution is used without isolation of the Nd-carboxylate and without further purification. The moisture containing solution is reacted with alkylaluminum or alkylaluminum hydride and organic halides under well-defined conditions (0–18 °C/> 30 min) [213–216].

In a similar method a well-defined excess of water ($n_{H2O}/n_{Nd} \geq 5/1$) is applied in combination with an excess of carboxylic acid. In this route either Nd oxide or Nd carbonate are used as Nd sources. The excess of water is necessary to start the reaction. In a patent given by Huang et al. the reaction is also kicked by the addition of diluted hydrochloric acid. The reaction is performed in a temperature and time range of 80–150 °C/2–4 h. At the end of the reaction the mixture separates easily into two clear phases which do not contain unreacted Nd_2O_3. Centrifugation is not necessary to achieve phase separation. Water is removed from the upper phase by azeotropic distillation. The water content of the dried Nd-carboxylate solution is below 2000 ppm [217, 218].

In the nitrate route $Nd(NO_3)_3$ is dissolved in water and the Nd-carboxylate is extracted from the aqueous phase by an organic solvent which contains the respective lithium-, sodium-, potassium- or ammonium-carboxylates. After the completion of the extraction Nd-carboxylate is present in the organic phase and the lithium-, sodium-, potassium- or ammonium nitrates are left in the aqueous phase. The two phases are separated and azeotropic distillation is applied to the organic phase in order to remove water [219, 220].

For the preparation of solid, non-sticky powders of Nd-carboxylates (2-ethylhexanoate, versatate and naphthenate) the solution of the respective Nd carboxylate is obtained by the nitrate route as described above. The organic solution is washed with water prior to the azeotropic removal of the latter. The powder is obtained by subsequent evaporation of the solvent either at normal or at reduced pressure [221, 222].

The viscosity of anhydrous Nd-carboxylates in organic solvents significantly increases with the concentration of the Nd-salt. At reasonable concentrations (10 wt. %) the viscosities are rather high [223]. A significant reduction of the solution viscosity of, for example, NdV is achieved by the addition of small amounts of aluminum alkyls [154, 155].

The Nd-carboxylates prepared by the described large-scale methods contain excess carboxylic acid and excess water. The viscosities of Nd carboxylates in organic solvents are significantly reduced by the presence of these two impurities which also have an impact on the course of the polymerizations and on various features of the resulting BR (Sect. 2.2.3).

2.1.1.3
Neodymium Alcoholates

Neodymium-alcoholates were mentioned in the patent literature prior to Nd-carboxylates [37, 38, 224–228]. The Nd-alcoholates most frequently mentioned in literature comprise $Nd(OBu)_3$ [224, 225, 229, 230], $Nd(O^iPr)_3$ [231–233], and Nd aryl oxides [185, 234, 235] (Scheme 3). Also adduct compounds

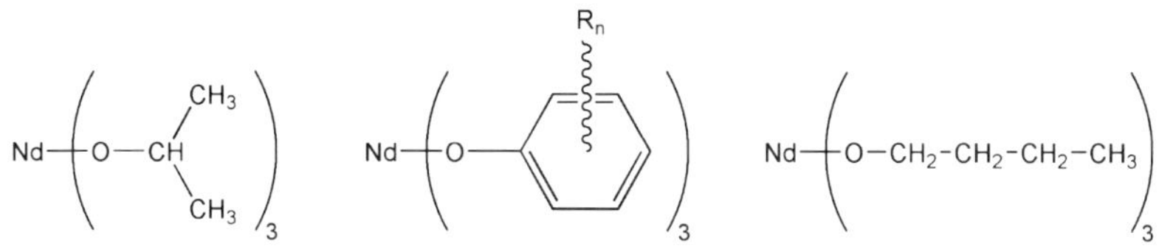

Nd—(O—CH(CH₃)₂)₃   Nd—(O—C₆H₄-Rₙ)₃   Nd—(O—CH₂-CH₂-CH₂-CH₃)₃

Scheme 3 Most commonly used neodymium alcoholates

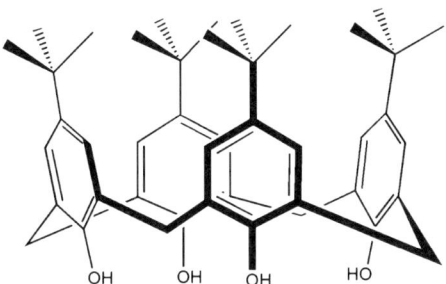

Scheme 4 Calixarene ligand (e.g. calix[4]arene)

such as $Nd(OMe)_3 \cdot (AlMe_3)_4$ [236, 237] and trinuclear Nd-complexes like $Nd_3(O^tBu)_9 \cdot THF$ [235] are described. Adducts of aluminum alkyls with Nd-alcoholates such as $Nd(O^tBu)_3 \cdot (AlMe_3)_3$ are obtained by the addition of Al-alkyls to the alcoholates at 0–10 °C. If such complexes are activated by perfluorinated triphenyl borane, poly(butadiene) with a *cis*-1,4-content in the range 0.7–73% is obtained [238, 239].

Various studies were also performed on sophisticated calixarene (Scheme 4) Nd-complexes. In combination with appropriate cocatalysts these systems yield poly(diene)s with high *cis*-1,4-contents (90–96%) [240–245].

Cocatalysts in Nd-Alcoholate-Based Systems

The usual cocatalysts for the activation of Nd-alcoholates comprise common aluminum alkyls, alumoxanes and magnesium alkyls which have already been described for the activation of the Nd halides (Sect. 2.1.1.1) and Nd carboxylates (Sect. 2.1.1.2): $AlMe_3$ (TMA) [185, 234], TIBA [224, 225, 229, 230], DIBAH [226, 227, 232], MAO [232, 246], modified methyl alumoxane (MMAO) [231] and MgR_2 [235]. The ratios of cocatalyst/Nd-alcoholate are comparable with those described for the activation of Nd carboxylates. Table 4 gives a selection of catalyst systems based on neodymium alcoholates.

In addition to the polymerization of conjugated dienes Nd-alcoholates are mainly used for the polymerization of cyclic polar monomers like lactones, lactides, e.g. see [247] and carbonates, e.g. see [248].

As for the preparation of Nd carboxylates, in the laboratory, the amide route is the most convenient route to various Nd alcoholates. In this synthesis

Table 4 Catalyst systems based on neodymium alcoholates

Neodymium alcoholate	Cocatalyst	Halide donor	cis-1,4-Content/%	Refs.
Nd(OBu)$_3$	TIBA	AlBr$_3$	98.1	[224, 225]
Nd(OBu)$_3$	DIBAH	EtAlCl$_2$	98.1	[226, 227]
Nd(OBu)$_3$	DIBAH	BzCl	98.5	[37, 38]
Nd(OBu)$_3$	DIBAH	B(C$_6$F$_5$)$_3$	92	[229, 230]
Nd(O^{i}Pr)$_3$	MMAO	–	82–93	[231]
Nd(O^{i}Pr)$_3$	MAO	–	> 90	[246]
Nd(O^{i}Pr)$_3$	MMAO	DEAC ^{t}BuCl Me$_3$SiCl	95	[231]
Nd(O^{i}Pr)$_3$	DIBAH	^{t}BuCl	96	[232]
Nd(O^{i}Pr)$_3$	MAO	^{t}BuCl	94	[232]
Nd(O^{i}Pr)$_3$	TIBA	[HNMe$_2$Ph]$^+$ [(C$_6$F$_5$)$_4$]$^-$	90	[233]
Nd$_3$(O^{t}Bu)$_9$·THF	MgR$_2$	–	> 5	[235]
Nd(OMe)$_3$(AlMe$_3$)$_4$	–	R$_x$Al$_y$Cl$_z$	96.8	[236, 237]
Nd(OAr)$_3$	TMA	DEAC	> 99	[185, 186, 234]

Nd(N(SiMe$_3$)$_2$)$_3$ is reacted with the respective alcohol to yield the anhydrous Nd alcoholate [211, 212].

2.1.1.4
Neodymium Phosphates and Phosphonates

Neodymium phosphate-based catalysts were used as early as in 1978 for the polymerization of IP by Monakov et al. [249, 250]. At a later stage neodymium phosphate-based catalyst systems were claimed by Asahi in a patent issued in the mid 1980s [251, 252]. A neodymium-phosphate which is predominantly mentioned in the context of diene polymerization is neodymium bis(2-ethylhexyl)phosphate (NdP). In Chinese scientific literature NdP (Scheme 5) is often abbreviated by its commercial name Nd(P$_{204}$)$_3$.

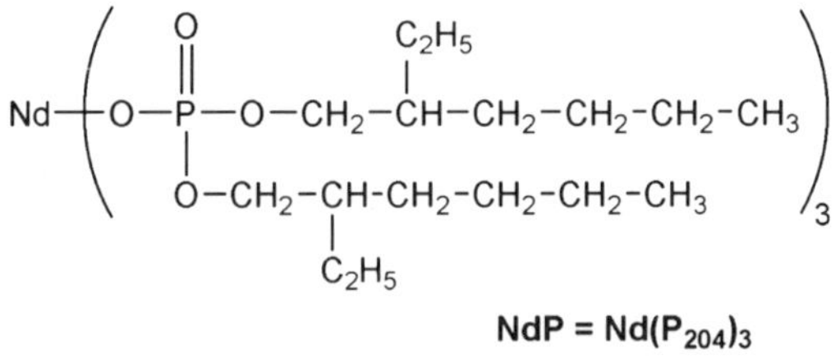

$$\text{Nd}\left(\text{O}-\overset{\overset{\displaystyle O}{\|}}{\underset{\underset{\displaystyle O-CH_2-CH-CH_2-CH_2-CH_2-CH_3}{|}}{P}}-O-CH_2-\overset{\overset{\displaystyle C_2H_5}{|}}{CH}-CH_2-CH_2-CH_2-CH_3\right)_3$$
$$\underset{C_2H_5}{}$$

NdP = Nd(P$_{204}$)$_3$

Scheme 5 Catalyst component neodymium bis(2-ethylhexyl)phosphate (NdP = Nd(P$_{204}$)$_3$)

In addition to the polymerization of dienes the versatility of NdP-based catalysts is exceptional regarding the number of different non-diene monomers which can be polymerized with these catalysts. Acetylene is polymerized by the binary catalyst system NdP/AlEt$_3$ [253, 254]. Lactides are polymerized by the ternary system NdP/AlEt$_3$/H$_2$O [255, 256]. NdP/TIBA systems are applied in the copolymerization of carbon dioxide and epichlorhydrine [257] as well as for the block copolymerization of IP and epichlorohydrin [258]. The ternary catalyst system NdP/MgBu$_2$/TMEDA allows for the homopolymerization of polar monomers such as acrylonitrile [259] and methylmethacrylate [260]. The quaternary system NdP/MgBu$_2$/AlEt$_3$/HMPTA is used for the polymerization of styrene [261].

Beside NdP (= Nd(P$_{204}$)$_3$) Xiu et al. and Liu et al. also describe the phosphorus-containing catalyst component Nd(P$_{507}$)$_3$ which is bis(2-ethylhexanol)phosphonate (Scheme 6) [262, 263]. Xu et al. applied Nd(P$_{507}$)$_3$ as well as Nd(P$_{204}$)$_3$ in combination with the cocatalyst TIBA for the homopolymerization of hexylisocyanate [262]. The Nd-phosphonate-based catalyst system Nd(P$_{507}$)$_3$/TIBA/H$_2$O is also used for the homopolymerization of styrene [263].

Though diene polymerization by NdP-based systems was already described by Monakov et al. in 1978 [249, 250] and by Shen et al. in 1980 [92] the high potential of NdP-based catalysts was not fully recognized until 2002 when Laubry applied a ternary NdP-based catalyst for the production of poly(isoprene) with a remarkably high *cis*-1,4-content (> 99%) [264–269]. Another interesting feature of the described catalyst is the selective polymerization of IP in the crude C5 cracking fraction [264, 265]. In a further patent Laubry also describes the preparation of random BD-IP-copolymers by NdP catalysis [270, 271]. Outstandingly remarkable for the reported catalyst system based on NdP/DIBAH/aluminum alkyl chloride are the low ratios of aluminum alkyl cocatalyst/NdP ($n_{DIBAH}/n_{NdP} = 1-5$) at which high catalyst activities are observed. In comparison, Nd-carboxylate systems are inactive at such low n_{Al}/n_{Nd}-ratios, e.g. [178, 272]. The reason for this unusual feature of NdP-based catalyst systems seems to be due to the fact that only very low amounts of Al-alkyls are consumed in the formation of the active catalyst. This result is of high economical relevance as the amount of Al-cocatalyst contributes as a major factor to overall catalyst costs.

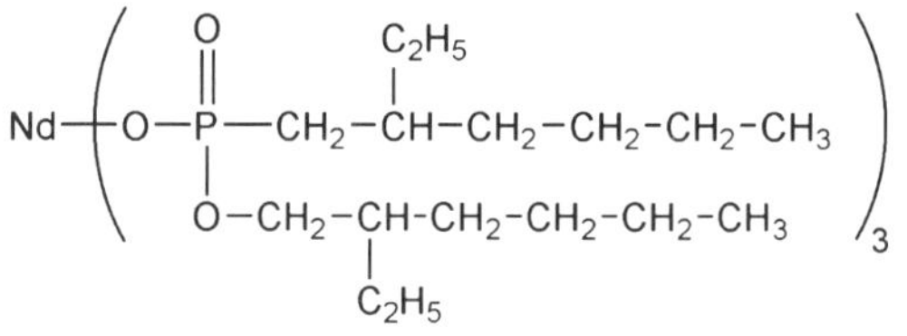

$$Nd \!-\!\!\left(\!\! O\!-\!\!\underset{\underset{O-CH_2-CH-CH_2-CH_2-CH_2-CH_3}{|}}{\overset{\overset{O}{\|}}{P}}\!\!-\!CH_2-\overset{\overset{C_2H_5}{|}}{CH}-CH_2-CH_2-CH_2-CH_3 \!\!\right)_{\!\!3}$$
$$\underset{C_2H_5}{}$$

Nd(P$_{507}$)$_3$

Scheme 6 Nd-phosphonate-based catalyst component Nd(P$_{507}$)$_3$

Preparation of Neodymium Bis(2-ethylhexyl)phosphate (NdP)
The preparation of NdP is described by Laubry [268, 269]. According to this patent Nd(III) chloride hexahydrate is reacted with the respective phosphoric acid in water/acetone. After several washing and drying steps anhydrous NdP is obtained.

2.1.1.5
Neodymium Allyl Compounds

In discussions about the nature of the active species in the polymerization of dienes by Ziegler/Natta catalyst systems allyl species have already been suggested in the 1960s [273–278]. This discussion has continued through the past decades [139, 279–283]. Today, it is widely accepted that Nd-allyl-groups are the key element in the insertion of dienes into the Nd carbon bond.

The anionic tetrakis allyl complex Li[Nd(C$_3$H$_5$)$_4$]·1.5 dioxane was the first Nd-allyl compound to be synthesized by Mazzei in 1981 [284]. In diene polymerization the catalytic activity of this complex is modest and yields poly(diene)s with a high *trans*-1,4-content (84%) and a surprisingly high 1,2-content (15%). The high 1,2-content is explained by a simultaneous anionic polymerization which is initiated by Li(C$_3$H$_5$). Li(C$_3$H$_5$) is formed in a side reaction in which Li[Nd(C$_3$H$_5$)$_4$]·1.5 dioxane decomposes [285, 286].

Since the mid-1990s, Taube and co-workers remarkably contributed to the chemistry of Nd-allyl compounds. These authors applied their expertise in nickel-catalyzed polymerizations to Nd chemistry [81, 285, 287]. At first, Taube et al. improved Mazzei's synthesis of Nd-allyl complexes [286]. By the abstraction of allyl lithium from Li[Nd(C$_3$H$_5$)$_4$]·1.5 dioxane by means of BEt$_3$ the first neutral Nd-allyl complex Nd(η^3–C$_3$H$_5$)$_3$·dioxane was obtained [288]. In the polymerization of dienes (in toluene), this complex shows a low catalytic activity and yields poly(diene)s with a high *trans*-1,4-content (94%). By the addition of cocatalysts such as DEAC or MAO the activity is increased 7-fold (with 2 eq. DEAC) or 16-fold to 60-fold (with 30 eq. MAO). As a consequence of cocatalyst addition the microstructure changes from high-*trans*-1,4 (94%) to high-*cis*-1,4 (94% *cis* with 2 eq. DEAC, 59–84% *cis* with 30 eq. MAO). The authors suggest that the cocatalyst abstracts one or two allyl anions from the complex Nd(η^3–C$_3$H$_5$)$_3$ resulting in a cationic Nd allyl species which has vacant sites at which diene monomers are coordinated [288]. Taube and his group performed several studies with catalyst systems based on Nd(η^3–C$_3$H$_5$)$_3$·dioxane. A comparison of the two alumoxanes MAO and HIBAO showed that MAO is the more efficient activator [289]. For the catalyst system Nd(η^3–C$_3$H$_5$)$_3$·dioxane/MAO one poly(butadiene) chain is formed per Nd-atom and no chain transfer is found. The highest *cis*-1,4-content that can be obtained with this system is 84%. In accordance with metallocene-catalyzed polymerizations of olefins in which cationic metallocene complexes are the active species,

$$(C_3H_5)_3Nd \cdot L_n + MAO \rightarrow [(C_3H_5)Nd^{2+} \cdot 2(C_3H_5)^- \cdot MAO \cdot L_n]$$

$$(C_3H_5)NdCl_2 \cdot L_n + MAO \rightarrow [(C_3H_5)Nd^{2+} \cdot 2\,Cl^- \cdot MAO \cdot L_n]$$

Scheme 7 Activation of $Nd(\eta^3-C_3H_5)_3 \cdot$ dioxane and $Nd(\eta^3-C_3H_5)Cl_2 \cdot 2$ THF by MAO according to Taube et al. (L = dioxane or THF) [289, 290]

Taube suggests a cationic Nd-allyl complex as the active catalyst species (Scheme 7) [289, 290].

Subsequent studies on BD polymerization by the solvent-free allyl compound $Nd(\eta^3-C_3H_5)_3$ (without cocatalysts added) showed that—depending on monomer concentration—2 to 3 poly(butadiene) chains are generated per Nd-atom [291]. By the addition of cocatalyst the number of chains generated per Nd-atom is reduced to one.

The role of halide donors in Nd-Ziegler/Natta-polymerization was elucidated for $Nd(\eta^3-C_3H_5)_2Cl \cdot$ THF and $Nd(\eta^3-C_3H_5)Cl_2 \cdot 2$ THF [290]. The solubility of these complexes in hydrocarbon solvents is very poor and so is their activity in diene polymerization. By the addition of 30–50 eq. of MAO, the solubility of these complexes is considerably increased and homogeneous catalyst systems are obtained. Thus, by the addition of MAO, polymerization activity and *cis*-1,4-content (98%) are increased. In Nd-carboxylate systems the authors assume that complete alkylation of all Nd-centers is unlikely [289]. The higher activity of Nd-allyl chloride complexes versus trisallyl Nd-complexes is explained by an easier abstraction of a chloride atom from the Nd-center as opposed to abstraction of an allyl group (Scheme 7) [290]. Ligand abstraction is considered to be the decisive step for the generation of a vacant coordination site on Nd [290]. Further studies on the polymerization kinetics initiated by the catalyst complexes $Nd(\eta^3-C_3H_5)_2Cl \cdot 1.5$ THF and $Nd(\eta^3-C_3H_5)Cl_2 \cdot 2$ THF, both activated by the cocatalysts HIBAO and MAO, revealed the following results [292]:

- The polymerizations show features of a living polymerization (proven by linear plots of M_n vs. $[M]_0/[Nd]_0$).
- The kinetic equations are:

  for $Nd(\eta^3-C_3H_5)_2Cl \cdot 1.5THF/HIBAO$: $\quad r_p = k_p \cdot [Nd] \cdot [BD]^{1.8}$

  for $Nd(\eta^3-C_3H_5)Cl_2 \cdot 2THF/MAO$: $\quad r_p = k_p \cdot [Nd] \cdot [BD]^2$

- Molar masses are independent of the amount of cocatalyst.
- One poly(butadiene) chain is generated per one Nd-atom.

In the most recent publication by Taube et al. these authors summarize their previous work on Nd-catalysis and report on the results of supplementary experiments [293]. In this detailed study it is shown that the dioxane complex of $Nd(\eta^3-C_3H_5)_2Cl$ is active by itself (without further cocatalysts) and that it catalyzes the polymerization of BD (*cis*-1,4-content = 85%). This is a strong

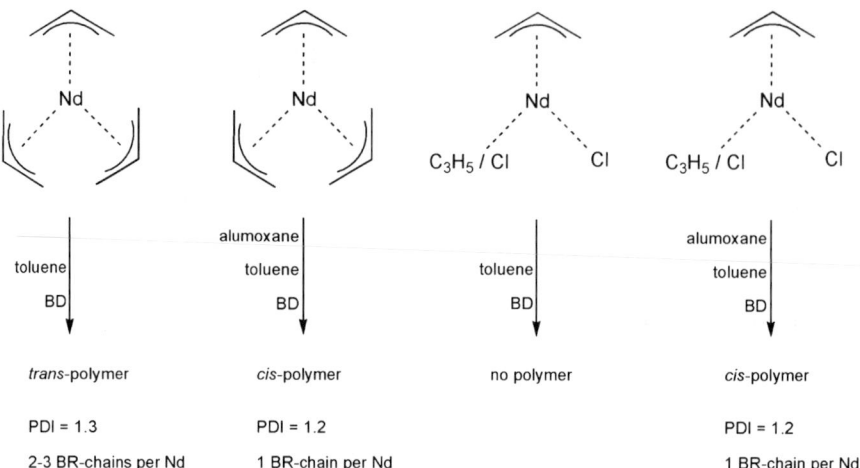

Scheme 8 Summary of results on BD polymerization with neodymium allyl (chloride) complexes by Taube et al. (solvent ligands are omitted for clarity). The activities of the three active complexes increase *from left to right*

indication that Nd(η^3–C$_3$H$_5$)$_2$Cl represents the key structural elements of an active Nd diene polymerization catalyst [293]. Scheme 8 gives the essentials of this study.

The activities of Nd(η^3–C$_3$H$_5$)Cl$_2$·2.5 THF and of Nd(η^3–C$_3$H$_5$)$_2$Cl·1,4-dioxane are substantially increased by the addition of 5–30 eq. of AlMe$_3$ and by the preformation of Nd(η^3–C$_3$H$_5$)$_3$ with 2Ph$_3$CCl, 2Ph$_3$CCl/5AlEt$_3$, 2AlMe$_2$Cl, 2AlMe$_2$Cl/30AlMe$_3$, 2AlEt$_2$Cl or 2AlEt$_2$Cl/10AlEt$_3$. Catalyst activities are significantly higher in aliphatic than in aromatic solvents [293]. This study also gives conclusions on the active catalytic species (Sect. 4.2).

It has to be mentioned that as early as in 1991 Porri et al. reported on the reaction of NdCl$_3$ with Mg(C$_3$H$_5$)Cl in THF yielding an undefined Nd allyl compound which was successfully tested in diene polymerizations [141, 167, 294]. For the polymerization experiments the cocatalysts TIBA, TMA, TIBAO and MAO were used and no halide donor was added. The undefined Nd allyl compound + TIBA yields a catalyst system that is reported to be at least three times more active than the system Nd^{i}O/TIBA/DEAC. The application of MAO with the Nd allyl compound increases catalytic activity 30-fold.

Soon after these studies Wu et al. succeeded to fully characterize Nd(C$_3$H$_5$)$_2$Cl·2 MgCl$_2$·2 TMEDA which was obtained by the reaction of NdCl$_3$ with Mg(C$_3$H$_5$)Cl in the presence of tetramethylethylene diamine (TMEDA). THF was used as the solvent [295].

Recently, Lorenz et al. reported on novel azaallyl Nd compounds which were obtained by the reaction of NdBr$_3$·(THF)$_{3.5}$ with dilithium hexa-1,5-diene-1,6-diamide [296]. In this study two η^3-azaallyl Nd complexes are

Scheme 9 Azaallyl ligands: different modes of coordination: **a** = enamide mode (η^1- or σ-azaallyl) and **b** = azaallyl mode (η^3- or π-azaallyl)

compared. The coordinatively less saturated η^3-azaallyl and bromide containing Nd complex exhibits higher polymerization activity (116.4 g(polymer) mmol^{-1}(Nd) h^{-1}) than the described coordinatively higher saturated η^3-azaallyl Nd complex (23.3 g(polymer) mmol^{-1}(Nd) h^{-1}). The η^3-azaallyl Nd bromide complex yields poly(butadiene) with a *cis*-1,4-content of 93.5%.

It is discussed that during polymerization catalysis the change of the coordination mode of the azaallyl ligand might be of importance as azaallyl ligands can either coordinate to Nd in an enamide mode (η^1- or σ-azaallyl) or in an azaallyl mode (η^3- or π-azaallyl) (Scheme 9).

2.1.1.6
Neodymium Cyclopentadienyl Complexes

There are various types of Nd-based cyclopentadienyl (Cp) derivatives which are in the focus of recent publications. The Cp derivatives comprise the following structural features:

- monocyclopentadienyl dichloro (CpNdCl$_2$);
- monocyclopentadienyl dialkyl (CpNdR$_2$);
- monocyclopentadienyl diallyl (CpNd(η^3–C$_3$H$_5$)$_2$);
- monoyclopentadienyl tris allyl anionic (Li$^+$ [CpNd(η^3–C$_3$H$_5$)$_3$]$^-$);
- dicyclopentadienyl monochloro (Cp$_2$NdCl);
- dicyclopentadienyl monoalkyl (Cp$_2$NdR);
- silylene bridged dicyclopentadienyl derivatives ([R$_2$Si(Cp)$_2$]Nd – Cl/R).

Most of these compounds show a limited solubility in non-polar solvents. In addition, the respective alkyl derivatives are rather unstable in solution and decompose easily [297]. The peculiarity about CpNd complexes is their ability to polymerize various alkenes such as α-olefines, styrene, α,ω-dienes as well as polar acrylates [298, 299].

The polymerization of BD with the binary system CpNdCl$_2$ · THF/AlR$_3$ yields poly(butadiene) with a *cis*-1,4-content as high as 98%. Among several tested aluminum cocatalysts DIBAH is the most effective activator for CpNdCl$_2$ · THF [300, 301]. Similar results were obtained with the respective indenyl neodymium dichloride systems IndNdCl$_2$ · THF/AlR$_3$ (Ind = indenyl) [302, 303]. With the bisindenyl systems Ind$_2$NdCl/TMA and

Ind$_2$NdCl/MAO poly(butadiene)s with modest cis-1,4-content (73%) were obtained [304].

According to Taube et al. the polymerization of BD by CpNd(η^3–C$_3$H$_5$)$_2$/MAO exhibits the features of a living polymerization. It is shown that each Nd-atom participates in the polymerization. As a result extremely narrow MMDs (PDI = 1.1) are obtained. In the complexes CpNd(η^3–C$_3$H$_5$)$_2$ and Cp*Nd(η^3–C$_3$H$_5$)$_2$ the presence of Cp and Cp* increases 1,2-contents to 5–10% [286, 305].

The dimethylsilyl-bridged dicyclopentadienyl complex [Me$_2$Si(3-Me$_3$SiC$_5$H$_3$)$_2$]NdCl (Scheme 10) was successfully applied for the preparation of a strongly alternating BD/ethylene-copolymer (Sect. 2.3.5) [306].

Recent work of Boisson et al. focuses on silylene-bridged Nd complexes of the type depicted in Scheme 10. The respective bis fluorenyl complex [Me$_2$Si(C$_{13}$H$_8$)$_2$]NdCl [307] and the mixed Cp/fluorenyl complex [Me$_2$Si(C$_5$H$_4$)–(C$_{13}$H$_8$)]NdCl [308, 309] have been described. The work on silylene-bridged Nd sandwich complexes performed by Boisson et al. is reviewed in [310].

Kaita and co-workers investigated pentamethylcyclopentadienyl (Cp*) neodymium complexes [311–315]. These authors harnessed [(Cp*)$_2$Nd(μ-Me)$_2$AlMe$_2$] and [(Cp*)$_2$Nd]$^+$ [B(C$_6$F$_5$)$_4$]$^-$ for the cis-specific polymerization of BD. For the activation of the complex [(Cp*)$_2$Nd]$^+$ [B(C$_6$F$_5$)$_4$]$^-$ the addition of 5 eq. of TIBA was necessary as the cationic complex on its own was not active in BD polymerization.

Neodymium cyclopentadienyl complexes are also obtained in-situ by the addition of Cp-derivatives (e.g. indene, cyclopentadiene, pentamethylcyclopentadiene, tetramethylcyclopentadiene, di-tert-butylcyclopentadiene, methylcyclopentadiene and fluorene) to standard Nd-catalyst systems such as NdV/MAO. It can be assumed that the respective cyclopentadienyl-anions are formed by proton abstraction from the Cp-derivatives. In the homopolymerization of BD the addition of Cp-derivatives results in an increase of the 1,2-content of about 4–10%. In addition, the in-situ formed Nd Cp-derivatives

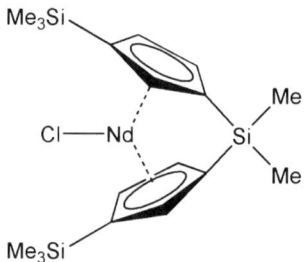

Scheme 10 Neodymium complex [Me$_2$Si(3-Me$_3$SiC$_5$H$_3$)$_2$]NdCl used for copolymerization of BD and ethylene [306], reprinted with permission of Wiley-VCH Verlag GmbH & Co. KGaA

allow for the copolymerization of BD and St and for the homopolymerization of ethylene or methacrylates [316, 317].

2.1.1.7
Neodymium Amides

Neodymium amides have already been mentioned in the context of the synthesis of Nd-carboxylates and Nd alcoholates (Sects. 2.1.1.2 and 2.1.1.3) [211, 212]. The use of neodymium amides as catalyst components in Ziegler/Natta polymerizations is a recent development and has only been investigated by a few groups, so far. In 1999 Boisson et al. tested $Nd(N(SiMe_3)_2)_3$ in BD polymerization [318]. For the catalyst system $Nd(N(SiMe_3)_2)_3$/TIBA/DEAC an optimum in activity was found at a molar ratio of 1/40/2 and the catalyst system yields BR with a *cis*-1,4-content of 97.6% which is comparable to the *cis*-1,4-contents obtained with Nd-carboxylate or Nd-alcoholate-based systems. The authors put emphasis on the monomeric nature of the Nd amide $Nd(N(SiMe_3)_2)_3$ as it does not form clusters [319]. This is considered to be a significant advantage for achieving high catalyst activities.

The polymerization of BD was also performed with $Nd(N(SiMe_3)_2)_3$ which was activated by MAO, TIBA, $B(C_6F_5)_3$, $[HNMe_2Ph]^+$ $[B(C_6F_5)_4]^-$ and $[CPh_3]^+$ $[B(C_6F_5)_4]^-$ [320]. The cationic complex $[Nd\{N(SiMe_3)_2\}_2(THF)_2]^+$ $[B(C_6F_5)_4]^-$ obtained by the reaction of $Nd(N(SiMe_3)_2)_3$ with $[HNMe_2Ph]^+$ $[B(C_6F_5)_4]^-$ is highly active in BD polymerization but not very *cis*-specific (< 90%).

Dow Chemical Company investigated homogeneous as well as heterogeneous Nd-amide-based catalysts with regard to their industrial applicability [321, 322].

2.1.1.8
Neodymium Acetylacetonates

Like carboxylates acetylacetonates are bidentate ligands for the Nd center. In the mid 1970s Monakov et al. started investigations on the use of Nd acetylacetonates for the polymerization of dienes [323, 324]. Nd-acetylacetonate and Nd-benzoylacetonate were again mentioned in 1980 by Shen et al. [92]. During the time of Nd-BR commercialization the influence of acetylacetone on Nd-based catalyst systems was intensely studied by JSR. The increase of the solubility of Nd-salts in hydrocarbon solvents by acetylacetone was claimed in 1983 [325, 326]. From this time onwards JSR filed numerous patents in which acetylacetone containing Nd catalyst systems were described [327–343].

Zhang et al. investigated a catalyst system comprising neodymium acetylacetonate, dibutyl magnesium and chloroform in the homopolymerization as well as in the copolymerization of the two monomers IP and styrene (St).

These authors succeeded in synthesizing the copolymer poly(IP-*co*-St) which predominantly contains *cis*-1,4-isoprene units (88%) [344].

2.1.1.9
Further Neodymium Compounds

Ligands which have recently been explored in "post-metallocene chemistry" are presently being adopted for Nd-catalysts. For example Thiele, Wilson and co-workers (Dow Global Technologies Inc.) recently filed a patent and published a paper on neodymium diimine and diiminopyridine complexes which yield poly(diene)s with high *cis*-1,4-contents (95–97%) [345–347].

Nd-boranate-based catalysts were also used in the polymerization of dienes. $Nd(BH_4)_3 \cdot (THF)_3$/TEA yields poly(butadiene) with a *trans/cis*-ratio $\sim 50/50$. If $Nd(BH_4)_3 \cdot (THF)_3$ is combined with a stoichiometric amount of $MgBu_2$ catalyst activities are increased, control of molar masses is improved and poly(diene)s with a *trans*-1,4-content of up to 99% are obtained. [348, 349].

Heteroleptic Nd complexes with three different ligands (Cp^*-type, diketiminate and boranate) are used for the preparation of BR with a high *trans*-1,4-content. If these Nd-precursors are combined with the cocatalyst $MgBu_2$ the *trans*-1,4-selectivity is increased to 98.4%. These effects are observed in the polymerization of IP [350].

The use of Nd-hydride-based catalyst systems is described in a patent filed by Phillips Petroleum Co. in 1987 [351, 352]. The hydride-containing catalyst system $NdH_3/AlBr_3/Et_2AlH$ yields BR with a high *cis*-1,4-content of 97%.

2.1.1.10
Comparative Studies with Different Neodymium Components

Though a vast number of studies on the characteristics of neodymium-mediated polymerizations were performed to the present day, only a few studies focus on the influence of the anion of the Nd precursor. As already mentioned in Sect. 2.1.1.2 Wilson systematically varied the structure of carboxylates and studied the influence on hydrocarbon solubility and on polymerization activity [183]. The dependence of polymerization activity on various halogenated Nd-carboxylates $Nd(OCOR)_3$ (R = CF_3, CCl_3, $CHCl_2$, CH_2Cl, CH_3) was the target of a study by Kobayashi et al. [177].

In addition to these two studies the polymerization kinetics of three different Nd-compounds which were activated by DIBAH and EASC were comparatively studied. In this investigation a Nd alcoholate [NdA = neodymium(III) neopentanolate], a Nd phosphate [NdP = neodymium(III) 2-ethyl-hexyl-phosphate] and a Nd carboxylate (NdV) were compared with a special focus on the variation of the molar ratios of n_{DIBAH}/n_{Nd} and n_{Cl}/n_{Nd} [272]. For each of these ternary catalyst systems the polymerization activities depend on the n_{Cl}/n_{Nd}-ratio and for each catalyst system the maximum activity is located at a specific n_{Cl}/n_{Nd}-ratio: for NdV at $n_{Cl}/n_{NdV} = 2$, for NdA at $n_{Cl}/n_{NdA} = 1$ and for NdP at $n_{Cl}/n_{NdP} = 2$. The activities of the three catalyst systems (expressed as turnover frequency = TOF) show different dependencies on the n_{DIBAH}/n_{Nd}-ratios. For NdV TOF exhibits a maximum at $n_{DIBAH}/n_{NdV} = 30$. For NdA TOF increases with an increasing ratio of n_{DIBAH}/n_{NdA} and for NdP TOF decreases with increasing n_{DIBAH}/n_{NdP}-ratios (Fig. 3).

In the whole study the highest catalyst activity (TOF = 808 200 h^{-1}) is observed for NdV/DIBAH/EASC at $n_{DIBAH}/n_{NdV} = 30$. It is interesting to note that the NdP-based catalyst system exhibits its highest activity (TOF = 722 200 h^{-1}) at a n_{DIBAH}/n_{NdP}-ratio as low as 5. This performance is unique as the NdV-based and the NdA-based systems are completely inactive at this low n_{DIBAH}/n_{Nd}-ratio. Since the costs for the aluminum alkyl cocatalyst are a major factor in total catalyst cost the high activity of the NdP system at $n_{DIBAH}/n_{NdP} = 5$ has to be particularly emphasized from an economic point of view [268, 269].

In Fig. 4 data on the microstructure are confined to cis-1,4-contents as 1,2-contents remain almost unchanged (0.2–1.3%). For the three catalyst systems the cis-1,4-contents decrease with increasing amounts of DIBAH as

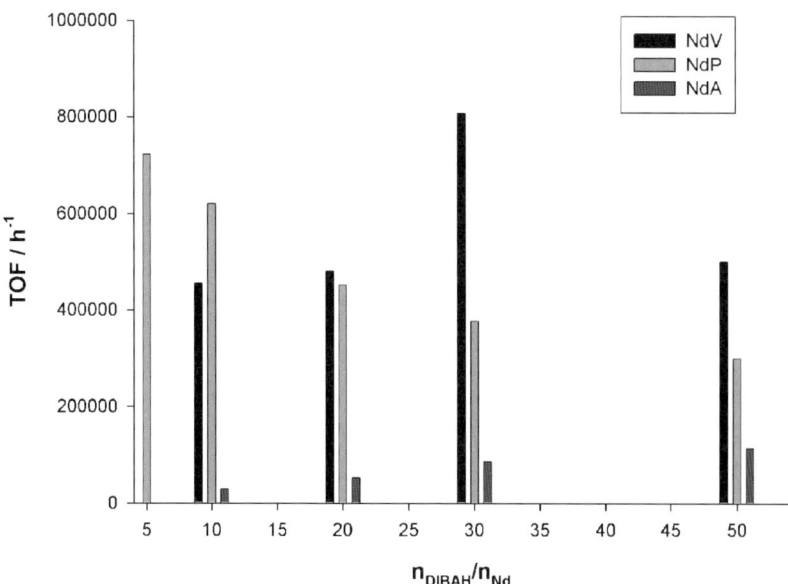

Fig. 3 Comparison of the turnover frequency (TOF) for the polymerization of BD with three catalyst systems: (1) NdV/DIBAH/EASC, (2) NdA/DIBAH/EASC and (3) NdP/DIBAH/EASC at various molar ratios n_{DIBAH}/n_{Nd} [272], reproduced by permission of Taylor & Francis Group, LLC., http://www.taylorandfrancis.com

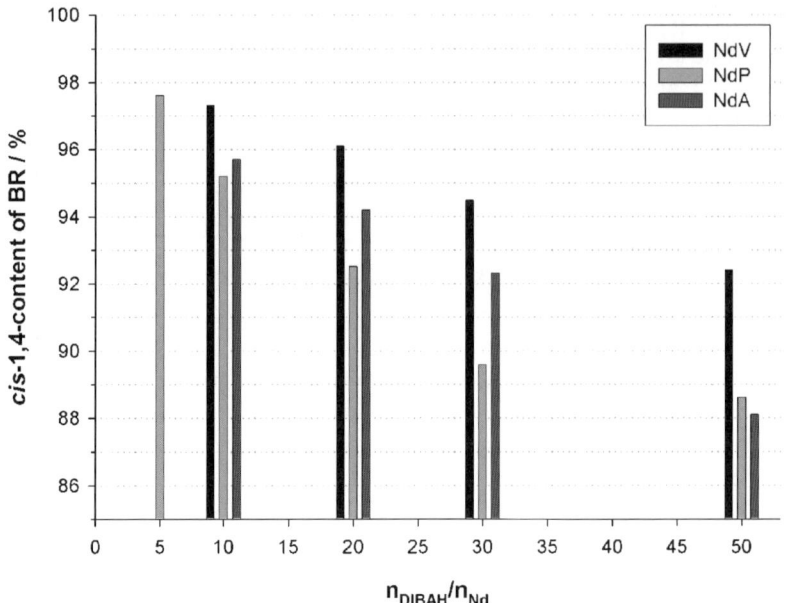

Fig. 4 Dependence of *cis*-1,4-contents on the molar ratio n_{DIBAH}/n_{Nd} for the catalyst systems (1) NdV/DIBAH/EASC, (2) NdA/DIBAH/EASC and (3) NdP/DIBAH/EASC [272], reproduced by permission of Taylor & Francis Group, LLC., http://www.taylorandfrancis.com

shown in Fig. 4. Since only NdP allows for a polymerization at a molar ratio as low as $n_{DIBAH}/n_{Nd} = 5$ the *cis*-1,4-content observed for this system is the highest within the whole study (97.6%). At the higher molar ratios of $n_{DIBAH}/n_{Nd} = 10, 20, 30$ and 50 NdV-based catalyst systems yield higher *cis*-1,4-contents than NdP- or NdA-based systems.

The decrease of *cis*-1,4 contents with increasing amounts of DIBAH observed for each of the three catalyst systems can be explained by a model which is discussed in Sect. 4.3.

2.1.2
Cocatalysts/Activators

In order to generate the active Nd-species the Nd precursors are activated by "cocatalysts" or "activators". Some aspects about cocatalysis were already addressed in the previous Sects. 2.1.1.1–2.1.1.10. In this work qualitative aspects on cocatalyst components and cocatalysis are summarized from a general point of view. Quantitative aspects and the impact of cocatalysts on polymerization rate and polymer properties are reviewed in Sect. 2.1.4.

The cocatalyst's prime function is the activation of the Nd-precursor in order to form the active catalyst. Secondly, the cocatalyst is often used for control of molar mass. This particularly applies to Al-, Mg- and Zn alkyls.

And thirdly, cocatalysts are scavengers for impurities such as moisture, excess carboxylic acids etc.

In the subsequent sections the use of the different classes of cocatalysts are reviewed. First, aluminum compounds of the type AlR_3 and AlR_2H (e.g. R = Me, Et, ^iBu, Oct) are addressed. These activators are amongst the most commonly used. Less often, alumoxanes like MAO, HIBAO and TIBAO are used. Cocatalysis with magnesium alkyls result in catalyst systems which are highly *trans*-1,4-specific. Finally, the few reports on the use of perfluorated arylboranes and arylborates of the type $B(C_6F_5)_3$ and $CPh_3^+ \ B(C_6F_5)_4^-$ as well as $[HNMe_2Ph]^+ \ [B(C_6F_5)_4]^-$ are addressed.

Aluminum-Based Cocatalysts

Catalyst systems which differ in the chemical structure of the Nd precursors (Nd chlorides, Nd carboxylates, Nd alcoholates, Nd phosphates etc.) do not require specific aluminum-based activators. As the differences in the interaction of various Nd-sources with the aluminum alkyl cocatalysts are rather of a quantitative than of a qualitative nature it can be concluded that the basic principles of activation are very similar for different Nd precursors. It is generally agreed that in the activation reaction the aluminum alkyls transfer alkyl or hydride groups to the Nd centers. In addition, aluminum alkyls act as Lewis acids which abstract halides or alkyl groups from Nd and create free coordination sites on Nd. The abstraction of halides and alkyl groups can also be accomplished by other strong acids or Lewis acids such as anilinium salts, trityl salts and perfluorinated arylboranes etc. During these reactions (which also occur in the presence of monomer) the active Nd-species is formed (details in Sect. 4.2). From the viewpoint of these mechanistic aspects it is obvious that the chain length of aluminum alkyls of the type AlR_3 and $AlHR_2$ also plays an important role. The impact of the chain length of the Al-alkyl cocatalyst on polymerization activities was studied by various authors. For example for the catalyst system $Nd^iO/AlR_3/DEAC$ polymerization activities decreased in the following order: $Al^iBu_3 > AlHBu_2 > AlEt_3$ [162, 165, 166]. In a later study Wilson systematically compared straight-chain aluminum alkyls of the type AlR_3 [174]. According to Wilson methyl substituents yield catalyst systems with low or even without any catalyst activity. Ethyl substituents result in increased polymerization activities. Maximum activities are achieved with propyl and butyl substituents. Further increases in the number of carbon atoms in the alkyl chains (> 4) do not lead to further improvements in catalyst activities. According to Wilson methyl and ethyl substituents do not provide sufficient solubility of the active species in the non-polar solvents which leads to partially heterogeneous systems. Wilson gives the following order for catalyst activities: $Al^nPr_3 > Al^nBu_3 = Al^nHex_3 = Al^nOct_3 = Al^nDodec_3 \gg AlEt_3 \gg AlMe_3$ [174]. He relates this order also to the decreasing extent of association of the aluminum alkyls. According to Wilson the limited activity of aluminum alkyl clusters can be overcome by an increase of temperature which leads to a reduction of association and accordingly results in a higher alkylating power of the cocatalysts.

Beside the activating effect aluminum alkyl cocatalysts are also efficient molar mass control agents. Control of molar mass is achieved by the adjustment of the molar ratio of n_{Al}/n_{Nd} (Sects. 2.1.4, 2.2.8 and 4.5). An increase in the amount of cocatalyst results in a decrease of molar mass. A change of the n_{Al}/n_{Nd}-ratio also influences the rate of the polymerization reaction which is a major shortcoming in the large-scale production of Nd-BR, particularly in continuous processes. Detailed discussions of this issue are found in Sect. 2.2.8. Because of this disadvantage research on Nd-BR still strikes out to find efficient non Al-based molar mass control agents which do not influence the rate of polymerization.

The influence of the n_{Al}/n_{Nd}-ratio on microstructure and especially on the *cis*-1,4-content is addressed in Sect. 2.1.4.

Activation of Nd carboxylates and Nd alcoholates by alumoxanes provides the option to omit halide donors. By the use of alumoxanes high activity levels are accomplished but *cis*-1,4-contents are considerably reduced as opposed to the use of AlR_3 with which very high *cis*-1,4-contents are accessible [175].

Magnesium-Based Cocatalysts

In Nd-catalyzed diene polymerizations aluminum alkyl cocatalysts can be replaced by magnesium alkyls. By this replacement the microstructure of the resulting polymer is completely changed and poly(diene)s with high *trans*-1,4-contents are formed [235, 353, 354]. Another interesting aspect brought about by magnesium alkyl cocatalysis is the preparation of poly(butadiene) block copolymers which comprise *cis*-1,4- and *trans*-1,4-building-blocks [355, 356]. According to Jenkins *trans*-1,4-specific polymerization is started by a MgR_2-containing Nd catalyst system. During the course of the polymerization a halide donor is added and diene insertion is switched to the *cis*-1,4-mode. This synthetic strategy was verified by the addition of EASC during a polymerization which was initiated by the catalyst system NdV/MgR_2 [157]. If the solvent 1,2-dichloroethane was used the catalyst system NdV/$MgBu_2$ yielded BR with a gradient composition of *cis*-1,4/*trans*-1,4-moieties. In this case 1,2-dichloroethane acts as a halide source from which chloride is successively released during the course of the polymerization. By the increasing availability of chloride *trans*-1,4-insertion of BD is slowly switched to the *cis*-1,4-polymerization mode [357].

The combination of Nd alcoholates with Mg alkyls also yields catalyst systems with a high *trans*-1,4-selectivity. In Nd-alcoholate-based catalyst systems magnesium alkyls are applied in considerably smaller amounts than aluminum alkyl cocatalysts ($n_{Mg}/n_{Nd} \geq 1$) [235].

Borane- and Borate-Based Cocatalysts

Perfluorinated phenylboranes and perfluorinated phenylborates are well-established activators in the metallocene-initiated polymerization of olefins. With the increasing commercial importance of metallocene technology for the polymerization of ethylene and the copolymerization of ethylene and 1-alkenes, perfluorinated phenylboranes and perfluorinated phenylborates became more readily accessible. As a consequence, a few studies on the influence of these highly fluorinated activators on Nd-catalysis are available in literature.

In the absence of halide donors activation of $Nd(O^tBu)_3 \cdot (AlMe_3)_3$ by $B(C_6F_5)_3$ yields BR with a *cis*-1,4-content below 75% [229, 230, 238, 239]. The activation of $Nd(N(SiMe_3)_2)_3$ by $B(C_6F_5)_3$), $[HNMe_2Ph]^+$ $[B(C_6F_5)_4]^-$ and $[CPh_3]^+$ $[B(C_6F_5)_4]^-$ has also been accomplished. With the boron activators higher *cis*-1,4-contents can be achieved than by activation with MAO [320].

The well-established perfluorinated phenylborate counter ion was also replaced by boron free cyclopentadienyl anions [358, 359]. By the sequential activation of NdV by Al-alkyls followed by $[C_6H_5N(CH_3)_2H]^+$ $[C_5(C_6F_5)_5]^-$ or by $[(C_6H_5)_3C]^+$ $[C_5(C_6F_5)_5]^-$ a cationic Nd species is generated which is stabilized by a non-coordinating boron-free cyclopentadienyl counter ion. As the cationic Nd-species is susceptible to side reactions with impurities the activation is performed in the presence of an excess of DIBAH. The replacement of the boron-containing tetrakis-pentafluorophenylborate anion by boron-free anions of the type $[C_5(C_6F_5)_5]^-$ yields BR with a *cis*-1,4-content in the range 74–91%.

2.1.3
Halide Donors

In the polymerization of dienes with Ziegler/Natta catalyst systems it is a well-established fact that the presence of halide donors is essential in order to achieve high catalytic activities and high *cis*-1,4-contents [360, 361]. The halide free catalyst system NdO/TIBA is a good example for a catalyst with a poor performance and a high *trans*-1,4-specificity [362, 363]. For various binary and ternary catalyst systems the qualitative impact of chlorides on the stereochemistry of BR is demonstrated in a series of fundamental experiments the results of which are summarized in (Table 5) [364].

Table 5 clearly demonstrates that *cis*-1,4-BR is only obtained if chloride is present in the catalyst system. In the absence of chloride BR with a high *trans*-1,4-content is obtained. These features were recently confirmed by Evans et al. who used well-defined Nd carboxylates for the polymerization of dienes [365]. According to Evans et al. halide atoms are transferred from the halide donor to Nd [366]. The role of halides for the achievement of high *cis*-1,4-contents was also demonstrated by Kwag et al. On the basis of density

Table 5 Influence of chloride in Nd catalyst systems on the stereochemistry of BR [364], copyright John Wiley & Sons Ltd., reproduced with permission

trans-1,4-Polymerization	cis-1,4-Polymerization
Nd(OR)$_3$/TIBA	Nd(OR)$_3$/DIBAC
Nd(COOR)$_3$/TIBA	Nd(COOR)$_3$/DIBAC
Nd(COOR)$_3$/Mg(Allyl)$_2$	Nd(COOR)$_3$/Mg(Allyl)$_2$/halide-donor
Nd(Ch$_2$Ph)$_3$/TIBA	Nd(Ch$_2$Ph)Cl$_2$/TIBA

functional calculations (B3LYP/CEP-31G) the authors suggest that by the coordination of chloride to Nd, 4f orbitals are induced, and the cis-coordination of dienes is enhanced. In addition, the coordination back-biting of the penultimate double bond of the poly(butadien)yl chain is favored. As a consequence anti-syn isomerization is blocked and cis-1,4-contents are increased (Sect. 4.3) [367].

According to Shen et al. fluorides are significantly less effective than chlorides for obtaining BR with a high cis-1,4-content [92]. The use of the halide donor BF$_3$·OEt$_2$ was examined by Kwag et al. In the catalyst system NdV/TIBA/BF$_3$·OEt$_2$ BR with a fairly high cis-1,4-content > 97% was obtained [368–370]. Contrary to Ziegler/Natta catalyst systems based on Co or Ni in Nd-mediated polymerizations of dienes the highest cis-1,4-contents are obtained with bromide and iodide donors (Table 6).

In Nd-based catalyst systems there are only small differences between bromide and chloride donors regarding polymerization activities and cis-1,4-contents. Chlorine-containing donors are preferred, especially in large-scale production, as chlorine donors are readily available and have a modest price.

Quite evidently, for binary catalyst systems of the type NdCl$_3$·D/AlR$_3$ an additional catalyst component which acts as a halide source is not required. NdCl$_3$ constitutes an appropriate halide source which provides halogen at a fixed molar ratio of $n_X/n_{Nd} = 3$.

Table 6 Influence of halides on the cis-1,4-content (in %) of Ziegler/Natta-BR [92], copyright Soc Chem Ind. Reproduced with permission. Permission is granted by John Wiley & Sons Ltd. on behalf of the SCI

Halide	Catalyst metal			
	Ti	Co	Ni	Nd
F	35	93	98	95.7
Cl	75	98	85	96.2
Br	87	91	80	96.8
I	93	50	10	96.7

For ternary catalyst systems a vast number of halide donors was investigated which renders a complete quotation impossible. It is important to note, however, that for a given halide, the actual halide source neither has a strong influence on catalyst activity nor on *cis*-1,4-contents. As halogen sources which are found in the literature cover the whole range from ionic halides to covalently bound halogen atoms the strength by which the halide is bound to the donor is not a critical factor.

For Nd-carboxylate-based systems the classes of halide donors comprise compounds with Al – X-bonds, C – X-bonds, Si – X-bonds and many others. Examples for often used alkyl aluminum chlorides are ^{i}Bu$_2$AlCl (DIBAC), EtAlCl$_2$ (EADC), Et$_2$AlCl (DEAC) and Et$_3$Al$_2$Cl$_3$ (EASC) [49, 174, 176, 178–180, 188, 205, 272]. One reason for the use of alkyl aluminum halides is their good solubility in hydrocarbon solvents. AlBr$_3$ which is not well soluble in these solvents was barely referenced during the past decade. Among the few papers on the use of AlBr$_3$ is a publication by Gehrke et al. [159].

In alkyl aluminum chlorides of the type R$_x$Al$_y$Cl$_z$ two different chemical moieties which cause alkylation as well as chlorination are present in one molecule. Therefore, R$_x$Al$_y$Cl$_z$-type activators do not require the separate addition of other halide donors in order to achieve high *cis*-1,4-contents. In Nd-based catalyst-systems the dual role of R$_x$Al$_y$Cl$_z$ compounds is demonstrated by Watanabe and Masuda [364]. These findings only hold true for Nd-based catalyst systems. For lanthanum-based catalyst systems Lee et al. found that the use of alkyl aluminum chlorides results in *trans*-1,4-polymerization (93–94%) [371]. However, usually, in Nd catalysts the alkylating power of R$_x$Al$_y$Cl$_z$ is not sufficient at the applied amounts of R$_x$Al$_y$Cl$_z$. Thus, an additional "standard" cocatalyst has to be added for the activation of the Nd precursor.

Examples of halide donors with a C – Cl-bond are ^{t}BuCl [175, 232], CCl$_4$ [157, 372–375] and CHCl$_3$ [376, 377]. Examples for Si-containing Cl-donors are SiCl$_4$ [157, 187] and RSiCl$_3$ [157, 378, 379]. TiCl$_4$ [157], PCl$_3$ and other halide donors can also be beneficially used but are less often referenced. Even tin halides have recently been mentioned by Kumho [380, 381].

In addition to the established halide sources JSR claims a series of unusual compounds such as: tin chloride, methylated chlorosilanes, beryllium halides, magnesium halides, calcium halides, barium halides, zinc halides, cadmium halides, mercury halides, manganese chloride, rhenium chloride, copper halides, silver halides and gold halides [382, 383]. In a subsequent JSR-patent the range of halide sources is extended to chlorine-, bromine- and iodine-containing organic compounds such as benzoyl chloride, benzyl chloride, xylene dichloride, benzoylbromide, benzylidene bromide, benzoyliodide etc. [384, 385]. Some of these halide donors particularly metal halides have a poor solubility in organic solvents. The solubility of these halides is claimed to be improved by appropriate additives which are identical or similar with the additives claimed by JSR to improve the solubility of

Nd-precursors. These additives comprise: alcohols, organic acids, esters of organic acids, lactones of organic acids, tributyl phosphate, tripenylphosphin, diethyl phosphinoethane, diphenyl phosphinoethane, acetylacetonate, triethylamine, N,N'-dimethyl acetoamide, tetrahydrofuran, diphenylether etc. In this context alkyl aluminum chloride which is prereacted with TBP is worth mentioning [386, 387].

Comparative studies on the impact of the different halides in ternary Nd-carboxylate catalyst systems are available from Wilson et al. [164, 174] and from Quirk et al. [187]. Wilson et al. evaluate the impact of the halide in Nd(carboxylate)$_3$/DIBAH/^{t}BuX. Polymerization activities decreased in the following order: ^{t}BuCl > ^{t}BuBr > ^{t}BuI [164]. A comparative study on the influence of various $R_xAl_yCl_z$ compounds was also performed by Wilson [174]. Polymerization activities decrease in the following order: AlMe$_2$Cl > AlMeCl$_2$ = Al$_2$Me$_3$Cl$_3$ and AlEt$_2$Cl > AlEtCl$_2$ = Al$_2$Et$_3$Cl$_3$ with all $R_xAl_yCl_z$ compounds being more active than ^{t}BuCl. Quirk et al. compared well-established halide donors in the ternary catalyst system with NdV/DIBAH/Cl donors at constant molar ratios $n_{Nd}/n_{Al}/n_{Cl} = 1/25/3$. Data on monomer conversion, M_w, M_w/M_n and cis-1,4-content are given in Table 7 [187]. According to Quirk the number of soluble catalytic centers is increased and average molar masses are reduced by the addition of TBP.

In Nd alcoholate-catalyzed polymerizations, as a rule of thumb, the same halide donors are applied at the same molar ratios as with Nd-carboxylate-based catalyst systems. In the literature, hydrocarbon soluble as well as hydrocarbon insoluble halide donors are combined with Nd-alcoholates. Examples are: benzyl chloride (BzCl) [37, 38], AlBr$_3$ [224, 225], AlEtCl$_2$ [226, 227], Et$_2$AlCl [231], ^{t}BuCl [231, 232] and Me$_3$SiCl [231].

Other neodymium precursors such as Nd phosphates and Nd allyls are most commonly applied in combination with alkyl aluminum chlorides. The principles outlined for the selection of the appropriate halide for Nd

Table 7 Influence of different chloride donors in the systems NdV/DIBAH/Cl-Donor [187], reprinted with permission of John Wiley & Sons, Inc.

Chloride source	Monomer conversion/%	M_w/ g·mol^{-1}	M_w/M_n	cis-1,4-Content/%
EASC[a]	84	240 000	4.3	97
DEAC[a]	83	220 000	5.0	96
DEAC[b]	95	190 000	3.7	97
^{t}BuCl[c]	77	380 000	5.9	97

[a] Without separate preformation of catalyst components prior to polymerization
[b] TBP-modified system [279]
[c] Preformation of the catalyst components at 20 °C for 24 °C

carboxylate- and Nd-alcoholate-based catalyst systems also apply for catalyst systems which are based on other Nd-precursors.

To our knowledge, only one example is known in which the addition of halide donors (Et$_2$AlCl, ^{t}BuCl and Me$_3$SiCl) was not beneficial for catalyst activity. Dong et al. studied the polymerization of IP with the catalyst system Nd(O^{i}Pr)$_3$/MMAO. Addition of the halide donors reduced catalyst activity in the following order: Et$_2$AlCl > ^{t}BuCl > Me$_3$SiCl [231].

2.1.4
Molar Ratio $n_{Cocatalyst}/n_{Nd}$

Variations of the amount of cocatalyst which are usually expressed by the molar ratio n_{Al}/n_{Nd} have a significant influence on polymerization rates, molar masses, MMDs and on the microstructures of the resulting polymers. These aspects are addressed in the following sections with a special emphasis on ternary catalyst systems. For ternary systems it has to be emphasized, however, that in many reports the ratio "n_{Al}/n_{Nd}" only accounts for the amount of aluminum alkyl cocatalyst and not for other Al-sources such as alkyl aluminum halides. Variations of the n_{Al}/n_{Nd}-ratios are also used for defined control of molar mass. This aspect is addressed in separate sections (Sects. 2.2.8 and 4.5).

Catalyst Activity
According to various studies binary catalyst systems of the type NdCl$_3 \cdot$ D$_n$/AlR$_3$ perform well when molar ratios are increased up to $n_{Al}/n_{Nd} = 40$. This performance for example applies to NdCl$_3 \cdot$ TBP/TIBA [133], NdCl$_3 \cdot$ 2THF/TEA [35], NdCl$_3 \cdot$ n 2-ethylhexanolate/TEA (n = 1.5 or 2.5) [114] and to NdCl$_3 \cdot$ 3 pentanolate/TEA [115]. For the catalyst system NdCl$_3 \cdot$ TBP/TIBA it is well-established that increases of n_{Al}/n_{Nd}-ratios > 40 do not result in further increases of catalyst activities. In this study by Iovu et al. n_{TIBA}/n_{Nd}-ratios had been varied in the range 20–100 [133].

In the context of ternary catalyst systems Throckmorton's pioneering work is worth mentioning although in this study Ce rather than Nd-based catalyst systems were used. For the two catalyst systems Ce octanoate/TIBA/DEAC and Ce octanoate/TIBA/EADC catalyst activities increased up to $n_{TIBA}/n_{Ce} = 20$. Further increases in the n_{TIBA}/n_{Ce}-ratios from 20–60 did not result in further activity improvements [34]. For a ternary didymium (Di)-based octanoate catalyst system Witte described a similar dependence. Within the n_{Al}/n_{Di}-range 20–40 significant increases of polymerization rates were reported. Further increases of the n_{Al}/n_{Di}-ratios did not have an additional effect [49].

Ternary Nd-carboxylate-based catalyst systems also show increasing catalyst activities with increasing n_{Al}/n_{Nd}-ratios. For NdV/TIBA/EASC the activity increase was described by a linear dependence in a double logarithmic plot (polymerization of BD at 45 °C) [204]. Also for NdO/TIBA/DEAC [188], NdV/TEA/SiCl$_4$ [187] and for NdV/DIBAH/EASC [178] increasing n_{Al}/n_{Nd}-ratios result in steady increases of polymerization activities. For the system NdV/DIBAH/EASC Quirk et al. report about a different dependence of reaction rates on n_{Al}/n_{Nd}-ratios. An increase of polymerization activities was only found in the n_{DIBAH}/n_{Nd}-range 10–40. Further increases of n_{DIBAH}/n_{Nd}-ratios from 40–60 left polymerization rates unaffected [191]. A leveling off of polymerization activities at high n_{Al}/n_{Nd}-ratios was also reported for Nd^{i}O/TEA/DEAC. Catalyst activities only increase in the n_{TEA}/n_{Nd}-range 15–20. Further increases of n_{TEA}/n_{Nd}-ratios from 20 to 60 result in only small increases of catalyst activities [168]. An activity maximum is found for Nd^{i}O/TEA/AlBr$_3$ at $n_{Al}/n_{Nd} = 40$ when n_{Al}/n_{Nd} is varied between 20–60 [159]. Maxima in the dependence of apparent rate constants were also reported for NdV/DIBAH/EASC at $n_{Al}/n_{Nd} = 40$. The location of the maximum was not affected by the n_{Cl}/n_{Nd}-ratios ($n_{Cl}/n_{Nd} = 2$ and 3) [179] but depends on the addition order of the different catalyst components, catalyst preformation and catalyst aging [169]. Increases of polymerization activities with increasing n_{TIBA}/n_{Nd}-ratios (60–120) were also reported for the gas-phase polymerization of BD. In this study the catalyst system NdN/TIBA/EASC was used [388].

The Nd-alcoholate-based binary catalyst systems show different dependencies on the variation of the n_{Al}/n_{Nd}-ratios. For Nd(O^{i}Pr)$_3$/MMAO [231] and also for Nd(O^{i}Pr)$_3$/[HNMe$_2$Ph]$^+$[B(C$_6$F$_5$)$_4$]$^-$/TIBA (1/1/30) [233] increasing polymerization rates were reported for the whole range of n_{Al}/n_{Nd}-ratios studied. The alcoholate-based binary catalyst system Nd(O^{i}Pr)$_3$/MAO exhibited an activity maximum at $n_{Al}/n_{Nd} = 80$ within the investigated range of n_{Al}/n_{Nd}-ratios = 10–80 [246] whereas the ternary alcoholate system Nd neopentanolate/DIBAH/EASC showed a steady increase of polymerization rates within the range of n_{Al}/n_{Nd}-ratios = 5–50 [272] (Sect. 2.1.1.10).

For the ternary neodymium phosphate-based catalyst systems (NdP/DIBAH/EASC) a totally different dependence of polymerization rates on n_{Al}/n_{Nd}-ratios was reported [264–269, 272]. According to these studies the catalyst system is highly active even at low n_{Al}/n_{Nd}-ratios < 5 and polymerization rates decrease with increasing n_{Al}/n_{Nd}-ratios within the n_{Al}/n_{Nd}-range = 5–50 (Sect. 2.1.1.10) [272].

For the Nd-amide-based system Nd(N(SiMe$_3$)$_2$)$_3$/TIBA/DEAC polymerization rates show a maximum at $n_{TIBA}/n_{Nd} = 40$ within the investigated range of n_{TIBA}/n_{Nd} ratios 0–40 [318].

The boranate system Nd(BH$_4$)$_3$(THF)$_3$/TEA exhibits a maximum at $n_{TEA}/n_{Nd} = 5$ in the investigated range of $n_{TEA}/n_{Nd} = 5$–50 [348].

Molar Mass and MMD

As variations in n_{Al}/n_{Nd}-ratios are used for the control of molar mass and the MMD this aspect is addressed in separate sections on molar mass regulation (Sect 2.2.8 and 4.5).

Microstructure

With a few exceptions, for binary as well as for ternary catalyst systems *cis*-1,4-contents decrease with increasing n_{Al}/n_{Nd}-ratios. This decrease of *cis*-1,4-contents can be best explained on the basis of a model given in Scheme 30 in Sect. 4.3. According to this model aluminum alkyls act as ligands competing for vacant Nd sites.

For binary catalyst systems the decrease of *cis*-1,4-contents on n_{Al}/n_{Nd}-ratios was confirmed for various NdX_3-based catalyst systems by Hsieh et al. [134, 139]. Also for $NdCl_3 \cdot$ n 2-ethylhexanolate/TEA (n = 1.5 or 2.5) increases of the n_{TEA}/n_{Nd}-ratios from 16.6 to 33 reduce *cis*-1,4-contents from ca. 99% to ca. 96% [114]. For the similar system $NdCl_3 \cdot$ 2THF/TEA *cis*-1,4-contents remain almost unaffected by changes of the n_{TEA}/n_{Nd}-ratio [35]. Contrary to this result Iovu et al. report about slight increases of *cis*-1,4-contents from ca. 92 to 96% with increasing concentrations of TIBA. In this study $NdCl_3 \cdot$ TBP/TIBA was used for the polymerization of IP [133].

To our knowledge, there is only one study in which the influence of magnesium alkyl cocatalysts on microstructure of poly(diene)s was investigated. Duvakina et al. used the catalyst system $NdCl_3$/TBP/Mg($^nC_4H_9$)($^iC_8H_{17}$) for BD polymerization and investigated the influence of the molar ratio n_{Mg}/n_{Nd} in the range of 6–60. This increase of n_{Mg}/n_{Nd} led to a decrease in *trans*-1,4-contents from 95 to 88–85% while the 1,2-content increased [135, 136].

For ternary catalyst systems the decrease in *cis*-1,4-contents with increasing n_{Al}/n_{Nd}-ratios is a common feature. The first report in which the decrease of *cis*-1,4-contents with increasing cocatalyst concentrations is addressed dates back to Throckmorton's study in which Ce-based catalyst systems were used [34]. The same dependence was confirmed in a subsequent study in which didymium octanoate-based catalysts were used [49]. In this study BD was polymerized and a decrease of the *cis*-1,4-content from 95% at $n_{Al}/n_{Di} = 20$ to 90% at $n_{Al}/n_{Di} = 60$ is reported.

Oehme et al. investigated the catalyst system Nd^iO/TEA/DEAC with respect to the ratio n_{TEA}/n_{Nd} which was varied in the range 15 to 60 [168]. The *cis*-1,4-content decreased from 94% at $n_{TEA}/n_{Nd} = 15$ to 86% at $n_{TEA}/n_{Nd} = 60$. Quirk et al. studied BD polymerization with the system NdV/DIBAH/EASC within the range of molar ratios $n_{DIBAH}/n_{Nd} = 10$–60 [191]. There was only a small dependence of *cis*-1,4-contents on the concentration of DIBAH. At low ratios $n_{DIBAH}/n_{Nd} = 10$–20 *cis*-1,4-contents were at 98–99%. An increase of n_{DIBAH}/n_{Nd}-ratios to 60 reduced *cis*-1,4-contents to 96%. For the same catalyst system NdV/DIBAH/EASC the influence of n_{Nd}/n_{DIBAH}-ratios on *cis*-1,4-content was confirmed [178]. In this study the decrease of the *cis*-

1,4-contents went along with the respective increase of *trans*-1,4-contents. The influence on the 1,2-content was much less pronounced.

For the two systems NdV/DIBAH/EASC and NdV/TIBA/EASC the influence of n_{Al}/n_{Nd}-ratios on BR microstructure were studied. For both systems the usual decrease of the *cis*-1,4-content with increasing n_{Al}/n_{Nd} was observed. At respective cocatalyst concentrations the use of TIBA results in higher *cis*-1,4-contents [205].

For the alcoholate-based system Nd(O^{i}Pr)$_3$/MAO the opposite trend was observed. The *cis*-1,4-content increased from 86.3% at $n_{Al}/n_{Nd} = 10$ to 91.3% at $n_{Al}/n_{Nd} = 80$ [246]. Also for Nd(O^{i}Pr)$_3$/MMAO variations of n_{Al}/n_{Nd}-ratios has a strong influence on *cis*-1,4 contents. At 30 °C the *cis*-1,4 content is only 82% at $n_{Al}/n_{Nd} = 50$, whereas an increase of n_{Al}/n_{Nd} to 100 increased the *cis*-1,4-content $\geq 91\%$ [231].

In the Nd amide system Nd(N(SiMe$_3$)$_2$)$_3$/TIBA/DEAC the percentage of *cis*-1,4-insertion decreases only from 98.2 to 97.6% when the ratio n_{TIBA}/n_{Nd} is increased from 10 to 40 [318].

2.1.5
Molar Ratio n_{Halide}/n_{Nd}

Halide donors constitute an essential component of ternary catalyst systems (Sect. 2.1.3). In these systems variations of the molar ratios n_X/n_{Nd} (X = halide) affect catalyst activities, molar masses, MMDs and the microstructures of the poly(diene)s.

For Nd-based catalyst systems the quantitative range of halide donors investigated falls between $0 < n_X/n_{Nd} \leq 10$. A maximum in catalyst activity is usually observed at molar n_X/n_{Nd}-ratios between 2 and 4, e.g. [49, 89, 168, 178, 187, 232, 272, 318]. Various factors such as addition order of the catalyst components, catalyst preformation and catalyst aging have an influence on the location of the optimum molar n_X/n_{Nd}-ratio (Sect. 2.1.6).

Catalyst Activity

There is unanimous agreement between various authors that there is a strong dependence of catalyst activity on the molar ratio n_X/n_{Nd}. All studies report about an activity maximum at a specific molar ratio n_X/n_{Nd}. This maximum results from the superposition of two opposing trends:

- Increase of the fraction of active Nd with increasing amounts of halide donor.
- Catalyst deactivation by the formation of insoluble NdX$_3$ with increasing amounts of halide donor.

In ternary catalyst systems halides play an important role in the activation process of the Nd-precursor (Sect. 4.2). At low n_X/n_{Nd}-ratios the contribution of the halide is small. Increasing n_X/n_{Nd}-ratios, on one hand result in catalyst activation. On the other hand, increasing n_X/n_{Nd}-ratios cause the formation of insoluble $NdCl_3$ which results in the reduction of catalyst activity. At very high n_X/n_{Nd}-ratios, however, catalyst activities are particularly low but not zero. Residual activities can be attributed to the fact that freshly precipitated NdX_3 is activated by aluminum alkyls and forms an active species. Definitive proof for the activation of NdX_3 by aluminum alkyls is the first generation of binary catalyst systems which are based on $NdCl_3/AlR_3$.

The first study in which the molar ratio n_X/n_{Ce} was systematically varied dates back to Throckmorton's pioneering work in which he used the catalyst system Ce octanoate/TIBA/halide donor [34]. Throckmorton compared the performance of the halide donors: Et_2AlF, Et_2AlBr, Et_2AlCl, $EtAlCl_2$ and Et_2AlI. For Et_2AlF the activity maximum was determined at $n_F/n_{Ce} \approx 45$. For Et_2AlBr a rather broad activity maximum was found at $n_{Br}/n_{Ce} = 1.5$ to 3. Et_2AlCl and $EtAlCl_2$ showed maximum catalyst activities at $n_X/n_{Ce} = 2$. Also for the ternary Di-based catalyst system (Di-octanoate/AlR_3/$AlBr_3$) an activity maximum was found. In the investigated range of molar ratios $n_{Br}/n_{Di} = 1.2-3$ catalyst activities showed a maximum at $n_{Br}/n_{Di} = 2.4$ [49].

For Nd-based ternary catalyst systems the dependence of polymerization rates on n_X/n_{Nd} ratios was in the focus of numerous studies. The results of these studies are summarized in Table 8.

As can be seen from the data in Table 8 the location of the activity maxima considerably differs between the various studies. At the first glance, the differences correlate with the chemical composition of the catalyst systems. It has to be emphasized, however, that these differences can be equally well attributed to differences in the experimental techniques between the various studies (catalyst preformation, catalyst aging, in-situ-preparation of catalysts etc.).

Table 8 Summary of studies on the dependence of the catalyst activity maxima on n_X/n_{Nd} ratios

Catalyst system	Molar ratios n_X/n_{Nd} at maximum of catalyst activity	Refs.
NdV/TIBA/EASC	2.2	[204]
NdV/DIBAH/SiCl$_4$	4.0	[187]
NdV/DIBAH/EASC	3.0	[191]
NdV/DIBAH/EASC	2.0–3.0	[178, 205]
Nd(RCOO)$_3$/DIBAH/^{t}BuCl	2.5–3.0	[89, 164]
Nd^{i}O/TEA/DEAC	2.5	[168]
Nd^{i}O/TEA/AlBr$_3$	2.4	[159]
Nd^{i}O/TIBA/DIBAC	2.5–3.0	[160]
Nd(N(SiMe$_3$)$_2$)$_3$/TIBA/DEAC	2.0–2.5	[318]
Nd(O^{i}Pr)$_3$/DIBAH/^{t}BuCl	3.0	[232]
Nd(O^{i}Pr)$_3$/MAO/^{t}BuCl	1.0–1.5	[232]

Molar Mass and Molar Mass Distribution

Various patterns regarding the dependence of molar mass on n_X/n_{Nd} ratios are observed:

- Molar mass goes through a maximum at a specific n_X/n_{Nd} ratio.
- Molar mass steadily decreases with increasing n_X/n_{Nd}-ratio.
- Molar mass depends on n_X/n_{Nd} ratios in a complex manner.

For catalyst systems for which a maximum in the dependence of molar mass is reported the molar mass maximum is often located at the same n_X/n_{Nd}-ratio as the activity maximum. This observation applies to Nd^{i}O/TIBA/DIBAC with maxima at $n_{Cl}/n_{Nd} = 2.5$–3.0 [160] and to Nd^{i}O/TEA/DEAC with a maximum at $n_{Cl}/n_{Nd} = 2.5$ [168]. For the catalyst system Nd(^{i}OPr)$_3$/DIBAH/^{t}BuCl a maximum molar mass was obtained at $n_{Cl}/n_{Nd} = 2$ whereas the activity maximum was at 3.0 [232]. For Nd(^{i}OPr)$_3$/MAO/^{t}BuCl the maximum of molar mass and catalyst activity coincide at a very low ratio of $n_{Cl}/n_{Nd} = 0.5$ [232].

With the MMAO-activated catalyst system Nd(^{i}OPr)$_3$/MMAO/halide donor a unique dependence of molar mass on the ratio of n_{Cl}/n_{Nd} was obtained. Molar mass steadily decreases with increasing n_{Cl}/n_{Nd}-ratios (between 0.5 to 2.0) regardless of the type of chlorine source (Et$_2$AlCl, ^{t}BuCl and Me$_3$SiCl) [231].

For the catalyst system NdV/DIBAH/EASC a rather complex dependence of molar mass on n_{Cl}/n_{Nd}-ratios was observed. Further insight comes from an

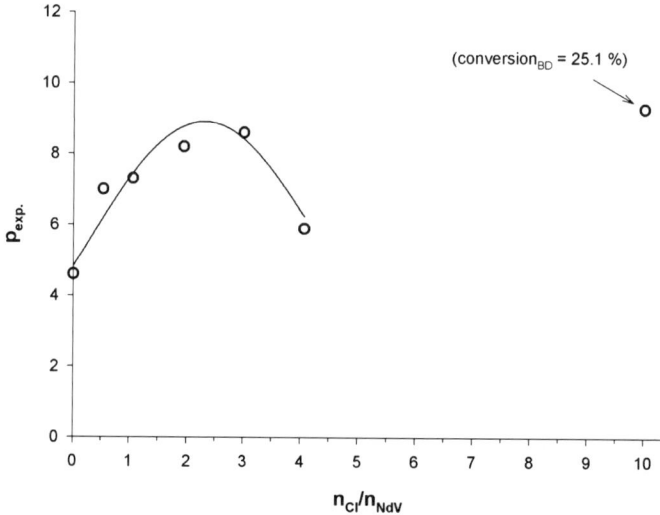

Fig. 5 Dependence of the formal number of chains per Nd atom ($p_{exp.}$) on the molar ratio n_{Cl}/n_{NdV} for the polymerization of BD (100% monomer conversion) with the catalyst system NdV/DIBAH/EASC [205]

analysis of the dependence of the formal number of polymer chains generated by one Nd-atom ($p_{exp.}$) (Fig. 5) [178, 205].

This dependence shows a maximum at the same molar ratio as the respective polymerization rates ($n_{Cl}/n_{NdV} = 2$–3). Because of the coincidence of the two maxima it was concluded that there is a correlation of $p_{exp.}$ and the number of active centers. In addition to the maximum of $p_{exp.}$ at $n_{Cl}/n_{NdV} = 2$–3 another high $p_{exp.}$ value is obtained at $n_{Cl}/n_{NdV} = 10$. Contrary to the maximum $n_{Cl}/n_{NdV} = 2$–3 the high $p_{exp.}$-value at $n_{Cl}/n_{NdV} = 10$ goes along with an extremely low polymerization rate. The discrepancy between catalytic activity and $p_{exp.}$ at $n_{Cl}/n_{NdV} = 10$ is explained by the presence of a second active species which results from the activation of $NdCl_3$ by DIBAH [205].

PDI and the modality of MMDs also strongly depend on n_{Cl}/n_{Nd}-ratios and on the catalyst systems used. Catalyst systems which yield monomodal distributions at low n_{Cl}/n_{Nd}-ratios and broad distributions which are often bimodal at high n_{Cl}/n_{Nd}-ratios are based on: NdV/DIBAH/SiCl$_4$ [187], NdV/DIBAH/EASC [191] and Nd(^{i}OPr)$_3$/DIBAH/^{t}BuCl [232].

A catalyst system for which PDI is almost unaffected by the molar ratio n_{Cl}/n_{Nd} is based on Nd(^{i}OPr)$_3$/MAO/^{t}BuCl [232], whereas a PDI minimum (PDI = 3.2 to 4.0 at $n_{Cl}/n_{Nd} = 2.5$) was reported for Nd(OCOR)$_3$/DIBAH/^{t}BuCl [89, 164].

For the catalyst system NdV/DIBAH/EASC the PDI increases with increasing molar ratios from PDI = 2.3 at $n_{Cl}/n_{NdV} = 1$ via PDI = 2.4 at $n_{Cl}/n_{NdV} = 2$ and PDI = 2.8 at $n_{Cl}/n_{NdV} = 3$ to PDI = 3.0 at $n_{Cl}/n_{NdV} = 4$ [205].

For the whole n_{Cl}/n_{Nd} range studied ($n_{Cl}/n_{NdV} = 2$ to 5) BR with a bimodal distribution is obtained with the catalyst system Nd(N(SiMe$_3$)$_2$)$_3$/TIBA/DEAC. The bimodality of the polymer is attributed to two active centers. Contrary to the previous studies a larger fraction of the high molar mass BR is found at low chlorine contents ($n_{Cl}/n_{Nd} = 2$) [318].

In summary, the studies on the impact of n_X/n_{NdV}-ratios on polymerization rate, molar mass, MMD and PDI support the view on the existence of several active species. The relative concentrations of the active species are governed by the molar ratios n_X/n_{Nd}. Particularly the occurrence of bimodal distributions provides strong evidence for the existence of two distinct active catalyst species.

Microstructure

The molar ratio n_X/n_{Nd} also influences the microstructure of poly(diene)s. Unfortunately, these studies are not consistent regarding the dependence of *cis*-1,4-content on n_X/n_{Nd} ratios.

None or only a negligible effect of the n_X/n_{Nd} ratio on *cis*-1,4-contents was observed for the catalyst systems NdV/DIBAH/SiCl$_4$ [187] and NdV/DIBAH/EASC [191]. In both studies the *cis*-1,4-contents only slightly vary between 96–97%. Also for the catalyst system Nd^{i}O/TEA/DEAC the impact on *cis*-1,4-contents is small (85–88%) [168]. Small effects were also reported for the catalyst systems NdA/DIBAH/EASC and NdP/DIBAH/EASC [178, 205, 272].

Maxima in *cis*-1,4-contents were observed for the catalyst systems NdV/DIBAH/EASC [178, 205, 272], Nd(O^{i}Pr)$_3$/DIBAH/^{t}BuCl [232], Nd(O^{i}Pr)$_3$/MAO/^{t}BuCl [232] and Nd(N(SiMe$_3$)$_2$)$_3$/TIBA/DEAC [318].

For the catalyst system NdV/DIBAH/EASC the impact of the halide concentration on *cis*-1,4-content is shown in Table 9. Within experimental error the optimum of the *cis*-1,4-contents at $n_X/n_{Nd} = 2-4$ coincides with the activity maximum at 2–3 (Table 8).

As can be seen from Table 9 an increase in the n_{Cl}/n_{Nd}-ratio from 0.5 to 3.0 results in an increase of the *cis*-1,4-content. A further increase of the n_{Cl}/n_{Nd}-ratio to 4.0 and 10.0 decreases the *cis*-1,4-content. BR which is obtained without halide donor (at a very low catalyst activity) exhibits a unique microstructural composition: 71.9% *cis*-1,4, 21.2% *trans*-1,4 and 6.9% 1,2. This observation corroborates the fact that high-*cis*-1,4-poly(butadiene) products only can be obtained in the presence of halide donors.

Table 9 Dependence of BR-microstructure on n_X/n_{Nd}-ratio for the catalyst systems NdV/DIBAH/EASC [178, 205, 272], reproduced by permission of Taylor & Francis Group, LLC., http://www.taylorandfrancis.com

Molar ratio n_X/n_{Nd}	*cis*-1,4/ %	*trans*-1,4/ %	1,2/ %
0.00	71.9	21.2	6.9
0.50	92.5	6.6	0.9
0.67	93.1	5.8	1.1
1.00	94.5	4.7	0.8
1.33	95.0	3.9	1.1
2.00	96.1	3.1	0.8
3.00	97.1	2.1	0.8
4.00	96.2	2.7	1.1
10.00	94.3	4.4	1.3

Table 10 Dependence of BR-microstructure on n_X/n_{Nd}-ratio for the catalyst system Nd neopentanolate/DIBAH/EASC [205, 272], reproduced by permission of Taylor & Francis Group, LLC., http://www.taylorandfrancis.com

Molar ratio n_X/n_{Nd}	*cis*-1,4/ %	*trans*-1,4/ %	1,2/ %
1.00	94.2	5.3	0.5
2.00	95.3	4.2	0.5
3.00	96.0	3.2	0.8

Table 11 Dependence of BR-microstructure on n_X/n_{Nd} ratio for the catalyst system NdP/DIBAH/EASC [205, 272], reproduced by permission of Taylor & Francis Group, LLC., http://www.taylorandfrancis.com

Molar ratio n_X/n_{Nd}	cis-1,4/ %	trans-1,4/ %	1,2/ %
1.00	93.4	6.3	0.3
2.00	92.5	7.2	0.3
3.00	gel	gel	gel

For the catalyst system Nd(O^{i}Pr)$_3$/DIBAH/^{t}BuCl the maximum cis-1,4-content ($\approx$ 96%) is at n_{Cl}/n_{Nd} = 3 within the n_{Cl}/n_{Nd}-range 0–8 investigated [232]. Also for Nd neopentanolate/DIBAH/EASC the cis-1,4-content is at $\approx$ 96% for n_{Cl}/n_{Nd} = 3 (Table 10) [205, 272].

MMAO activation of the ternary Nd alcoholate catalyst system Nd(O^{i}Pr)$_3$/MMAO/halide donor results in comparable cis-1,4-contents in the range of 91% to 96%, irrespective of the chloride source (Et$_2$AlCl, ^{t}BuCl and Me$_3$SiCl) [231]. For the catalyst system Nd(O^{i}Pr)$_3$/MAO/^{t}BuCl the cis-1,4-content is reduced to $\approx$ 93% at n_{Cl}/n_{Nd} = 2–2.5 [232].

For the Nd-phosphate-based catalyst system NdP/DIBAH/EASC cis-1,4-contents decrease with increasing n_{Cl}/n_{Nd}-ratios 1.0–2.0 from 93.4 to 92.5. For even higher n_{Cl}/n_{Nd}-ratios gelled polymer is obtained (Table 11) [205, 272].

For the Nd-amide-based catalyst system Nd(N(SiMe$_3$)$_2$)$_3$/TIBA/DEAC a maximum in cis-1,4-content is found at n_{Cl}/n_{Nd} = 2 (97.6%) [318].

2.1.6
Addition Order of Catalyst Components, Catalyst Preformation and Catalyst Aging

Nd-based catalyst systems comprise a variety of catalyst components. Therefore, the order of addition of these components to the monomer solution is not trivial. The situation is even more complicated as some or even all of the catalyst components can be "prereacted" prior to the addition to the monomer solution. In "catalyst prereaction" which is also referred to as "catalyst preformation" or "catalyst aging" the catalyst components are mixed and reacted prior to the addition to the monomer solution. Reaction time, reaction temperature and the presence of small amounts of monomer play decisive roles. Alternatively, the catalyst components are separately added to the monomer solution. In this approach the active catalyst is generated "in-situ".

During catalyst preformation the concentrations of the catalyst components are usually higher than during the in-situ-preparation of catalysts. Therefore, the build-up-reaction of the active catalyst species occurs more rapidly [162, 165, 166]. The parameters associated with catalyst preformation and catalyst aging have a significant impact on catalyst homogeneity, polymerization activities, molar masses and MMDs. By appropriate adjustments of catalyst preformation and catalyst aging PDI can be varied from relatively narrow ($M_w/M_n < 3$) to broad ($M_w/M_n > 7$). As these variations in PDI are caused by relatively small and simple changes careful attention to the details of the experimental procedures are required. In this context it is worth mentioning that some aspects of catalyst preformation were already acknowledged in the first patent filed by Enoxy on the use of neodymium catalysts for diene polymerization [37, 38].

The literature referring to catalyst preformation of binary catalyst systems is not addressed in this work. In this review the available studies on ternary catalyst systems are summarized. The available literature is discussed in the following two subsections "*Preformation without Monomer*" and "*Preformation in the Presence of Diene Monomers*".

Preformation without Monomer

A unique variant of catalyst preformation in the absence of dienes was described by Enichem in an early patent on Nd-BR. The active Nd catalyst was prepared by the reaction of neodymium oxide with carboxylic acid and ^{t}BuCl in vaseline at 80 °C. Subsequently, aqueous HCl was added at 80 °C. Finally, the addition of the aluminum alkyl cocatalyst yielded the active Nd catalyst [389, 390].

For the preparation of BR with a narrow MMD the influence of catalyst addition order and catalyst preformation/aging was first recognized by Jenkins et al. [391, 392]. Catalyst preformation at − 15 °C to − 60 °C and catalyst aging at − 20 °C to − 40 °C were essential to obtain BR with a relatively narrow MMD ($M_w/M_n < 3.1$) and low intrinsic viscosities. These results were obtained with the catalyst systems NdV or NdN/TEA/^{t}BuCl. The sequence of addition occurred in the following order: TEA + Nd component + ^{t}BuCl.

Systematic studies on the addition order of catalyst components focused on the catalyst system NdV/DIBAH/^{t}BuCl. The variations included all six conceivable addition orders. It was found that the order in which the halide donor ^{t}BuCl was added plays a decisive role concerning the heterogeneity of the catalyst and its activity [158]:

1. DIBAH + NdV + ^{t}BuCl
2. NdV + DIBAH + ^{t}BuCl
3. NdV + ^{t}BuCl + DIBAH
4. ^{t}BuCl + NdV + DIBAH
5. DIBAH + ^{t}BuCl + NdV
6. ^{t}BuCl + DIBAH + NdV

Catalyst solutions which were homogeneous to the naked eye were only obtained by the routes (1) and (2) in which ^{t}BuCl was added at the end of the sequence. Routes (3) and (4) in which ^{t}BuCl was added at an earlier stage resulted in catalyst mixtures which formed a precipitate after 20 h of aging time. Addition orders (5) and (6) resulted in the immediate formation of heterogeneous catalyst slurries. The initial catalyst activities followed the order of catalyst heterogeneity: (5) > (3) > (1). The heterogeneous catalysts showed a high initial activity and a reduced activity at later stages of the polymerization. The catalyst activity after longer polymerization periods followed the order of catalyst homogeneity: (1) > (3) > (5). Wilson and Jenkins explained these observations by the presence of two active catalyst species. Initial polymerization was assigned to insoluble particles mainly consisting of $NdCl_3$. The second slow type of chain growth which occurred in a quasi-living manner was attributed to soluble catalyst species. The distribution of molar masses was broader for heterogeneous catalysts and followed the orders of addition: (6) > (4) > (1). Route (1) resulted in a catalyst which was stable for a number of weeks [89]. For the addition orders (1) and (2) no difference in reaction kinetics and MMD was observed.

Oehme et al. investigated the influence of addition orders for the catalyst system $Nd(OCOR)_3$/TIBA/EASC [163]. The catalyst components were mixed at 25 °C and aged for 30 min at the same temperature. Catalyst activities and molar masses decreased in the order (1) > (2) > (3):

1. $Nd(OCOR)_3$ + EASC + TIBA
2. TIBA + $Nd(OCOR)_3$ + EASC
3. TIBA + EASC + $Nd(OCOR)_3$

The cis-1,4-content was almost unaffected by the addition order. At low catalyst concentrations the addition order had a stronger influence on catalyst activity. An impact of the addition order on the modality of MMD was found, too. According to the authors these observations gave an indication for the formation of different catalyst centers. Further investigations by Gehrke et al. focused on the influence of catalyst aging on molar mass. The catalyst system Nd^iO/TIBA/DEAC was used in this study. An increase of the preformation temperature resulted in an increase of molar mass. This increase was assumed to be caused by catalyst deactivation. For the achievement of low molar masses catalyst aging at < 40 °C/5 min was recommended [160]. In another study by Gehrke et al. the impact of catalyst preformation on the modality of the MMD was investigated. For this study the catalyst system $Nd(OCOR)_3$/TIBA/EASC (n_{TIBA}/n_{Nd} = 25 and n_{Cl}/n_{Nd} = 3) was used [169]. In the first step $Nd(OCOR)_3$ and EASC were prereacted; subsequently, TIBA was added. An increase of catalyst aging time from 0 min to 1080 min reduced catalyst activity and changed the modality of MMD from bimodal to unimodal. These changes were attributed to the existence of two catalyst species. The species which produced the low molar mass BR was supposedly deactivated by the extension of preformation time. The second species which produced the BR fraction with the higher molar mass was less sensitive to an increase of

alyst preformation systematically increases catalyst activities irrespective of the preformation temperature.

For the catalyst system NdV/DIBAH/EASC Quirk et al. studied the influence of the order of catalyst addition on monomer conversion, M_w, M_w/M_n and cis-1,4-content without prior preformation of the catalyst system [191]. The results of this study are given in Table 14.

As can be seen from Table 14 the addition order affects reaction rate, molar mass and PDI. The authors suggested that in-situ activation results in the formation of two types of active species the relative concentrations of which were governed by the addition sequence. Addition order (1) EASC + NdV + DIBAH promotes the formation of insoluble species which produced polymer with a broad, bimodal MMD (PDI = 7.5) at low catalyst activity. Addition order (3) DIBAH + NdV + EASC leads to the formation of more soluble catalyst species which exhibit increased catalyst activity and produce BR with a monomodal MMD (PDI = 3.4). The influence of the addition order on cis-1,4-content is negligible.

Quirk et al. also studied the impact of catalyst aging times between 24–72 h for the catalyst system NdV/DIBAH/SiCl$_4$ (1/25/1). Catalyst preformation was carried out by mixing the three components in the order DIBAH + NdV + SiCl$_4$ at 20 °C. Catalyst activities are increased by an extension of aging times and by the presence of IP monomer ($n_{IP}/n_{Nd} = 5$) during the preformation step. Activities of preformed and aged catalysts are superior over in-situ prepared catalysts [187].

In the patent literature catalyst preformation in the presence of dienes is exploited for the production of BR with a low PDI. PDIs < 3.1 are obtained for the catalyst system NdN/DIBAH/SiX$_4$ or R$_n$-SiX$_{4-n}$ ($n = 1, 2, 3$). The preferred order of addition is NdN + DIBAH + diene + SiX$_4$ or R$_n$-SiX$_{4-n}$ ($n = 1, 2, 3$). Catalyst preformation at 0–30 °C/5 h–6 d is recommended [393, 394].

Catalyst preformation in the presence of dienes is also favorable for the preparation of Nd-BR with a low solution viscosity and a low cold flow. These properties result from a high degree of branching and a low PDI. Nd-BR which meets these requirements is prepared with the catalyst system NdV/DIBAH/EASC. Catalyst preformation in the presence of IP and dimethyl-di-2,4-pentadienyl-(E,E)-silane (DMDPS, Scheme 11) are essential features. The optimized addition order for the catalyst components is: DIBAH + IP + NdV. This catalyst mixture was aged at 50 °C for 90 min. Prior to the addition of the Cl-donor it is cooled to 5 °C. The catalyst was subse-

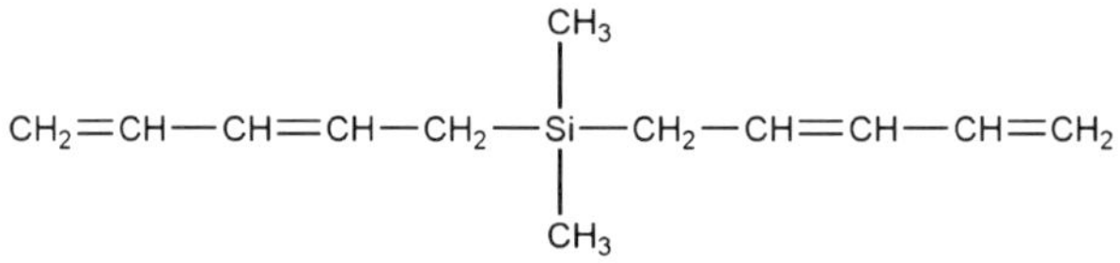

$$CH_2=CH-CH=CH-CH_2-\underset{\underset{CH_3}{|}}{\overset{\overset{CH_3}{|}}{Si}}-CH_2-CH=CH-CH=CH_2$$

Scheme 11 Dimethyl-di-2,4-pentadienyl-(E,E)-silane (DMDPS)

Table 15 Influence of catalyst preformation and DMDPS on Nd-BR properties [395, 396]

Preformation of catalyst components	Monomers present during polymerization	Nd-BR Properties		
		Mooney viscosity [7] (ML 1 + 4)/ 100 °C MU	Solution viscosity (5.43 wt. % in toluene at 23 °C)/ mPas	Cold flow (50 °C)/ mg·min^{-1}
No	BD	46	770	9.1
No	BD + DMDPS	41	330	4.9
Yes [1]	BD	45	290	22.2
Yes [1]	BD + DMDPS [2]	39	85	6.7
Yes [1]	BD + DMDPS	80	190	1.2

[1] Catalyst preformation in the presence of isoprene
[2] Addition of DIBAH prior to polymerization

quently aged for 12 h at ambient temperature. The polymerization of BD in the presence of DMDPS yields Nd-BR with the requested property profile (Table 15) [395, 396].

BR with narrow MMDs ($M_w/M_n \geq 3.5$) and a low solution viscosity can also be obtained by the use of a multi-component catalyst system which comprises the following six components: (1) Nd-salt, (2) additive for the improvement of Nd-solubility, (3) aluminum-based halide donor, (4) alumoxane, (5) aluminum (hydrido) alkyl, and (6) diene. The solubility of the Nd-salt is improved by acetylacetone, tetrahydrofuran, pyridine, N,N-dimethylformamide, thiophene, diphenylether, triethylamine, organo-phosphoric compounds and mono- or bivalent alcohols (component 2). The catalyst components are prereacted for at least 30 seconds at 20–80 °C. Catalyst aging is preferably performed in the presence of a small amount of diene [397, 398]. As the additives employed for the increase of the solubility of Nd salts exhibit electron-donating properties it can be equally well speculated that poisoning of selective catalyst sites favors the formation of polymers with a low PDI.

From the wealth of observations on catalyst preformation and catalyst aging reported in the literature it is extremely difficult to draw definitive conclusions about the nature of the reactions which occur during this process step. Nevertheless, it is rather evident that contrary to metallocene catalysts, the majority of Nd-based catalyst systems do not exhibit the features of single site catalysts (SSC). As bimodal and multimodal MMDs are observed for Nd-based catalyst systems various active species (or active sites) are present

---

[7] Mooney viscosity is the torque of a melt of rubber determined in a Mooney rheometer. The torque is given in Mooney units (MU). It is determined on pure rubber or rubber compounds at elevated temperature (usually between 100 °C to 150 °C) after a well-defined heating procedure (usually 1 min without application of shear and subsequently 4 to 8 min with the application of shear). The method is described in detail in DIN 53523; ASTM D 1646; ISO 289

during polymerization. PDI and the modality of MMDs depend on the relative concentration of the different active species and on the lifetime of theses species during the course of monomer conversion. Heterogeneous as well as soluble species are involved which exhibit different activities and yield different molar masses. The homogeneity of the catalyst systems greatly depends on the conditions applied during catalyst preformation and catalyst aging. Reaction conditions which favor the formation of $NdCl_3$ lead to reaction patterns that can be attributed to heterogeneous species. Catalyst heterogeneity depends on the point of addition of the halide donor. The formation of homogeneous species is favored in the presence of monomer during catalyst preformation. Also the degree of catalyst homogeneity is important for the livingness of the polymerization. In combination with the optimization of catalyst preformation and catalyst aging selective poisoning of specific catalyst species or catalyst sites is another important aspect regarding the preparation of narrow or even monomodal MMD.

Further aspects of catalyst preformation and catalyst aging are covered in the context of supported catalysts in Sects. 2.1.7 and 3.2.

2.1.7
Supported Catalysts

The use of supported Nd catalyst systems is described for the polymerization of BD and IP in solution, gas phase and slurry. In this section the use of supported catalyst is reviewed for solution processes, only. The use of supported catalysts for gas-phase polymerization (Sect. 3.2) and for slurry polymerization (Sect. 3.1) is addressed in the cited sections.

As early as in 1985, supported catalysts were described for the use in a solution process [399]. Bergbreiter et al. used catalyst supports on the basis of divinylbenzene-styrene-copolymers as well as on polyethylene. These authors found that the use of supported catalysts has no influence on the stereospecificity of diene polymerization.

Barbotin et al. tested various silica supports [400]. Special emphasis was given to the chemical fixation of the Nd-component. The activity of the supported catalysts was optimized for the polymerization of BD and IP in toluene. As a result of these optimizations, the supported catalysts performed better than the respective unsupported catalysts [401, 402]. An increase of polymerization activities was also observed for $MgCl_2$ supports [403, 404]. In this study $NdCl_3$ was attached to $MgCl_2$ in refluxing THF. The $MgCl_2/NdCl_3$-based reaction product was isolated and purified prior to the preformation with DEAC in heptane. The precipitate obtained in the preformation reaction was used for the polymerization of BD after TIBA was added. BR with a high cis-1,4-content (98.2%) was obtained. Catalyst supports on the basis of poly(styrene-co-acrylic acid) and poly(ethylene-co-acrylic acid) were also used for the polymerization of BD [405]. Nd was bound by the acrylic acid moieties. According to Yu et al. catalyst activities depend on the acrylic acid content, the morphology of the polymer supports, the n_{Nd}/n_{COOH}-ratio as well as on the distribution of Nd on the polymer support.

The immobilization of Nd tetramethyl aluminate complexes on porous silica supports (MCM-48) was reported by Fischbach et al. [406, 407]. In this study activation was accomplished by DEAC. The polymerization of IP in hexane yielded IR with high *cis*-1,4-contents (> 99%).

Tris-allyl-neodymium Nd(η^3– C_3H_5)·dioxane which performs as a single site catalyst in solution polymerization was heterogenized on various silica supports which differed in specific surface area and pore volume. The catalyst was activated by MAO. In the solution polymerization the best of the supported catalysts was 100 times more active (determined by the rate constant) than the respective unsupported catalyst [408].

2.1.8
Other Additives

Literature describes a series of additives which are deliberately added to Nd catalyst systems. These additives aim at an increase of the solubility of catalyst components in organic solvents and at an increase of polymerization activity. In addition, additives are used to influence the width and the modality of MMDs. By the use of additives the number of catalyst components is increased from two (binary catalyst systems) or three (ternary catalyst systems) to multi-component systems which comprise up to seven or even eight different components. Additives reported in the literature and the influence of these additives is discussed in the following section.

Electron Donors

The impact of electron donor ligands in NdX_3-systems has already been discussed in Sect. 2.1.1.1. Ligands such as alcohols, trialkyl phosphates, alkyl sulfoxides, alkyl amides, THF, *N*-oxides, pyridine etc. are added in order to facilitate water removal from $NdCl_3 \cdot 6H_2O$ by azeotropic distillation and in order to increase solubility and activity of $NdCl_3$-based catalyst systems in organic solvents.

The solubility of neodymium carboxylates in organic solvents is also improved by the addition of electron donors such as acetylacetone, tetrahydrofuran, *N,N'*-dimethylformamide, thiophene, diphenylether, triethylamine, pyridine, organic phosphorus compounds etc. Also the storage stability of neodymium carboxylates in organic solutions (reduction of sediment formation) is increased by these additives. Mixtures of the Nd-precursor and the respective additives are reacted in the temperature range 0–80 °C. The sequential addition of Al-compound and halide donor yield the active polymerization catalysts [409, 410].

The solubility of rare earth hydrides in organic solvents is increased by appropriate additives, too. For this purpose the hydrides are reacted with electron-donor ligands such as alkyl benzoates, alkyl propionates, alkyl toluates, dialkylethers, cyclic ethers, alkylated amines, N,N'-dimethylacetamide, N-methyl-2-pyrrolidone, trialkyl and triaryl phosphines, trialkyl phosphates and triaryl phosphates, trialkyl phosphates, hexamethylphosphoric triamide, dimethyl sulfoxide, etc. Prior to use as a polymerization catalyst the prereacted mixture of the rare earth hydride plus the additive is prereacted with Al-alkyl-based Lewis acids in the temperature range of 60–100 °C for 10 min to 24 h [351, 352].

For the preparation of Nd-BR with narrow MMDs a catalyst system which comprises five components is applied: Nd-salt/electron donor additive/aluminum based halide donor/alumoxane/aluminum (hydrido) alkyl. The following additives are recommended in order to improve the solubility of the Nd salt: acetylacetone, tetrahydrofuran, pyridine, N,N-dimethylformamide, thiophene, diphenylether, triethylamine, organo-phosphoric compounds and mono- or bivalent alcohols. The catalyst components are prereacted for at least 30 seconds in the temperature range of 20–80 °C and catalyst aging is preferably performed in the presence of a small amount of diene [397, 398].

Additives are also used to improve the solubility of halide donors [382, 383]. Metal(II) halides such as magnesium chloride, calcium chloride, barium chloride, manganese chloride, zinc chloride and copper chloride etc. are used as halide sources. In order to increase the solubility of the halides they are reacted with electron donors which have been previously described for the increase of solubility of Nd-components [338, 339]. The number of catalyst components is further increased if two Al-compounds (alumoxane + aluminum (hydrido) alkyl) are used. In addition, a small amount of diene can also be present during the preformation of the different catalyst components as described by JSR. In some catalyst systems the total number of components reaches up to eight [338, 339]. Such complex catalyst systems are also referred to in other JSR patents [384, 385] (Sect. 2.2.6).

Water

In standard Ziegler/Natta-systems traces of water are considered an impurity, even though water is known to have a beneficial effect in the stereospecific polymerization by Co-based catalyst systems [411]. So far, in Nd-catalyzed polymerizations the influence of water has received only little attention. For the system NdV/EASC/TIBA it was mentioned that water had no impact as long as the total water content in the polymerization mixture stayed below 5 ppm. In this study water was added as a byproduct of the organic solution of Nd versatate. The water content of different lots varied in the range 25–700 ppm [412]. Other authors also recommended a low water content in the organic solutions of Nd carboxylates (< 2000 ppm, preferably

Table 16 Influence of water content in the BD polymerization with the catalyst system NdV/DIBAH/EASC [191], reprinted with permission from Elsevier

n_{H_2O}/n_{Nd}	Monomer conversion/%	M_w/ g·mol^{-1}	M_w/M_n	cis-1,4-Content/%
0.008	55	560 000	6.7	99
0.030	69	470 000	5.5	98
0.051	84	210 000	3.8	98
0.11	86	230 000	3.5	99
0.76	77	250 000	4.7	98
1.51	68	300 000	4.2	98

< 1000 ppm) [413, 414]. Organic solutions of Nd neododecanoates (NdV) with considerably higher water contents up to 20 000 ppm were also prepared. The impact of water, however, was not studied in detail [213–216]. There is only one study available in which limited amounts of water are claimed to be beneficial for catalytic activities ($n_{Al}/n_{H2O} = 2/1$) [415, 416].

For the catalyst system NdV/EASC/DIBAH the impact of water on monomer conversion, M_w, polydispersity and cis-1,4-content was systematically studied (Table 16) [191]. With increasing amounts of water catalyst activity passes through a maximum whereas M_w and M_w/M_n pass through a minimum. It has to be mentioned, however, that the overall effect of water on reaction rate and polymer properties are relatively small. In this study it is also shown that water has no influence on cis-1,4-contents [191].

It might be speculated that water reacts with the aluminum alkyl cocatalyst and forms alumoxanes which might also contribute to overall catalyst activity.

Carboxylic Acids

Quirk et al. studied the impact of excess versatic acid [191]. According to this study an excess of versatic acid $n_{excess\ acid}/n_{Nd} > 0.22$ reduces monomer conversion at a fixed polymerization time (1 h) and leads to an increase in M_w as well as PDI and has no effect on cis-1,4-contents (Table 17).

As shown by Kwag et al. the stoichiometric ratio of one molecule of carboxylic acid per one molecule of Nd-carboxylate leads to the fragmentation of Nd carboxylate oligomers. The fragmentation renders every Nd center accessible for catalyst activation and results in an exceptionally high catalytic activity [156, 417–419]. The monomeric Nd precursor is composed of Nd(neodecanoate)$_3$·(neodecanoic acid). The non-associated monomeric species Nd(neodecanoate)$_3$·(neodecanoic acid) is obtained by ligand exchange starting from Nd acetate and neodecanoic acid in boiling chlorobenzene [156]. Beside the described strategies, appropriate amounts of water and carboxylic acids are deliberately added to organic solutions of Nd carboxylates in order to decrease the solution viscosity [420, 421].

Table 17 Influence of excess versatic acid in the BD polymerization with the catalyst system NdV/DIBAH/EASC [191], reprinted with permission from Elsevier

$n_{\text{excess acid}}/n_{\text{Nd}}$	Monomer conversion/%	M_w/ g·mol^{-1}	M_w/M_n	cis-1,4- Content/%
0.22	88	220 000	3.4	97
0.54	63	330 000	4.7	97
0.91	59	310 000	5.4	98
1.43	53	380 000	5.8	97
1.66	55	370 000	5.2	97

These studies on the influence of water and carboxylic acids show that Nd-based catalyst systems are relatively robust to water and excess versatic acid. From an industrial point of view this is an extremely important aspect.

Ethers
Ethers (e.g. THF and diethyl ether) in high quantities are known to be strong catalyst poisons which interact with vacant Nd sites. This was proven by Wilson and Hsieh et al. who used ethers as solvents in Nd-catalyzed diene polymerization [134, 175]. On the other hand ethers are applied beneficially in small amounts in Nd-halide-based catalyst systems in order to solubilize active Nd species (Sect 2.1.1.1).

Acetylenes and Allenes
Acetylenes and/or allenes that can be present as impurities in BD have a detrimental effect on catalyst activities. In order to obtain reproducible results these impurities are removed by the pretreatment of the monomer with a catalyst suspension [165, 166].

Other Impurities
To the best of our knowledge, studies that focus on the impact of other additives or impurities such as vinylcyclohexene (Sect. 2.2.5) are not available. A large number of additives was also tested for molar mass control. In Sect. 2.2.8 this aspect is addressed in detail.

2.2
Technological Aspects of the Polymerization in Solution

In Sect. 2.2 important technological aspects of the neodymium-catalyzed solution processes are addressed. In this context a book chapter on large-scale BR production in solution by Taube and Sylvester is recommended [81].

2.2.1
Solvents

For the large-scale production of Nd-BR hexane, cyclohexane, pentane and technical mixtures of aliphatic hydrocarbons are preferred over aromatic solvents [82]. The use of aliphatic and cycloaliphatic solvents goes along with higher polymerization activities and advantages in the process steps after completion of the polymerization (stripping of solvent from polymer solution) and solvent recycling (purification and drying) [89]. Other reasons in favor of choosing aliphatic solvents are their lower price and their lower toxicity.

The first report on higher polymerization activities in hexane than in benzene dates back to Throckmorton's pioneering work on ternary Ce-based catalyst systems [34]. To the present day there are several reports available in the literature in which the influence of solvents on polymerization activities, microstructure of the resulting polymers and molar masses are compared. The relevant results of these investigations are summarized in this section.

Binary Systems

For binary catalyst systems there are four studies available that focus on the influence of solvents. For catalyst systems of the type $NdX_3 \cdot n$ D/AlR_3 the following order in catalytic activities was established: cyclohexane $\sim n$-hexane $\geq n$-pentane $>$ chlorinated arenes $>$ toluene $>$ 1-hexane $>$ tetrachloroethylene $\geq$ styrene [134]. The influence of these solvents on the *cis*-1,4-content is negligible and molar masses are lower in toluene than in aliphatic solvents. According to Hsieh et al. solvents with electron-donating properties compete with the monomer for vacant sites at the Nd centers. Ethers (e.g. THF) exhibit a particularly strong interaction with vacant Nd sites and are effective catalyst poisons. For the *trans*-1,4-specific binary catalyst system $NdCl_3 \cdot TBP/Mg(^nC_4H_9)(^iC_8H_{17})$ the *trans*-1,4-content decreases slightly in the following order: chlorobenzene $>$ toluene $>$ cyclohexane [135, 136]. For binary $Nd(O^iPr)_3$-based catalyst systems the impact of the solvents heptane, cyclohexane, toluene, dichloromethane was determined. In these studies IP was polymerized. For $Nd(O^iPr)_3$/MAO polymer yield and M_n are higher in heptane and cyclohexane than in toluene [246]. The reduced polymer yield in toluene is attributed to competitive coordination of the aromatic solvent with IP. In dichloromethane the polymer yield is the lowest among the solvents tested. In addition, the polymer obtained in dichloromethane is insoluble. *Cis*-1,4-contents increase in the order: heptane (91.1%) $<$ cyclohexane (91.9%) $<$ toluene (92.9%). Also for the catalyst system $Nd(O^iPr)_3$/MMAO the use of aliphatic solvents results in higher polymer yield and higher molar masses than toluene [231]. According to the authors the catalyst system $Nd(O^iPr)_3$/MMAO is homogeneous in heptane providing for polymers with a high molar mass, narrow MMD, and a high *cis*-1,4-content. The use of dichloromethane results in cyclization of poly(isoprene).

Ternary Systems

For ternary catalyst systems polymerization activities are also higher in aliphatic and cycloaliphatic hydrocarbons than in aromatic solvents. These features were observed for various catalysts systems. Detailed studies are available for Nd^{i}O/TIBA/DEAC [161, 165, 166], NdV/DIBAH/EASC [205, 422], Nd(N(SiMe$_3$)$_2$)$_3$/TIBA/DEAC [318, 320], NdV/DIBAH/^{t}BuCl [423] and NdV/MAO/^{t}BuCl [175].

According to Cabassi et al. the reduction of polymerization rates by aromatic compounds is caused by competitive coordination of monomer and arenes to vacant Nd-sites. The following coordination equilibrium was put forward in order to account for the observed effects (Scheme 12) [165, 166].

Scheme 12 Competitive coordination to Nd between dienes and arenes [161], reproduced with permission from Wiley-VCH Verlag GmbH & Co. KGaA

In this equilibrium the Nd-species to which a diene is coordinated is active in polymerization, whereas the Nd-species to which an arene is coordinated is inactive. According to the authors the experimentally determined ranking of activities: toluene > mesitylene > toluene (+ 7% hexamethylbenzene) correlates with the electron richness (i.e. Lewis basicity) of the aromatic compounds. The polymerization activity decreases with increasing Lewis basicity of the aromatic compound as the equilibrium is shifted and the concentration of the active species is reduced. These considerations were supported by the following experimental results (Table 18).

In addition to the experimental evidence published, the reaction given in Scheme 12 is supported by the fact that η^6-toluene complexes of Nd could be isolated, characterized and used for the initiation of diene polymerization [424].

Table 18 Polymerization of IP in different aromatic solvents with Nd^{i}O/DEAC/TIBA at 0 °C/1 h, $n_{IP}/n_{Nd} = 9000$, $c_{Nd} = 2.16 \times 10^{-4}$ mol·L^{-1} [165, 166]

Solvent	Monomer conversion/%
Toluene	18
Mesitylene	11
Toluene + 7% Hexamethylbenzene	negligible

In the already quoted publication by Cabassi et al. the use of alkenes as potential solvents was also investigated. As alkenes are not incorporated into the polymer chain the authors suggested the use of alkenes, such as *cis*-butene, 1-butene and isobutene as appropriate solvents for the polymerization of BD. Also, the technical potential of the C_4 cracking fraction as a suitable solvent for the industrial production of BR was recognized. This suggestion was later followed up by Laubry who succeeded to selectively polymerize IP in the C_5 cracking fraction [264, 265].

If proton transfer from appropriate substrates leads to stabilized Nd allyl or Nd benzyl species allyl or benzyl proton donors allow for the control of molar mass (Sect. 2.2.8). On the basis of this consideration hexane, toluene and *tert*-butylbenzene (TBB) were comparatively tested in the polymerization of BD. In this study the catalyst system NdV/DIBAH/EASC was used. The rate of polymerization decreases in the order hexane > TBB > toluene. Only in hexane does the monomer conversion proceed to full completion (Fig. 6) [422].

Linear first-order plots ($-\ln(1-x)$ vs. time) are only obtained for hexane. The use of TBB and toluene results in curved dependencies. The deviations from the respective hexane reference are more pronounced for toluene than for TBB (Fig. 7).

The low catalyst activity in toluene and the found non-linear M_n-conversion characteristics are explained by chain-transfer reactions. A reaction scheme that accounts for the abstraction of a proton from toluene by the allyl-end of the growing poly(butadiene) chain is given in Scheme 13.

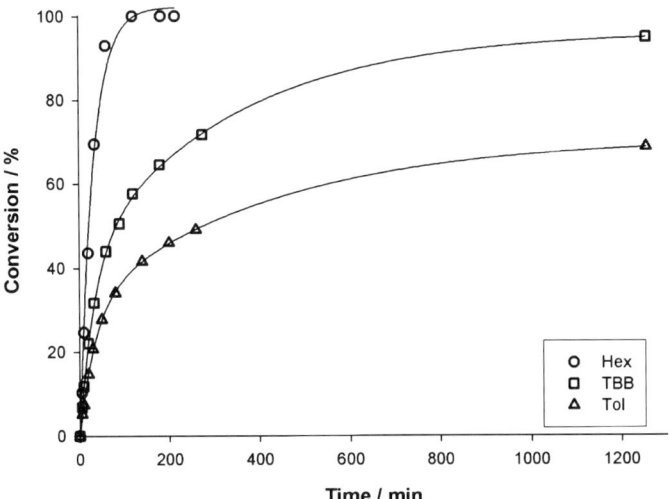

Fig. 6 Time-Conversion plots for the polymerization of BD with the catalyst system NdV/DIBAH/EASC in the solvents *n*-hexane (Hex), *tert*-butylbenzene (TBB) and toluene (Tol) [422], reproduced by permission of Taylor & Francis Group, LLC., http://www.taylorandfrancis.com

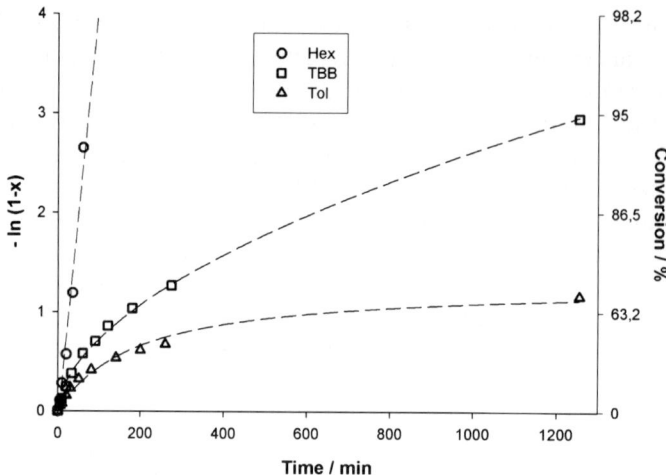

Fig. 7 Plot of $-\ln(1-x)$ vs. polymerization time (based on conversion time curves in Fig. 6) for BD polymerization with the catalyst system NdV/DIBAH/EASC in different solvents [422], reproduced by permission of Taylor & Francis Group, LLC., http://www.taylorandfrancis.com

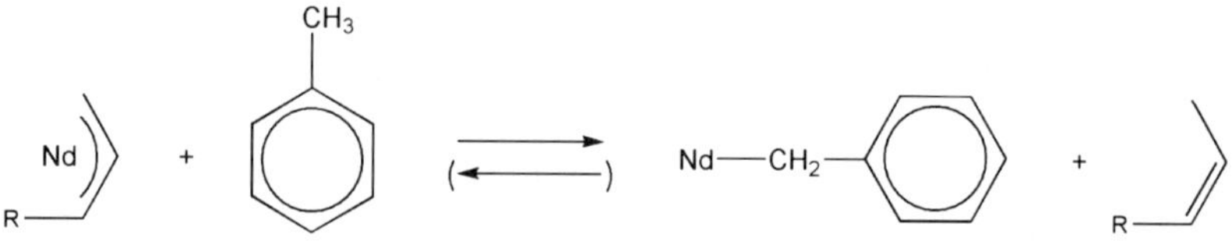

Scheme 13 Transfer of a benzyl-H-atom from toluene (solvent) to the allyl-end of a poly(butadien)yl chain [R = poly(butadien)yl] (ligands and charges are omitted for clarity) [422], reproduced by permission of Taylor & Francis Group, LLC., http://www.taylorandfrancis.com

According to Scheme 13 a benzyl derivative of Nd is formed. At a polymerization temperature of 60 °C the benzyl Nd intermediate once formed decomposes rapidly as Nd(benzyl)$_3$ is reported to be stable only below -15 °C [425, 426]. As a consequence of the low thermal stability of the Nd benzyl species proton transfer from toluene is irreversible and the overall rate of polymerization is reduced by the decrease of the amount of the active catalyst species. As TBB lacks benzyl protons it can only act as a π-donor. Therefore, TBB reduces the polymerization rate to a lower extent than toluene. Beside the interpretations given, the study also presents detailed investigations on the evolution of the MMDs with monomer conversion in the three solvents n-hexane, TBB, toluene [422]. In the two aromatic solvents a high molar mass fraction is more pronounced than in n-hexane.

A broad range of solvents (cyclopentane, hexane, cyclohexane, methylcyclohexane, toluene, xylene, dichloromethane, chloroform, carbon tetrachloride, 1,2-dichlorobenzene and THF) was compared by Wilson in the polymerization of BD with the catalyst system NdV/MAO/^{t}BuCl (Nd/MAO/Cl = 1/264/3) [175]. The highest polymerization activities are obtained with cyclopentane, hexane, cyclohexane and methylcyclohexane. No difference regarding the rate of polymerization and the characteristics of the polymer is observed with these solvents. The use of toluene and xylene significantly reduces the rate of polymerization. By the use of 1,2-dichlorobenzene, however, catalyst activity is not affected negatively. According to Wilson the chlorine groups on the aromatic ring reduce Lewis basicity and as a consequence the ability of the arene to compete with the monomer for free coordination sites on Nd is decreased. Insoluble polymers are obtained in dichloromethane and chloroform. No polymerization occurs when carbon tetrachloride and THF are used.

Boisson et al. investigated $Nd(N(SiMe_3)_2)_3$-based catalyst systems in heptane and toluene. At an optimized composition of $Nd(N(SiMe_3)_2)_3$/TIBA/DEAC = 1/40/2 the catalyst system was more active in heptane than in toluene and yields the higher cis-1,4-content in heptane (97.6%) than in toluene (93.1%) [318, 320].

A recent study performed by Mello et al. focuses on a comparison of n-hexane and cyclohexane in the polymerization of BD with the catalyst system NdV/DIBAH/^{t}BuCl. In this study Mello et al. use the pure solvents and mixtures of n-hexane and cyclohexane [423]. Cyclohexane yields BR with a significantly lower molar mass than n-hexane. According to Mello et al. this effect is due to the thermodynamically better solvent quality of cyclohexane. The authors found no strong influence of the cyclohexane/n-hexane ratio neither on catalyst activity nor on microstructure.

In the most recent study by Dong et al. the impact of solvents was determined for the ternary catalyst system $Nd(O^iPr)_3/[HNMe_2Ph]^+[B(C_6F_5)_4]^-$/TIBA (1/1/30) [233]. In heptane and cyclohexane polymer yields are particularly high. Polymerization in heptane provided a polymer with a relatively narrow MMD whereas the use of cyclohexane resulted in a bimodal MMD. Cis-1,4 contents are slightly higher in cyclohexane (91.3%) than in heptane (90.4%). Polymerization in dichloromethane results in a very low polymer yield and an increased cis-1,4-content (93.4%). In toluene no polymer is obtained, at all.

2.2.2
Monomer Concentration

In the large scale Nd-BR production monomer concentrations range between 10–23 wt. %. In order to keep solution viscosities at a manageable level, in the first approximation a very simple approach is applicable: The higher the aimed molar mass the lower one has to set the monomer concentration, and vice versa. Apart from molar mass, MMD, branching and monomer conversion, the upper limit of polymer concentration also depends on engineering factors. From an engineering point of view the ability of the equipment to handle highly viscous solutions and the efficiency of heat removal are important issues [49, 82].

Polymer solutions with a particularly high solid concentration are obtained by the application of Mooney jump technology. Mooney jump simply means the increase of molar mass by a chemical reaction. Mooney jump technology comprises a first reaction step during which a low molar mass BR- or IR-feedstock is prepared in the usual way. Because of the low molar mass of the feedstock high polymer concentrations can be achieved at the end of the polymerization. The Mooney jump reaction is performed subsequent to completion of the polymerization when heat removal is no longer an issue. Oil extension is usually performed simultaneously with the Mooney jump reaction in order to keep solution viscosities at moderate levels. As Mooney jump technology allows for the handling of high polymer concentrations the amounts of solvent which have to be removed are relatively low. Therefore energy costs for the production of Mooney jumped BR or IR are particularly favorable [427, 428] (Sect. 2.2.6).

2.2.3
Moisture and Impurities

In large-scale production the presence of impurities cannot be avoided. Impurities are brought into the system as byproducts of the raw materials: monomers, solvents, catalyst components, additives etc. The presence of impurities in monomers and solvents is mainly due to insufficient removal of moisture in their recycling. The most important byproducts of Nd carboxylates (dissolved in an organic solvent) are Nd compounds such as Nd oxide, Nd hydroxide, carboxylic acids, e.g. versatic acid which can be present in NdV, and water. The impact of some of these relevant impurities is mainly available from scientific reports. The according literature is summarized in Sect. 2.1.8.

2.2.4
Monomer Conversion, Shortstop and Stabilization of Polymers

In the polymerization of BD by Ti-, Co- and Ni-based catalyst systems the polymerization has to be shortstopped at a specific monomer conversion in order to avoid the formation of gel. In contrast, polymerization catalysis by Nd catalysts does not need control of monomer conversion. As gel formation is particularly low with Nd catalysts full monomer conversion can be accomplished [427, 428].

If the polymerization is shortstopped at a specific monomer conversion water, acids and alcohols are used. Water has to be properly mixed with the polymer solution. Otherwise water is not efficient in quenching the polymerization. Preferred acids are long-chain organic acids, e.g. stearic acid. It is important that the acids dissolve well in the organic medium and that the acids have high boiling points in order to avoid contamination of recycled monomer and solvent. Also quenching by alcohols is limited to alcohols with high boiling points. For this purpose phenols, especially sterically hindered phenols, are preferred. These phenols also stabilize the dried polymer against oxidation. In this context 2,6-di-*tert*-butyl-*p*-kresol (BHT) and 2,2-methylene-bis-(4-methyl-6-*tert*-butylphenol) (BPH) deserve special attention. The shortstop of the polymerization is equally achieved by postpolymerization modification (Sect. 2.2.6). Quenching by the addition of modifiers either occurs by the reaction of the modifiers with the living polymer chains or by the poisoning of active catalyst sites by the modifiers used (e.g. PCl_3, SCl_2, S_2Cl_2, $SOCl_2$, S_2Br_2 etc.).

Prior to the removal of solvent, residual monomer, and BD dimers, stabilizers are added to the polymer solution in order to protect the polymer during the subsequent process steps of drying and storage. As the use of Nd-BR so far is limited to applications for which the color of Nd-BR is not an issue staining antioxidants are preferred over non-staining antioxidants. Important staining antioxidants belong to the class of phenylenediamines, e.g. *N*-isopropyl-*N'*-phenyl-*p*-phenylendiamine (IPPD), *N*-1,3-dimethylbutyl-*N'*-phenyl-*p*-phenylendiamine (6PPD), *N*-1,4-dimethylpentyl-*N'*-phenyl-*p*-phenylendiamine (7PPD) and *N,N'*-bis-1,4-(1,4-dimethylpentyl)-*p*-phenylendiamine (77PD).

Organic solvents, residual monomer and dimers (Sect. 2.2.5) are usually removed by vapor stripping during which an aqueous slurry of Nd-BR is obtained. BR is isolated from the slurry by appropriate dewatering screws prior to drying by hot air [82]. The final process steps of Nd-BR-solutions are not unique to Nd-BR. Therefore, these aspects are not addressed in this review.

2.2.5
Formation of Dimers

Polymerization and dimerization of dienes are competing reactions that occur simultaneously during the course of the polymerization. The dimerization of BD occurs by a Diels–Alder reaction which results in the formation of vinylcyclohexene (VCH), a non-conjugated diene (Scheme 14).

Dimerization is particularly pronounced at high monomer concentrations, long reaction times and at elevated temperatures. Ziegler/Natta-catalysts are

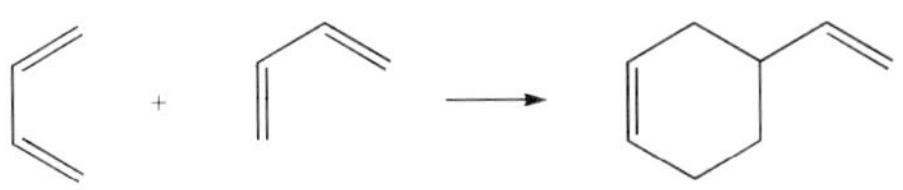

Scheme 14 Diels–Alder-dimerization of BD to vinylcyclohexene (VCH)

known to catalyze the dimerization of BD [429, 430]. The rate of dimerization significantly depends on the polymerization catalyst used. Unfortunately, in this respect there is no comparison of different Ziegler/Natta-catalysts available. According to our experience, the formation of dimers is comparatively high with Ti- and Ni-based catalyst systems. For Nd-catalysts this side reaction is particularly low. Regarding the standard catalyst systems which are used for the large-scale production of high-cis-BR dimer formation increases in the following order: Nd < Co < Ni ~ Ti.

As dimerization competes with polymerization, dimer formation very slightly reduces polymer yield. The toxicity and the smell of VCH are much more relevant. VCH is removed from the rubber solution together with residual monomer and solvent prior to the isolation of the rubber from the polymer solution. As VCH has a higher boiling point than BD and hexane an efficient stripping process has to be used in order to reduce VCH contents to environmentally friendly levels.

2.2.6
Post-Polymerization Modifications

Standard binary and ternary Nd-based catalyst systems yield high *cis*-1,4-BR with high linearity and without specific end groups. Branching, end group functionalization and cyclization can be achieved by post-polymerization reactions. Branching is beneficial for the reduction of polymer solution viscosities and allows for high polymer concentrations. Reduction of cold flow of unvulcanized rubbers is another positive feature of branching. Thus, branching is equally important for the preparation of rubber compounds and rubber solutions with reduced compound viscosities.

Because of the (partially) living character of the Nd-catalyzed diene polymerization the incorporation of functional end groups is also achieved by post-polymerization reactions. By the incorporation of specific functionalities carbon black or silica fillers are bound better to the rubber. As a consequence vulcanizate properties (tensile strength, rebound and resistance to wear) are improved.

In post-polymerization modification for branching bi- or multifunctional additives are applied at high monomer conversion usually before the shortstop of the polymerization. Enichem describes the use of PCl_3 [431, 432]. The beneficial effect of the Mooney jump with PCl_3 on black incorporation time (BIT) and cold flow is illustrated by the examples given in this patent (Table 19). In the preparation of rubber compounds the incorporation of carbon black is time and energy intensive. By the reduction of the BIT total mixing time and energy cost are reduced.

Branching by bifunctional sulfur derivatives such as SCl_2, S_2Cl_2, $SOCl_2$ and S_2Br_2 is claimed by Bayer [427, 428]. Especially Nd-BR technology is useful for this post-polymerization modification as it allows for the required low

Table 19 Addition of PCl$_3$ for branching of Nd-BR [431, 432]

	Reference without Mooney jump	Mooney jump with PCl$_3$		
Mooney viscosity (ML 1 + 4)/MU	41.5	40	45	50
BIT/s	385	150	120	180
Cold flow/mg · h^{-1}	37	11	–	–

concentrations of residual monomers and dimers. In this way side reactions of low molar mass products with sulfur compounds are avoided and the formation of undesirable and smelly products is reduced.

In a series of patents JSR describes branching and end group functionalization by post polymerization modification which is performed prior to quenching of the polymerization. For the modification tin tetrachoride as well as alkylated and arylated tin chlorides, silicon tetrachloride, diphenylmethanediisocyanate (MDI), carbon disulfide, styrene epoxide, 2,4,6-trichloro-1,3,5-triazine, organic diacids, diesters of organic diacids, acid anhydrides, diphenylcarbonate, divinylbenzene, dioctyltin carboxylates etc. are used [338, 339]. In a subsequent patent another ten classes of products are claimed for post-polymerization modification: quinones, thiazoles, sulfenamides, dithiocarbamates, thiuram disulfides, thioimides, epoxide containing amines, imides, amino containing aldehydes or thioaldehydes [433, 434]. The simultaneous use of two modification agents is also described [435, 436]. As a consequence of the performed modifications a reduction of cold flow and improvements of vulcanizate properties regarding moduli, tensile strength, rebound at 25 °C, heat-build-up, abrasion resistance etc. are observed. These improvements can be attributed to branching as well as to an improvement of interactions with carbon black.

In the above-mentioned patents post-polymerization modification is performed in a single step. An alternative is the performance of two sequential modification steps. In the first step the polymer is reacted with tin or germanium derivatives (triphenyl tin, tributyl tin, diphenyl tin dichloride, dioctyl tin dichloride, phenyl tin trichlorde etc.). In the subsequent modification step heterocumulene compounds (ketenes, thioketenes, isocyanates, thioisocyanates and carbodiimides) are applied [437, 438]. Specific interaction with silica is obtained by end group functionalization with vinyl monomers which contain hydroxyl or epoxy groups. The effects are observed if the PDI is below 5 [439, 440].

Control of cold flow without significant increases in Mooney viscosity is achieved by the addition of organoboranes during the course of the polymerization [441, 442]. For this purpose trialkylboranes are used (Scheme 15).

Cyclization can be also achieved in a separate reaction step after polymerization. Cyclization of Nd-based diene rubbers has been investigated since the

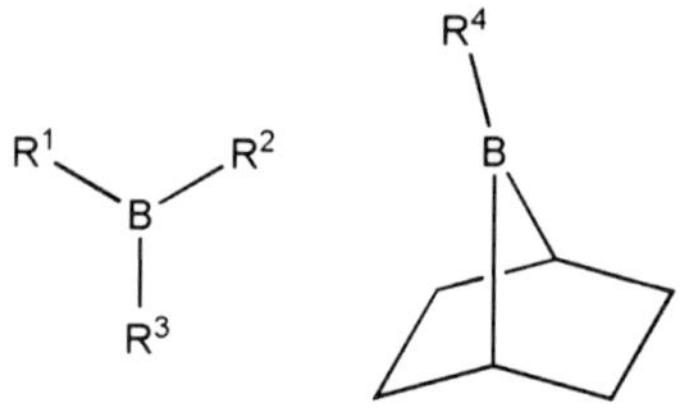

Scheme 15 Organoborane additives used for the control of cold flow in Nd-BR (R_1, R_2, R_3 = alkyl groups with 1–5 C-atoms; R_4 = alkoxy or alkyl groups with 1–5 C-atoms) [441, 442]

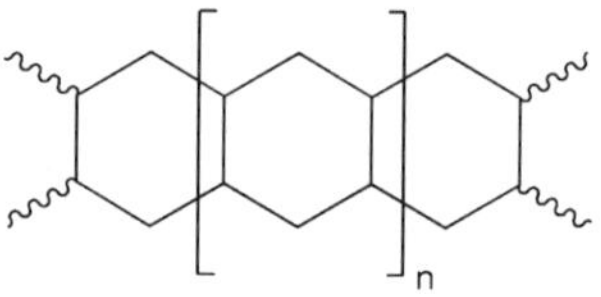

Scheme 16 Cyclized poly(butadiene)

mid 1980s [443, 444]. In 2005 Wang reported on in-situ cyclization of Nd-BR polymerized with the catalyst system $NdCl_3/3^i PrOH/AlEt_3$ [445]. Cyclization is achieved by the addition of allyl chloride to the reactive solution. By this means the formed BR cyclizes and also unreacted BD is cyclopolymerized. The obtained cyclized poly(butadiene) has a low intrinsic viscosity. The yields of the cyclized poly(butadiene) increase with increasing amounts of allyl chloride. A cyclization mechanism is suggested by Wang. A similar approach to cyclization was followed up in order to cyclize a BR/IR-copolymer. For the preparation of the BR/IR-copolymer the catalyst system NdN/TEA/^{t}BuCl was used [446, 447]. According to Wang, cyclized diene rubbers (Scheme 16) show beneficial properties in the formulation of adhesives, inks, paints, photoresists and tires [445].

2.2.7
Polymerization Temperature

Polymerization temperature has an influence on polymerization rate, molar mass, MMD and *cis*-1,4-content. As molar mass regulation by adjustment of polymerization temperature is important from an industrial point of view this aspect is given special emphasis in this subsection.

Ranges of polymerization temperatures in Nd-catalyzed diene polymerization which can be found in patents are summarized in Table 20.

The patents quoted in Table 20 give rather broad temperature ranges. An explanation for this feature is the polymerization mode which is used for the large-scale polymerization of Nd-BR. To the best of our knowledge, adiabatic rather than isothermic modes are used. In the adiabatic mode polymerization heat is neither removed by external nor by evaporation cooling. Therefore,

Table 20 Ranges of polymerization temperatures in Nd-catalyzed diene polymerization given in patents

Temperature range/ °C	Preferred temperature range/°C	Refs.
0–200	20–100	[448, 449]
25–150	60–90	[213, 214]
0–120	40–90	[450, 451]
–20–150	–	[452, 453]

the polymerization temperature is not constant and increases during the course of the reaction. The temperature ranges given in the general part of patent disclosures are principally consistent with adiabatic polymerization modes even though real temperature ranges are smaller.

The patents quoted in Table 20 do not give information on the effect of polymerization temperature on reaction rates, molar mass, MMD and cis-1,4-contents. This information is scattered in various scientific reports. The reports which are valuable in this context are summarized in Table 21 for binary Nd catalyst systems and in Table 22 for carboxylate-based systems.

Influence of Polymerization Temperature on Polymerization Rate

According to Porri et al. the largest effect of polymerization temperature is on reaction rates [50]. Information on the dependence of reaction rates is contained in most of the reports quoted in Tables 21 and 22. Unfortunately, most of these studies do not provide activation energies.

For only two binary catalyst systems are activation energies available. These activation energies are based on the polymerization of IP by $NdCl_3 \cdot TBP/TIBA$ ($E_a = 25.6$ kJ·mol^{-1}) [133] and on the polymerization of BD or IP by $NdCl_3 \cdot n$ ROH/AlR$_3$ ($E_a(BD) = 34.1$ kJ·mol^{-1} and $E_a(IP) = 34.3$ kJ·mol^{-1}) [134]. Activation energies for Nd-carboxylate-based ternary catalyst systems are significantly higher ($E_a = 46$–53 kJ·mol^{-1}) than for binary NdCl$_3$-based catalyst systems. Activation energies for the ternary catalyst systems are available for the copolymerization of BD and IP with a NdN-based catalyst system ($E_a = 46$ kJ·mol^{-1}) [93, 94] and for the polymerization of BD by NdV/DIBAH/EASC ($E_a = 53.0$ kJ·mol^{-1}). In the latter study the preexponential factor $k_0 = 10^{10}$ L·mol^{-1}·min^{-1} was also determined [205].

The activation energies available for binary and ternary catalyst systems are in a range from 20 to 70 kJ·mol^{-1}, which according to Odian is characteristic for polymerizations mediated by Ziegler/Natta-catalysts [456]. It is interesting to note that the polymerization of dienes catalyzed by NdCl$_3$-based catalyst systems show lower activation energies than Nd-carboxylate-based catalyst systems. The influence of halide donors on the temperature depen-

Table 21 Influence of polymerization temperature in binary Nd catalyst systems

Catalyst system	Temperature range/ °C	Molar mass or molar mass equivalent	Refs.
$NdCl_3 \cdot TBP/TIBA$	20–80	M_n	[133]
$NdCl_3 \cdot n$ 2-ethylhexanolate/TEA	25–75	Intrinsic Viscosity	[114]
$NdCl_3 \cdot 3$ pentanolate/TEA	25–75	Intrinsic Viscosity	[115]
$NdCl_3 \cdot n\ L/AlR_3$	30–70	Intrinsic Viscosity	[134]

Table 22 Influence of polymerization temperature in Nd-carboxylate-based catalyst systems

Catalyst system	Temperature range/ °C	Molar mass or molar mass equivalent	Refs.
NdV/EASC/DIBAH	0–80	M_v	[88]
NdV/EASC/DIBAH	0–60	DSV	[454]
NdV/MAO/^{t}BuCl and NdV/MAO	20–60	M_{peak}, PDI	[175]
Nd^{i}O/TIBA/^{i}Bu$_2$AlCl	25–75	M_v	[160]
Nd(OCOR)$_3$/TIBA/EASC	–20–80	M_n	[169]
NdV/DIBAH/^{t}BuCl	50–100	Mooney Viscosity, M_w, MMD	[455]
NdN/TIBA/EASC	30–70	M_w, PDI	[388]
NdV/DIBAH/EASC	20–80	M_n	[205]

dence of polymerization activities is reflected by a study in which the two catalyst systems NdV/MAO/^{t}BuCl and NdV/MAO are compared [175]. The ^{t}BuCl-containing catalyst system is more active at low temperatures (20 °C) and shows a lower temperature dependence between 20 to 60 °C.

Reports on steady increases of polymerization rates with increasing polymerization temperatures usually refer to an upper limit of polymerization temperature of around 60 °C. At temperatures > 60 °C catalyst deactivation becomes more prominent and overall catalyst activities decrease. There are two reports which point in this direction. A decrease of catalyst activities at elevated temperatures was observed for NdV/DIBAH/^{t}BuCl [455] and for NdN/TIBA/EASC [388]. Pires et al. studied the solution polymerization of BD whereas Ni et al. studied the polymerization of BD in the gas phase. The rate maximum observed by Pires et al. was at ∼ 80 °C whereas the reaction maximum in the gas-phase polymerization was at ∼ 50 °C. The reduction of polymerization rates at elevated temperatures can be explained by the decay of the number of active species. In gas-phase polymerization deactivation becomes evident at lower temperatures (50 °C) compared to the solution process (80 °C). This observation can be explained by the stabilization of active catalyst-species by monomer. As monomer concentration is lower in the polymer granules during gas-phase polymerization deactivation is more strongly pronounced at lower temperatures than in the solution process.

The influence of polymerization temperature on reaction rates was also studied for the two Nd-alcoholate-based catalyst systems $Nd(O^iPr)_3$/MAO [246] and $Nd(O^iPr)_3$/[HNMe$_2$Ph]$^+$[B(C$_6$F$_5$)$_4$]$^-$/TIBA [233]. With $Nd(O^iPr)_3$/MAO a remarkable increase of polymerization rates between 0 °C and 60 °C is observed. At 0 °C the polymer yield is < 12% even after 12 h, whereas polymer yield at 60 °C is ∼ 98% within 1 h reaction time. The polymerization of IP with the catalyst system $Nd(O^iPr)_3$/[HNMe$_2$Ph]$^+$[B(C$_6$F$_5$)$_4$]$^-$/TIBA (1/1/30) showed a moderate increase of polymerization rates between 30 and 60 °C. At 30 °C the polymerization went to completion after 5 h and at 60 °C 3 h were needed for 100% conversion [233].

Influence of Polymerization Temperature on Molar Mass and Molar Mass Distribution

For Nd-based catalyst systems increasing polymerization temperatures result in a considerable reduction of molar mass. As a representative example, the dependence of molar mass (M_v) on polymerization temperature is given for the ternary catalyst system NdV/DIBAH/EASC (Fig. 8).

The dependence of molar mass on polymerization temperature is exploited for the regulation of molar mass in the large-scale production of BR [81, 82].

Quantitative studies on the dependence of molar mass on polymerization temperature date back to Throckmorton (1969) who determined the influ-

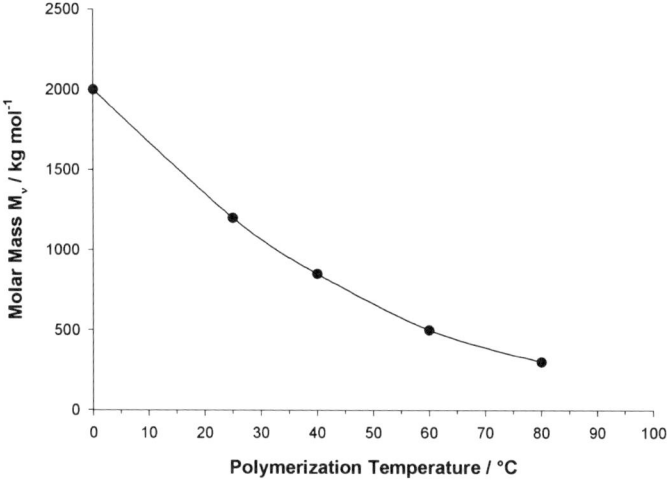

Fig. 8 Dependence of molar mass (M_v) on polymerization temperature for the catalyst system NdV/DIBAH/EASC [88]

ence of polymerization temperature on dilute solution viscosities in the temperature range 5–70 °C [34]. Witte also described the dependence of polymer solution viscosity on polymerization temperature for ternary Di-octanoate-based catalyst systems [49]. More recent studies on the dependence of molar mass on polymerization temperature exist for binary (Table 21) as well as for ternary carboxylate-based catalyst systems (Table 22).

For binary catalyst systems the decrease of molar masses with increasing polymerization temperatures is documented for: $NdCl_3 \cdot TBP/TIBA$ [133], $NdCl_3 \cdot n$ 2-ethylhexanolate/TEA (n = 1.5 or 2.5) [114] and $NdCl_3 \cdot n$ L/AlR$_3$ [134]. For ternary Nd-carboxylate-based catalyst systems the respective studies were performed with the following catalyst systems: NdV/DIBAH/EASC [88, 205], NdV/DIBAH/EASC [454], NdV/MAO/^{t}BuCl and NdV/MAO [175], Nd^{i}O/TIBA/^{i}Bu$_2$AlCl [160], Nd(OCOR)$_3$/TIBA/EASC [169], and NdV/DIBAH/^{t}BuCl [455]. There is agreement among the majority of authors that the reduction of molar mass with increasing polymerization temperature is caused by an increase of irreversible chain transfer reactions mainly with aluminum alkyl. In contrast, Friebe et al. interpret their results on the basis of a reversible transfer reaction (Sect. 4.5).

The dependence of M_n on monomer conversion for the system NdV/DIBAH/EASC is more complex than one would expect from the M_v/temperature plot given in Fig. 8. As shown in Fig. 9 straight M_n-conversion dependencies are only obtained at polymerization temperatures ≥ 40 °C. At low polymerization temperatures (20 °C) and at low monomer conversions (< 50%) a high molar mass fraction causes significant deviations from the straight line which is expected for a living polymerization.

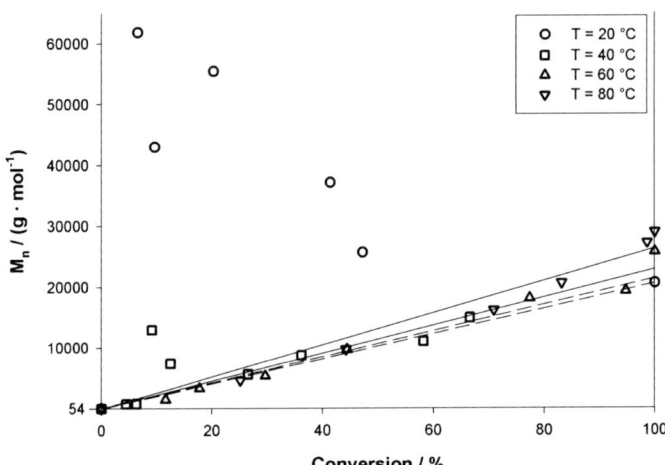

Fig. 9 Dependence of M_n on monomer conversion at different polymerization temperatures for the polymerization of BD with the catalyst system NdV/DIBAH/EASC [205]

According to Fig. 9 the increase of polymerization temperature results in a higher linearity of the respective M_n-conversion plots. This dependence indicates that the concentration of active catalyst sites (or active catalyst species) which form the high molar mass polymer is reduced with increasing polymerization temperature. As a result, PDI decreases with increasing polymerization temperature (Fig. 10) [205].

Figure 10 shows that the share of the high molar mass fraction decreases with increasing polymerization temperature.

The dependence of MMD and PDI on polymerization temperature is confirmed by studies reported by Oehme et al. ($Nd(OCOR)_3$/TIBA/EASC) [169] and Pires et al. (NdV/DIBAH/^{t}BuCl) [455]. Oehme et al. obtain a unimodal MMD at − 20 °C whereas at 80 °C two peaks are present. This is evidence for the presence of two active catalyst sites. According to Oehme et al. the active center which produces the low-molar-mass peak is more sensitive to increased temperatures than the second species. It is also possible that the latter species does not exist at low polymerization temperature. Pires et al. also explain their results on the basis of two different active catalyst sites. According to Pires et al. the less stable catalyst site which produces the low molar mass polymer gradually deteriorates in the temperature range 50–80 °C. At polymerization temperatures > 80 °C, however, an increase of PDI as well as of molar mass is observed. This effect is explained by polymer branching.

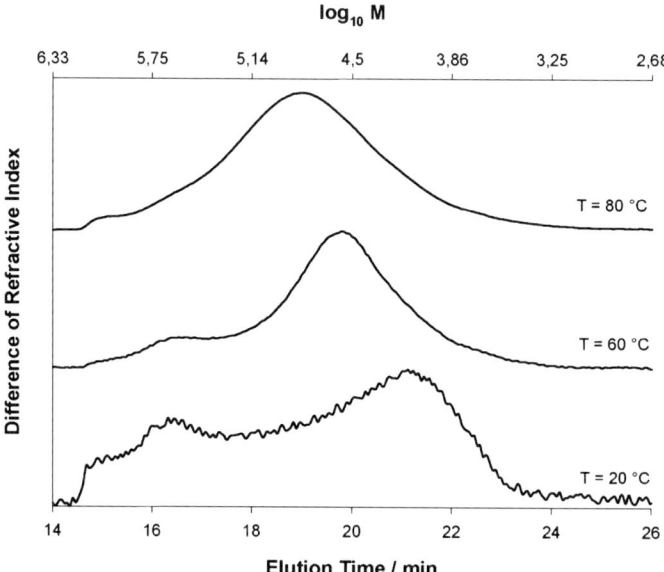

Fig. 10 MMDs of BR (conversion$_{BD}$ = 100%) polymerized with the catalyst system NdV/DIBAH/EASC at T = 20, 60 and 80 °C, PDI(T = 20 °C) = 6.08, PDI(60) = 3.34, PDI(80) = 3.63 [205]

Also for the two Nd-alcoholate-based catalyst systems $Nd(O^iPr)_3$/MAO [246] and $Nd(O^iPr)_3$/MMAO [231] M_n as well as PDI decrease with increasing polymerization temperature (30–60 °C).

Influence of Polymerization Temperature on Microstructure

The available studies unambiguously agree on the decrease of *cis*-1,4-content with increasing polymerization temperature. According to Throckmorton *cis*-1,4-contents decrease from 3 °C: 99.4%, 50 °C: 97.2%, 70 °C: 96.8% for ternary Ln-based catalysts [34]. The same trend was confirmed for a ternary Di-octanoate-based catalyst system [49]. Witte gives the following data: 0 °C: 97.5% and 60 °C: 95%. For binary as well as for ternary Nd-based catalyst systems the same trend is established.

For binary NdX_3-based catalyst systems reports on the decrease of *cis*-1,4-contents with increasing polymerization temperatures are available for $NdCl_3$/ROH/AlR_3 [134, 139], $NdCl_3 \cdot$ n 2-ethylhexanolate/TEA (n = 1.5 or 2.5) [114] and $NdCl_3 \cdot 3$ pentanolate/TEA catalyst systems [115].

For ternary catalyst systems, details on the dependence of *cis*-1,4-contents on polymerization temperature are available for: Nd(OCOR)$_3$/TIBA/EASC, which was preformed and aged prior to use (*cis*-1,4-contents: – 20 °C: 99%, 80 °C: ~ 94%) [169] for NdV/DIBAH/EASC (*cis*-1,4-contents: 0 °C: 98.8%, 80 °C: 91.7%) [88], NdV/DIBAH/EASC (*cis*-1,4-contents: 20 °C: 95.8%, 40 °C: 94.2%, 60 °C: 92.9%, 80 °C: 94.0%) [205] and for NdV/DIBAH/^{t}BuCl (*cis*-1,4-contents: 50 °C: 98.4%, 60 °C: 98.6%, 70 °C: 98.9%, 80 °C: 98.9%, 90 °C: 98.7%, 100 °C: 98.2%) [455].

Also for the Nd-alcoholate-based catalyst systems $Nd(O^iPr)_3$/MAO [246], $Nd(O^iPr)_3$/MMAO [231] and $Nd(O^iPr)_3$/[HNMe$_2$Ph]$^+$[B(C$_6$F$_5$)$_4$]$^-$/TIBA [233] *cis*-1,4 contents decrease with increasing polymerization temperature. The decrease is less pronounced for $Nd(O^iPr)_3$/MAO than for $Nd(O^iPr)_3$/MMAO. For $Nd(O^iPr)_3$/[HNMe$_2$Ph]$^+$[B(C$_6$F$_5$)$_4$]$^-$/TIBA the *cis*-1,4-contents are not significantly affected by changes in polymerization temperature.

In the gas-phase polymerization of BD with the catalyst system NdN/TIBA/EASC *cis*-1,4-contents decrease slightly with increasing polymerization temperature [388].

2.2.8
Control of Molar Mass

Molar mass strongly influences the performance of raw (unvulcanized) rubbers during the preparation of rubber compounds, e.g. addition of fillers and other ingredients. Also the processing characteristics of the compounded rubbers as well as the physical properties of the vulcanized rubbers significantly depend on the molar mass of the unvulcanized rubbers. In order to better meet various requirements, there is not only one BR grade available but a whole range of different product lines that mainly differ in molar mass (or Mooney viscosity). As a consequence, in large-scale BR-production, control of molar mass is an important issue. In Nd-catalyzed diene polymerization the most important means to control molar mass are:

- Variation of the monomer/catalyst-ratio (n_M/n_{Nd}).
- Variation of the amount of cocatalyst ($n_{cocatalyst}/n_{Nd}$).
- Variation of the polymerization temperature (Sect. 2.2.7).
- Additives.

As the influence of polymerization temperature on molar mass and MMD is addressed in Sect. 2.2.7 this issue is not discussed in this context. The residual aspects are reviewed in the following subsections.

Molar Mass Control by Variation of the Monomer/Catalyst-Ratio (n_M/n_{Nd})

The control of molar mass by variation of the n_M/n_{Nd}-ratio is directly linked with the (partially) living character of Nd-catalyzed diene polymerization. Studies which focus on n_M/n_{Nd}-ratios in the context of living polymerizations are addressed in Sect. 4.4. In this subsection only the influence of n_M/n_{Nd}-ratios on molar mass is discussed.

For binary NdCl$_3$-based catalyst systems there are two studies available for which the impact of n_M/n_{Nd}-ratios on intrinsic viscosities was investigated. These studies deal with NdCl$_3$ · 2THF/TEA [35] and with NdCl$_3$ · n 2-ethylhexanolate/TEA (n = 1.5 or 2.5) [114]. For both systems increasing n_M/n_{Nd}-ratios result in increases of molar mass.

For ternary Nd-carboxylate-based catalyst systems only two studies seem to be available for which n_M/n_{Nd}-ratios were investigated. Wilson compared the two Nd precursors neodecanoate and neodecanoate/neononadecanoate in the ternary catalyst system Nd(carboxylate)$_3$/DIBAH/^{t}BuCl. An increase of $m_{Nd}/100g(BD)$ in the range 0.12 to 0.19 resulted in the reduction of Mooney viscosities from 60 to 25 MU [183]. Sylvester et al. also varied the amount of Nd at constant BD concentrations. An increase of $m_{Nd}/100g(BD)$ from 0.025 to 0.30 reduced M_v from 750 kg · mol^{-1} to 300 kg · mol^{-1} [88].

For allyl Nd compounds a series of studies on the variation of n_M/n_{Nd}-ratios are available [288]. The allyl Nd compounds studied showed an influence of n_M/n_{Nd}-ratios on molar mass which is theoretically predicted for a living polymerization.

Molar Mass Control by Variation of the Amount of Cocatalyst ($n_{cocatalyst}/n_{Nd}$)

The variation of $n_{cocatalyst}/n_{Nd}$-ratios (mostly n_{Al}/n_{Nd}-ratios) is the most efficient way to control molar mass [81, 457, 458].

Already in his pioneering study Throckmorton investigated the impact of n_{Al}/n_{Ce}-ratios for different Al-alkyls on dilute solution viscosities (DSV) [34]. Throckmorton found that for the two ternary catalyst systems Ce octanoate/AlR$_3$/DEAC and Ce octanoate/AlR$_3$/EtAlCl$_2$ dilute solution viscosities decrease in the order $Al^nHex_3 > Al^nBu_3 > Al^iHex_3 > Et_2AlH$. Throckmorton explained the strong regulating effect of Al–H moieties by the high "mobility" of H atoms attached to Al. In a subsequent study in this area Witte systematically varied n_{Al}/n_{Nd}-ratios in order to prepare poly(diene)s with a defined molar mass [49]. Witte studied the polymerization of BD and determined the dependence of molar mass on the amount of aluminum alkyl cocatalysts in ternary Di-octanoate-based catalyst systems. Witte found a 50% decrease of intrinsic viscosities as a result of increasing n_{Al}/n_{Di}-ratios from 20 to 60 [49].

A major drawback of molar mass control by changing n_{Al}/n_{Nd}-ratios is the simultaneous alteration of polymerization rates. As shown for the system NdV/DIBAH/EASC, an increase in n_{DIBAH}/n_{NdV}-ratios from 10 to 30 reduces molar mass by 73% but also doubles the rate of polymerization [178, 179]. For NdV/TIBA/EASC the variation of n_{TIBA}/n_{NdV} from 10 to 30 reduces molar masses by 78% but increases the polymerization rate even 27-fold (Fig. 11) [179]. As shown by these two examples, on one hand, variations of n_{Al}/n_{Nd}-ratios have a considerable effect on molar mass, and on the other hand, lead to an undesired side effect regarding reaction rates. Because of these interdependencies, in the large-scale continuous production of Nd-BR, adjustments of the n_{Al}/n_{NdV}-ratios have to be counteracted by adaptations of residence time in order to keep monomer conversion per reactor and fi-

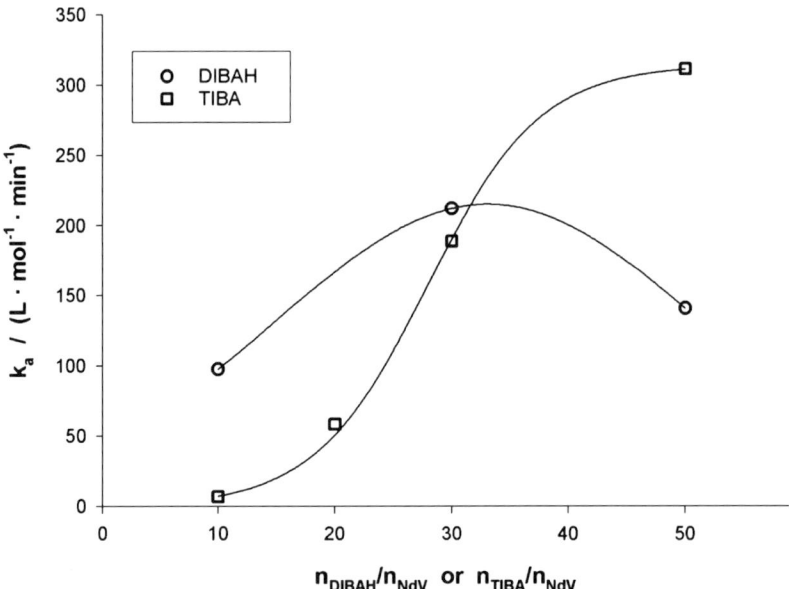

Fig. 11 Dependence of apparent polymerization rate constant k_a on n_{Al}/n_{NdV} in BD polymerization with the catalyst systems (1) NdV/DIBAH/EASC and (2) NdV/TIBA/EASC [179], reproduced by permission of Taylor & Francis Group, LLC., http://www.taylorandfrancis.com

nal monomer conversion at the end of the reactor line constant. From these considerations it is evident that an important requirement for an appropriate molar mass control agent is its ability to regulate molar mass without influencing the respective polymerization rates. The efforts to meet this target are summarized later in this Sect. 2.2.8 (Subsection "Additives").

In the following binary and ternary Nd catalyst systems are reviewed regarding the influence of the ratio n_{Al}/n_{Nd} on molar mass.

Influence of n_{Al}/n_{Nd} in Binary Nd Catalyst Systems

In Nd-halide-based binary catalyst systems the dependence of molar mass on n_{Al}/n_{Nd}-ratios was studied for various catalyst compositions. In these studies the following catalyst systems were used: $NdCl_3 \cdot 3ROH$ [92], $NdCl_3 \cdot n$ L/AlR_3 [134], $NdCl_3 \cdot TBP/TIBA$ [133], $NdCl_3 \cdot n$ 2-ethylhexanolate/TEA (n = 1.5 or 2.5) [114] and $NdCl_3 \cdot 2THF/TEA$ [35].

Shen et al. established linear relationships between the viscosity average molar mass and monomer conversion for the n_{Al}/n_{Nd}-ratios 17.5, 30 and 50. According to Shen et al. the slopes of the straight lines decrease with increasing n_{Al}/n_{Nd}-ratios [92]. For $NdCl_3 \cdot n$ L/AlR_3 inherent viscosities decreased with increasing n_{Al}/n_{Nd}-ratios (10–45) at various polymerization temperatures (30, 50 and 70 °C) [134]. For $NdCl_3 \cdot TBP/TIBA$ a reciprocal correlation between M_n and the amount of TIBA could be established. Also PDI decreased with increasing cocatalyst concentrations [133]. For $NdCl_3 \cdot n$ 2-ethylhexanolate/TEA (n = 1.5 or 2.5) increasing n_{TEA}/n_{Nd}-ratios resulted in a decrease of intrinsic viscosities. This effect was attributed to transfer reactions with TEA [114]. The same dependencies were confirmed when n_{TEA}/n_{Nd}-ratios were varied with the catalyst system $NdCl_3 \cdot 3$ pentanolate/TEA [115] and with $NdCl_3 \cdot 2THF/TEA$ [35]. For NdX_3-based catalyst systems the importance of Al – H groups for molar mass control was demonstrated by NMR-studies [459, 460]. The superiority of DIBAH over TIBA regarding molar mass control was also confirmed by Hsieh et al. for $NdCl_3$-based catalyst systems [134].

Influence of n_{Al}/n_{Nd} in Ternary Nd Catalyst Systems

The dependence of molar mass or molar mass equivalents on n_{Al}/n_{Nd}-ratios was investigated for the following ternary catalyst systems listed in Table 23.

In each of the studies quoted in Table 23 increasing n_{Al}/n_{Nd}-ratios result in decreases of molar mass. Between these studies there is unanimous agreement that molar mass reduction is caused by chain transfer with the cocatalyst. Most of the studies quoted in Table 23 consider the transfer reaction as irreversible. Only Friebe et al. explain their results on the basis of a reversible transfer of living polybutadienyl chains between Nd and Al [178, 179]. A comparison of chain transfer efficiencies between DIBAH and TIBA reveals that chain transfer is much less pronounced for TIBA (Sect. 4.5). For DIBAH chain transfer efficiency is 8-fold over that of TIBA and the substitution probability

Table 23 Studies on the dependence of molar mass on the n_{Al}/n_{Nd}-ratios in ternary Nd-carboxylate-based catalyst systems

Catalyst system	Range of n_{Al}/n_{Nd}-ratios	Molar mass or molar mass equivalent	Refs.
NdV/DIBAH/EASC	10–60	M_n	[191]
NdV/DIBAH/EASC	10–30	M_n	[178]
NdV/TIBA/EASC	10–30	M_n	[179]
NdV/TEA/SiCl$_4$	25–60	M_n	[187]
NdO/TIBA/DEAC	15–60	M_w and M_n	[188]
Nd^{i}O/TIBA/DIBAC	5–30	M_v	[160]
Nd^{i}O/TEA/DEAC	15–60	M_v	[168]
Nd^{i}O/TEA/DEAC	60–120	M_w, PDI	[388]
NdV/TIBA/EASC	20–176	Solution viscosity	[204]
NdV/MgBu$_2$/DEAC	3.8–22.5 (n_{Mg}/n_{Nd})	Solution viscosity	[157]

of Al – H is ∼ 22-fold over that of an isobutyl moiety attached to Al [179]. This huge difference reflects the great importance of Al-based cocatalysts with Al – H moieties. Mechanistic conclusions made from these investigations are described in Sect. 4.5. Although DIBAH exhibits comparatively high molar mass control activity only ca. 1/3 of the total amount of DIBAH participates in molar mass control [178, 179]. A possible explanation is the aggregation of DIBAH which reportedly forms trimers [461].

For Nd-alcoholate systems variations of n_{Al}/n_{Nd}-ratios were only studied for Nd(O^{i}Pr)$_3$. The catalyst system Nd(O^{i}Pr)$_3$/MMAO was used for the polymerization of IP. It was found that molar mass decreased with increasing n_{Al}/n_{Nd} ratios [231]. For the catalyst system in which MMAO was replaced by MAO (Nd(O^{i}Pr)$_3$/MAO) variations of n_{Al}/n_{Nd}-ratios resulted in a rather unusual dependence of M_n on n_{Al}/n_{Nd}. Dong et al. explained this effect by the superposition of catalyst activation and chain transfer [246]. In addition, high n_{Al}/n_{Nd}-ratios result in the formation of gel. At comparatively low n_{Al}/n_{Nd} ratios of 50 and 100 the resulting polymers dissolve well in common organic solvents (e.g. toluene, chloroform, tetrahydrofuran and carbon disulfide), whereas the polymer obtained at a high n_{Al}/n_{Nd}-ratio of 300 contained insoluble polymer fractions [231].

Taniguchi et al. investigated the IP polymerization with the catalyst system Nd(O^{i}Pr)$_3$/[HNMe$_2$Ph]$^+$ [B(C$_6$F$_5$)$_4$]$^-$/TIBA (1/1/30) with respect to the molar ratio n_{TIBA}/n_{Nd} in the range 10–50 [233]. Taniguchi et al. found that molar mass decreases by the factor 4.5 when n_{TIBA}/n_{Nd} is increased from 20 to 50.

In contrast, to the observations quoted so far, variations of n_{Al}/n_{Nd}-ratios had no effect on molar mass for the two allyl Nd systems Nd(η^3– C$_3$H$_5$)$_2$Cl · 1.5 THF and Nd(η^3– C$_3$H$_5$)Cl$_2$ · 2 THF [292].

Nd-cyclopentadienyl derivatives fall again in the known pattern. For $CpNdCl_2 \cdot THF/AlR_3$-based catalyst systems increases of n_{Al}/n_{Nd}-ratios result in decreases of molar mass. Molar mass regulation efficiencies increased in the following order: TEA < TIBA < DIBAH [300, 301]. The high molar mass regulating effect of Al – H containing compounds are exploited by Bridgestone for the production of Nd-BR with a low molar mass and a distinctly bimodal MMD [462, 463]. Bridgestone uses a quaternary catalyst system of the type Nd-component/AlR_3 or alumoxane/AlR_2H/halide component.

Beside the study given by Jenkins in 1985 [157] (Table 23) for molar mass control by magnesium dialkyl cocatalysts a molar mass regulating effect was recently described for the silylene-bridged bisfluorenyl neodymium chloride complex $[Me_2Si(C_{13}H_8)_2]NdCl$ in BD/ethylene copolymerization [307]. In this study the amount of cocatalyst was varied and with an increasing amount of MgR_2 a drop in polymer molar mass was observed. But in contrast to studies on the fast chain transfer of aluminum alkyl cocatalysts the chain transfer between Mg and Nd in the present system is described to be slow compared to chain propagation.

Molar Mass Control by Variation of Polymerization Temperature
The aspects of molar mass control by adjustments of the polymerization temperature are covered in Sect. 2.2.7. Therefore this issue is not addressed in this context.

Additives
With the well-established Ti-, Ni- and Co-based catalyst systems molar mass regulation is achieved by the addition of appropriate amounts of hydrogen, 1,2-butadiene or cyclooctadiene. In Nd-catalyzed BD polymerizations these molar mass control agents are not effective [82, 206, 207].

To the present day there is an ongoing search for the "magic additive" which allows molar mass control of Nd-catalyzed polymerizations without a detrimental effect on polymerization activities. This search is documented in the scientific as well as in the patent literature. In this context ethanol, dihydronaphthaline, chloroform, diethyl aniline, triphenylmethane, octanoic acid, allyl iodide and diallylether were unsuccessfully evaluated [464, 465]. Also propylene, oxygen, 1,5-hexadiene, ethyltrichloroacetate and n-butanol resulted in the deactivation of the catalyst system without the desired reduction of molar mass [157].

To the best of our knowledge, beside aluminum alkyls and hydridoaluminumalkyls only vinyl chloride [206, 207] and benzyl-H containing compounds such as toluene [157, 384, 385, 409, 410] are unambiguously effective in molar mass regulation. The reports on molar mass control by diethyl zinc are controversial [157, 180–182, 466, 467].

Molar mass control by vinyl chloride is described for the catalyst system $Nd^iO/TIBA/EtAlCl_2$ [206, 207]. According to this report vinyl chloride is effective in the range 1000–2500 ppm based on BD. Details on the reaction mechanism of molar mass control by vinyl chloride are not revealed in the patent.

The use of benzyl-H-containing compounds for molar mass control is described in a series of patents filed by JSR and in one scientific publication by Jenkins [157, 384, 385, 409, 410]. JSR claims the use of reagents with an active hydrogen in the benzylic position such as toluene, ethylbenzene, propylbenzene, isopropylbenzene, mesitylene or 2,4-dihydronaphthalene. In inert solvents these compounds allow for the control of molar mass as well as for the adjustment of MMDs (PDI < 4). In combination with the JSR catalyst also aluminum (hydrido) alkyls and silanes are active in molar mass control when added at later stages of the polymerization [384, 385, 409, 410]. Jenkins described the use of toluene as a molar mass regulating agent in BD polymerization with the system NdV/MgBu$_2$/DEAC. Jenkins evaluated the use of toluene for molar mass control in various hexane/toluene solvent mixtures. He found that toluene only has a marked effect when used as a pure solvent. But even in pure toluene molar mass is rather high [157].

According to Jenkins diethylzinc has no effect on molar mass [157]. In contrast to the negative result published by Jenkins there are reports from two other sources on the successful use of diethyl zinc [180–182, 466, 467]. These differences are either due to different catalyst systems or are due to differences in the addition order of catalyst components. Strong evidence in favor of molar mass control by diethyl zinc was provided by Lynch who used NdV/MgR$_2$-based catalyst systems [466, 467]. In combination with NdV/DIBAH/EASC the use of ZnEt$_2$ also resulted in a reduction of molar mass [180–182]. A careful study revealed that the formal number of polymer chains (p_{exp}) formed per Nd atom increases with increasing n_{ZnEt2}/n_{NdV}-ratios (Table 24).

From these observations the similarities of molar mass control by AlR$_3$ and ZnR$_2$ become evident. For both molar mass control agents the M_n-conversion plots are linear, the slopes of which decrease with increasing molar ratios of

Table 24 Influence of the amount of added ZnEt$_2$ (n_{ZnEt2}/n_{NdV}) on molar mass control (formal chain number $p_{exp.}$) in the catalyst system NdV/DIBAH/EASC [180], reproduced by permission of Taylor & Francis Group, LLC., http://www.taylorandfrancis.com

Molar ratio n_{ZnEt2}/n_{NdV}	Formal chain number $p_{exp.}$	PDI	cis-1,4-Content/%
0	7.6	2.63	94.5
5	8.6	2.43	90.2
10	9.3	2.15	88.6
30	10.2	1.82	84.3

n_{AlR3}/n_{NdV} and n_{ZnEt2}/n_{NdV}, respectively. It is concluded that there is a reversible exchange of alkyl groups between $ZnEt_2$ or AlR_3 on one hand and between living alkyl-poly(butadienyl) chains attached to Nd on the other hand (Sect. 4.5) [179, 180]. The addition of $ZnEt_2$ also results in the elimination of the high molar mass fraction which is formed at the start of the polymerization. As a consequence PDI decreases from 2.63 (without $ZnEt_2$) to 1.82 (30 eq. of $ZnEt_2$). The addition of $ZnEt_2$ only has a negligible influence on polymerization rates but significantly reduces cis-1,4-contents (Table 24). A comparison of the molar mass control efficiencies of $ZnEt_2$, and DIBAH and TIBA resulted in the following ranking: DIBAH > $ZnEt_2$ > TIBA [180].

2.2.9
Miscellaneous

Polymerization in the Presence of Carbon Black
According to a JSR patent Nd-catalysis allows for the polymerization of BD in the presence of carbon black. By this method a BR/carbon black master batch is obtained. The master batch technology eliminates time and energy consuming mixing of carbon black and rubber [340, 341]. According to this patent the content of bound rubber (amount of rubber which can be extracted from a carbon black/rubber mixture with an appropriate solvent, e.g. toluene) is increased and vulcanizate properties are improved.

Microwave Irradiation
For the two catalyst systems $Nd(BH_4)_3 \cdot (THF)_3/MgBu_2$ and $Nd(BH_4)_3 \cdot (THF)_3/$TEA microwave irradiation accelerates IP polymerization compared to conventional heating. Extended exposure to microwave irradiation (120 °C, > 30 min), however, results in the depolymerization of IR. Both catalyst systems yield trans-1,4-poly(isoprene). The microstructure is barely influenced by the mode of heating. According to the authors microwave irradiation increases the polarity of the Nd-alkyl-bonds which is considered as a major reason for the increase of polymerization activities at temperatures < 120 °C [468].

2.3
Homo- and Copolymerization in Solution

Beside the homopolymerization of BD Nd-based Ziegler/Natta catalyst systems are also applied for the homopolymerization of IP, the copolymerization of the dienes BD and IP, the homopolymerization and copolymerization of substituted dienes as well as for the copolymerization of BD with alkenes such as styrene, ethylene and other ethylene derivatives.

Because of the extraordinarily high cis-1,4-content of Nd-based IR and the industrial potential of this polymer this aspect is addressed in some detail.

From an industrial point of view the other aspects dealt with in this section are less exciting. Therefore, the homopolymerization of substituted dienes other than IP, the copolymerization of BD and styrene and the copolymerization of BD with ethylene and higher 1-alkenes are only briefly summarized. The Nd-catalyzed homo- and copolymerization of monomers with polar entities are not dealt with in this review.

2.3.1
Homopolymerization of Isoprene

So far, Li- and Ti-catalysts are applied for the production of commercially available *cis*-1,4-poly(isoprene) (IR = isoprene rubber). The *cis*-1,4-content of these synthetic IR-grades is lower or close to that of natural rubber (NR) (98%). For Li-based IR the *cis*-1,4 content is 92 to 94% ($\sim$ 93%, e.g. Cariflex$^{\copyright}$ IR-309, Shell) and for Ti-based IR the *cis*-1,4-content is 96 to 98% ($\sim$ 97%, e.g. Natsyn$^{\copyright}$ 200, Goodyear) [469, 470]. In comparison to NR, the lower *cis*-1,4-content of these synthetic IR-grades results in reduced strain-induced crystallization. As a consequence rubber compounds which are based on synthetic IR exhibit reduced building tack and the respective vulcanizates show inferior mechanical properties compared to NR-based products. In the production of synthetic IR (Li- and Ti-grades) molar mass is adjusted during polymerization. Therefore, IR does not have to be masticated[8] prior to compounding and processing. This is a significant advantage of IR over NR which has to be masticated prior to use. In addition, synthetic *cis*-1,4-poly(isoprene) does not contain any byproducts such as proteins [471]. In NR these impurities vary in content and cause unreproducibilies in scorch (premature vulcanization), speed of vulcanization, state of cure and vulcanizate properties [80]. Therefore, synthetically prepared IR shows a much greater consistency than NR does.

In order to gain independence from NR imports the former Soviet Union produced synthetic IR in a large scale whereas in the western world the production of synthetic IR always stayed below these volumes. Two reasons account for this: Firstly, in a free market IR faces strong competition with NR and secondly the monomer IP has never been readily available at a price which has allowed competitive IR production.

The preparation of IR by the use of Nd catalysts was already mentioned in the early patents on Nd-catalyzed diene polymerization [154, 155, 325, 326, 472–477]. Because of the limited availability of IP monomer in the western world and due to non-competitive prices of IP the option of manufacturing IR by the application of Nd-catalysis has not been the focus of the rubber producing industry during the past decades. But as a consequence of short NR supplies and strong increases of NR prices the application of Nd catalysis to the polymerization of IP has attracted new interest. In this context the

---

[8] Mechanical molar mass degradation.

most recent activities of Michelin have to be particularly mentioned [264–267, 270, 271].

Concerning scientific literature Shen et al. reported in 1980 on the use of Nd catalysts for IP polymerization. These authors apply the catalyst system $NdCl_3 \cdot 3ROH/AlR_3$ for the synthesis of IR ($\sim$ 95% cis-1,4-content; $T_g = -72\,°C$) [92]. Subsequently the application of Nd-based catalysts to the polymerization of IP was investigated in many studies, e.g. by Hsieh et al. in 1985 [134]. However, due to the higher technological importance of BR the number of studies on the polymerization of IP is considerably outnumbered by those on BD. The vast majority of papers which deal with the polymerization of IP does not reveal a significant difference of the features and dependencies observed for the polymerization of BD. In comparative studies in which BD is replaced by IP usually a slightly reduced rate of polymerization is observed for the monomer IP. This trend is also observed for the copolymerization parameters of BD and IP (Sect. 2.3.2) where r_{BD} is consistently higher than r_{IP}.

In some recent patents Michelin reveals great progress in the preparation of IR by Nd catalysis (Sect. 2.1.1.4). In these patents IR with an exceptionally high cis-1,4-content of 99.0–99.6% is described [266, 267]. In order to obtain such a high cis-1,4-content a Nd-phosphate-based catalyst is combined with low n_{Al}/n_{Nd}-ratios of 1 to 4. Contrary to other Nd catalysts, such as NdV-based systems, NdP-based catalyst systems exhibit high activities even at low n_{Al}/n_{Nd} ratios. These low n_{Al}/n_{Nd} ratios are essential for achieving such outstandingly high cis-1,4-contents. In addition, Michelin reports on the selective polymerization of IP in the crude C_5 cracking fraction [478]. In this way Michelin successfully avoids the expensive isolation of IP from the crude C_5 cracking fraction and the subsequent IP purification prior to its polymerization [264, 265].

Beside this work of Michelin, the application of Nd-catalyzed gas-phase polymerization to IR by Zhang et al. confirm the features elaborated for the gas-phase polymerization of BD (Sect. 3.2) [479, 480].

In addition to the work on the optimization of the cis-1,4-content of IR some work has been devoted to obtaining poly(isoprene) with a high trans-1,4-content. Bonnet et al. report on the use of the complex $Cp^{*\prime}Nd(BH_4)_2(THF)_2$ ($Cp^{*\prime} = C_5Me_4{}^nPr$) which exhibits a rather high activity (up to 37 300 g(polymer) mol^{-1}(Nd) h^{-1}) after activation with magnesium dialkyls [349]. This catalyst system is highly trans-specific (98.5%) and yields the highest trans-1,4-content for poly(isoprene) yet described for a homogeneous organometallic catalyst. Furthermore, by application of this half-sandwich complex monomodal IR with a very narrow distribution of molar masses ($M_w/M_n = 1.15$) is produced in a quasi-living manner. The high activity of the catalyst system is explained by the presence of the electron-rich $Cp^{*\prime}$-ligand which favors high trans-1,4-contents by $\eta^3 - \eta^1$-allylic rearrangements. In addition, the catalyst system exhibits single-site character which is considered to be the essential reason for the narrow distribution of molar masses.

2.3.2
Copolymerization of Butadiene and Isoprene

The similarity in chemical structure of BD and IP allows the copolymerization by many Ziegler/Natta catalysts. To our knowledge, the first study on Nd-catalyzed copolymerization of BD and IP was reported by Monakov et al. [152, 153]. Theses authors copolymerized the two dienes with the catalyst system Nd stearate/TIBA/DEAC in toluene at 25 °C. The respective copolymerization parameters reported in this study are: $r_{BD} = 1.94$ and $r_{IP} = 0.62$.

Shen et al. determined the BD/IP copolymerization parameters for the polymerization with the ternary catalyst system NdN/TIBA/EASC at 50 °C: $r_{BD} = 1.4$ and $r_{IP} = 0.6$ [92]. Over a wide range of BD/IP copolymer compositions the experimentally determined T_g values significantly deviate from the theoretical curve which was calculated by the Fox equation for random copolymers. Only for IP-contents < 10 wt. % does the experimentally determined data coincide with the theoretical curve. Shen et al. also succeeded to synthesize block copolymers comprising poly(butadiene) and poly(isoprene) building blocks [92].

Hsieh et al. used NdCl$_3$-based catalysts for the copolymerization of BD and IP [134]. They determined the following copolymerization parameters: $r_{BD} = 1$ and $r_{IP} = 1$. For a copolymer with a 50/50 composition (wt. % based) a single $T_g = -85$ °C was found and for the dependence of $1/T_g$ on copolymer composition a linear relationship was obtained.

Oehme et al. used the catalyst NdO/TEA/DEAC for the BD/IP-copolymerization and determined the reactivity ratio of BD and IP by the method of Kelen–Tüdös: $r_{BD} = 1.09$ and $r_{IP} = 1.32$. This is the only study in which $r_{IP} > r_{BD}$ [168]. Studies on the copolymerization of BD and IP with the catalyst system NdO/TIBA/EASC performed by the same group showed that the cis-1,4-content of the BD units decreases with increasing content of incorporated IP [163]. A higher content of incorporated IP results in a lower PDI of the copolymer. The dependence of T_g on copolymer composition indicates a random distribution of BD and IP units in the copolymer.

For the catalyst system Nd^{i}O/TIBA/DIBAC (1/2.5/15) Gehrke et al. determined the copolymerization parameters for BD and IP at 60 °C according to the method of Kelen and Tüdös: $r_{BD} = 2.28$ and $r_{IP} = 1.79$. The catalyst system was preformed for 30 min at 25 °C prior to polymerization [481].

The catalyst system Nd(OCOCCl$_3$)$_3$/TIBA/DEAC was used for the copolymerization of BD and IP by Kobayashi et al. [176]. The copolymerization parameters were determined at 0 °C: $r_{BD} = 1.22$ and $r_{IP} = 1.14$. This study also revealed that the microstructures of the poly(butadiene) and poly(isoprene) moieties were not influenced by polymerization temperature.

The use of ternary NdP-based catalyst systems for the copolymerization of BD and IP was reported by Laubry (Michelin) [270, 271]. Also gas-phase polymerization was recently applied for the preparation of IP/BD block copoly-

Table 25 Copolymerization parameters r_{BD} and r_{IP} for the copolymerization of butadiene and isoprene with Nd catalyst systems

Catalyst system	r_{BD}	r_{IP}	Refs.
Nd stearate/TIBA/DEAC	1.94	0.62	[152, 153]
NdN/TIBA/EASC	1.4	0.6	[92]
$NdCl_3$/AlR_3	1.0	1.0	[134]
NdO/TEA/DEAC	1.09	1.32	[168]
Nd^iO/TIBA/DIBAC	2.28	1.79	[481]
$Nd(OCOCCl_3)_3$/TIBA/DEAC	1.22	1.14	[176]

mers. In this study the authors used a silica-supported NdN-based catalyst system [482, 483].

On the basis of the available information it can be concluded that the copolymerization of BD and IP occurs in a close to random fashion. In spite of small differences in the copolymerization parameters reported it may be summarized, that the homopolymerization of BD is slightly favored over the homopolymerization of IP. There is just one exception in the literature for which $r_{IP} > r_{BD}$ [168]. In the other publications available r_{BD} is consistently higher than r_{IP} (Table 25).

By the hydrogenation of BD/IP-copolymer feedstocks fully and partially hydrogenated rubbers with low crystallinities and low glass transition temperatures are obtained. For the preparation of highly amorphous hydrogenated rubbers a random copolymer composition is essential. In this respect feedstock preparation by the use of the Nd-carboxylate-based system NdV/DIBAH/EASC yielded better results than BD/IP copolymer feedstocks prepared by anionic polymerization with BuLi [484, 485].

2.3.3
Homopolymerization and Copolymerization of Substituted Butadienes (other than Isoprene)

Substituted conjugated 1,3-butadienes are especially interesting with regard to conclusions about the mechanism of the polyinsertion reaction. From the substitution pattern of the monomers and the microstructure of the formed polymers mechanistic conclusions can be drawn. These issues are addressed in detail in review articles by Porri et al. [50, 141, 486].

In this subsection some studies on the homo-and copolymerization of substituted conjugated dienes (Scheme 17) are summarized.

Monomers such as E-1,3-pentadiene (piperylene) and E-2-methyl-1,3-pentadiene give 1,4-polymers with an asymmetric carbon atom. Therefore, these monomers in principle will result in polymers with an isotactic, syndiotactic or atactic structure.

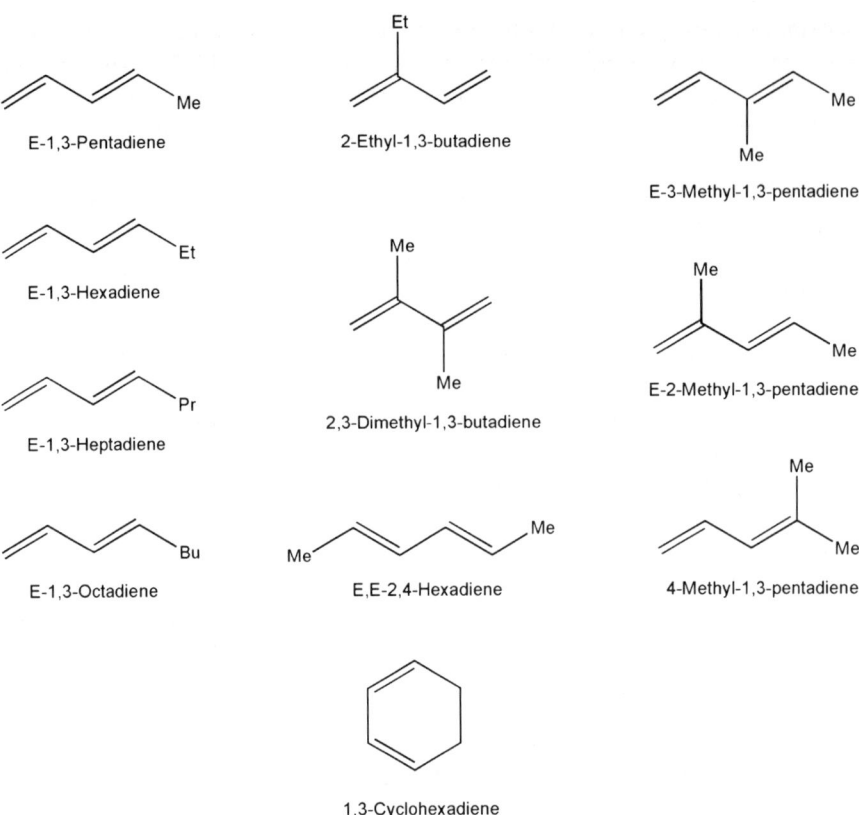

Scheme 17 Substituted conjugated 1,3-butadienes which have been homo- and/or copolymerized by Nd-catalysts

Early studies on the homopolymerization of E-1,3-pentadiene yielded polymers with a high *cis*-1,4-content and an isotactic structure, whereas E-2-methyl-1,3-pentadiene resulted in a polymer with a mixed *cis*-1,4/*trans*-1,4-structure [487–492]. Investigations on the polymerization of E-1,3-pentadiene with the system NdN/TIBA/DEAC partially support these findings as a poly(1,3-pentadiene) with a *cis*-1,4-threo-disyndiotactic structure was obtained [492]. A somewhat lower *cis*-1,4-content of 70% was obtained when the polymerization of E-1,3-pentadiene was catalyzed by $(CF_3COO)_2NdCl/TEA$ [493, 494]. When 2,3-Dimethyl-1,3-butadiene is polymerized with the catalyst $NdN/TIBA/EtAlCl_2$ the resulting poly(2,3-dimethylbutadiene) predominantly contains *cis*-1,4-units [495, 496].

A study in which BD and E-1,3-pentadiene were homopolymerized by the application of the same ternary Nd-carboxylate-based system revealed a significant drop in *cis*-1,4-structures from 95% (BD) to 80% (E-1,3-pentadiene) [490, 491].

An even more significant difference in microstructures was found for polymers obtained by the Nd catalyzed polymerization of BD, IP, E-1,3-pentadiene and E,E-2,4-hexadiene. The first three monomers yield polymers with high cis-1,4-contents whereas the latter monomer yields a polymer with 100% trans-1,4-configuration [495, 496].

E-3-Methyl-1,3-pentadiene was polymerized using NdO/TIBA/DEAC and isotactic polymers with a high crystallinity were obtained. The polymers essentially consisted of cis-1,4 units ($\geq 80\%$). Isotactic cis-1,4-poly(3-methyl-1,3-pentadiene) co-exists in two morphological structures [497].

A study on the homo- and copolymerization of a variety of dienes such as 2,3-dimethyl-1,3-butadiene, 2-ethyl-1,3-butadiene, E-1,3-pentadiene, E-1,3-hexadiene, E-1,3-heptadiene, E-1,3-octadiene, E,E-2,4-hexadiene, E-2-methyl-1,3-pentadiene, 1,3-cyclohexadiene mainly focused on mechanistic aspects [139]. It was shown that 1,4-disubstituted butadienes yield trans-1,4-polymers, whereas 2,3-disubstituted butadienes mainly resulted in cis-1,4-polymers. Polymers obtained by the polymerization of 1,3-disubstituted butadienes showed a mixed trans-1,4/cis-1,4 structure (60/40). The microstructures of the investigated polymers are summarized in Table 26.

In later studies on the homopolymerization of E-1,3-pentadiene with NdO/TIBA/DEAC crystalline polymers with cis-1,4-contents in the range 84–99% and a high isotacticity were obtained. It was found that the cis-1,4-content increases when the polymerization temperature is decreased from room temperature to $-30\,°C$. The polymerization of E-2-methyl-1,3-pentadiene resulted in polymers which almost exclusively comprised cis-1,4-units and no dependence of the cis-1,4-content on polymerization temperature was observed. The obtained poly(2-methyl-1,3-pentadiene) was composed of various polymer fractions with different stereo regularities [165, 166].

A comprehensive study on the homopolymerization of various substituted poly(butadiene)s was performed by Porri et al. [167, 294]. In this study the homopolymerization of the monomers listed in Table 27 was compared and

Table 26 Microstructure of poly(diene)s obtained by Nd catalysis [139], reproduced with permission of the American Chemical Society

Polymer	cis-1,4-Content/%	trans-1,4-Content/%
Poly(butadiene)	70–99	0.1–30
Poly(isoprene)	≤ 99	< 1
Poly(2,3-dimethylbutadiene)	> 97	–
Poly(E-1,3-pentadiene)	60–95	0–5
Poly(E-1,3-hexadiene)	30–60	10–30
Poly(E-2-methyl-1,3-pentadiene)	40	60
Poly(E,E-2,4-hexadiene)	–	98

Table 27 Polymerization rate and polymer properties of substituted butadiene derivatives [167], reproduced with permission from Wiley-VCH Verlag GmbH & Co. KGaA

Monomer	Polymerization		Microstructure		$[\eta]$(Tol/ 25 °C)/ dL·g^{-1}	m.p./ °C	i.p.*/ Å
	Time/ h	Conv./ %	cis-1,4/ %	trans-1,4/ %			
1,3-Butadiene	0.25	53	98–99		> 10		
2-Methyl-1,3-Butadiene	1	75.8	91.8		4.6		
2-Ethyl-1,3-Butadiene	5	31.4	93		2.9		
2,3-Dimethyl-1,3-Butadiene	23	7.5	100			194.7	
E-2-Methyl-1,3-Pentadiene	30	10.2	100			161	7.9
1,3-Pentadiene	6	30	87		5.2	45	8.15
E-1,3-Hexadiene	5	17.7	85		1.3	85.9	8.0
E-3-Methyl-1,3-Pentadiene	5	28.2	79.4		2.0	68.9	8.02
E,E-2,4-Hexadiene	100	8.3		100		91.1	8.24

* identity period

the respective polymers were characterized. For this study the catalyst system Nd carboxylate/TIBA/DEAC was used.

In this study it is demonstrated that polymerization rates as well as molar masses decrease with increasing substitution at the butadiene moiety. It is remarkable that in all cases except for E,E-2,4-hexadiene high cis-1,4-contents are obtained. Most of the polymers based on substituted butadienes of the type $CH_2 = CH - CH = CHR$ are highly crystalline and highly isotactic.

In very recent years Nd-catalysis in the homo- and copolymerization of substituted conjugated dienes was no longer in the focus of research work.

2.3.4
Copolymerization of Butadiene and Styrene

It is interesting to note, that there are publications on the Nd-catalyzed copolymerization of BD and styrene (St) as well as on the selective polymerization of BD which is performed in the presence of St as the solvent. The copolymerization of BD and St cannot be achieved with standard binary or ternary catalyst systems which yield BR with a high cis-1,4-content. This is best demonstrated by the fact that the ternary "high-*cis*" catalyst system NdV/DIBAH/EASC exclusively yields the BR homopolymer even when the polymerization of BD is carried out in a large excess of styrene being used as a solvent (Sect. 3.3). In order to achieve the copolymerization of BD and St catalyst systems have to be used which are less stereospecific and yield BR with a significantly lower *cis*-1,4- and a higher *trans*-1,4-content.

One catalyst system which allows for the copolymerization of BD and St is based on NdV and is activated by MAO. It is important to note that this catalyst system does not contain a halide donor. By the addition of cyclopentadienyl derivatives (e.g. cyclopentadiene, indene, anthracene etc.) St incorporation is increased. It may be speculated that by the influence of MAO a proton is abstracted from the cyclopentadienyl derivatives added and that the resulting cyclopentadienyl-type anions coordinate to the active Nd sites [498, 499]. In this way the activity pattern of Nd is changed and copolymerization of BD and St is made possible. As shown in the attached Table 28 increases of St incorporation occur simultaneously with decreases of *cis*-1,4-contents.

Also Nd(η^3- C_3H_5)$_3$-based catalyst systems have been applied successfully for the copolymerization of BD and St. The *cis*-1,4-content of incorporated BD moieties is around 10% [500, 501].

Another catalyst system which allows for the copolymerization of BD and St comprises the catalyst components Nd(OCOCCl$_3$)$_3$/TIBA/DEAC [177]. The catalyst yields low *cis*-1,4-contents and high contents of *trans*-1,4- and 1,2-structures in the BD units. The content of incorporated St increases with increasing polymerization temperature.

Table 28 Increase of styrene incorporation by the addition of cyclopentadienyl (C) derivatives (catalyst system: NdV/MAO) [498, 499]

Patent example	1	4	5	6	7	9	10	15	16	18
Type of Cp derivative added		1H-Indene	1,2,3,4,5-Pentamethyl-cyclopenta-1-H-diene		1H-Indene		1H-Indene		no Cp type additive	
Content of incorporated St/wt.-%	14.4	23.5	7.9	18.8	18.6	19.1	14.4	13.3	4.4	5.5
Content of incorporated BD/wt.-%	85.6	76.5	92.7	81.2	81.4	80.9	95.6	86.4	95.6	94.5
cis-1,4/%	37.8	41.0	36.4	27.0	50.9	51.8	57.9	67.9	61.6	58.2
trans-1,4/%	54.8	51.5	52.9	62.8	41.7	40.8	34.5	24.2	32.1	35.3
1,2/%	7.4	7.5	10.7	10.3	7.4	7.4	7.6	7.9	6.3	6.5

For the copolymerization of BD and St Zhang et al. use various Nd precursors in combination with TIBA and DIBAC [502]. These authors obtain copolymers with a content of incorporated styrene of ca. 10–20 mol %. At this level of incorporated styrene the BD moieties exhibit cis-1,4-contents between 90 to 97%. The amount of incorporated St increases with increasing polymerization temperature and with increasing St content in the monomer feed. It is demonstrated that the cis-1,4-content of the incorporated BD moieties decreases with increasing content of incorporated St. In this study styrene incorporation depends on the composition of the catalyst system. The highest amount of styrene is incorporated at the molar ratios of $n_{TIBA}/n_{Nd} = 20$ and $n_{Cl}/n_{Nd} = 3$.

Further studies by Zhang et al. extend the range of Nd-based catalyst systems which allow for the copolymerization of BD and St [377]. In this study Nd phosphates, Nd carboxylates and Nd acetylacetonate were tested with several cocatalysts and halide donors. The catalyst components were added in a special sequence and catalyst aging was applied. It is shown that the use of the cocatalyst $MgBu_2$ yields a higher content of incorporated St than the application of the cocatalyst TIBA. On the other hand the use of TIBA leads to a higher cis-1,4-content (94.7%) than $MgBu_2$ (85.4%). In a comparative study in which various Nd-components were activated with $MgBu_2/CHCl_3$ it turned out that NdP yields the highest cis-1,4-content but allows for only 9 mol % of St incorporation. On the other hand the system $Nd(OCOPh)_3/MgBu_2/CHCl_3$ yields a cis-1,4-content as low as 64% but incorporates 30 mol % St. In general, an increase of the amount of $MgBu_2$ enhances trans-1,4-polymerization and favors the incorporation of St. The same trends are also observed when polymerization temperature is increased from 5 °C to 64 °C and when the content of styrene in the monomer feed is increased. The authors give an estimate for the copolymerization parameters of BD and St: $r_{BD} = 36$ and $r_{St} = 0.36$ [377].

For the random copolymerization of BD and St Gromada et al. evaluated various dialkylmagnesium cocatalysts in combination with the well-defined complexes $Nd_3(O^tBu)_9(THF)_2$, $Nd(O-2,6-^tBu-4-Me-Ph)_3(THF)$ and $Nd(O-2,6-^tBu-4-Me-Ph)_3$ [235]. It was shown that the preparation of BD/St copolymers was only possible by the use of aryloxide-based Nd systems. BD moieties in the copolymer predominantly exhibit trans-1,4-structure and the St content of the copolymers ranges between 3–13.5 mol %. The copolymers have a molar mass $M_n \leq 42\,800\,g \cdot mol^{-1}$. Aliphatic Nd precursors such as $Nd_3(O^tBu)_9(THF)_2$ are shown to be inactive in the copolymerization of BD and St.

The impact of solvent on the copolymerization of BD and St was studied for the catalyst system $Nd(N(SiMe_3)_2)_3/[HNMe_2Ph]^+ [B(C_6F_5)_4]^-/$ TIBA [320]. With aromatic solvents such as toluene copolymers with a St content above 12% were obtained. With cyclohexane the St content of the copolymers remained below 12%. Again, the cis-1,4-content of incorporated

BD considerably decreases with increasing St incorporation (from 83% *cis* at 8 mol % styrene to 54% *cis* at 15 mol % styrene) and catalyst activity decreases with increasing content of St in the monomer feed. With this system maximum St contents of ca. 15 mol % were obtained when toluene was used as the solvent.

BD/St-copolymers were also prepared by the use of the catalyst system Nd(OCOCCl$_3$)$_3$/TIBA/DEAC [503]. The copolymer exhibited 79% *cis*-1,4-structure in the BD units at a content of incorporated styrene of 23 mol %. In this study diades were also determined. According to the authors BD moieties which are adjacent to St moieties predominantly exhibit a *trans*-1,4-configuration whereas BD moieties in BD-BD-diades exhibit a *cis*-1,4-structure [503]. It therefore can be concluded that the microstructure of an entering BD monomer is controlled by a penultimate effect. This effect can be best described by a model in which backbiting coordination of a penultimate BD unit to Nd is involved [177, 367].

Unique effects in the copolymerization of BD and St were reported by Oehme et al. who apply a special addition and aging procedure for the catalyst system NdO/TIBA/CCl$_4$ (molar composition: 1/29/24) [504]. Exceptionally high *cis*-1,4-contents go along with high contents of incorporated St. Even at St contents as high as 80 mol % the microstructure of the BD units still is ca. 90% *cis*-1,4. The only disadvantage of these catalyst systems are low activities when high amounts of St are incorporated. The authors also report on the copolymerization parameters for this system: r_{BD} = 4.3 and r_{St} = 0.5.

The application of Nd catalysis to the preparation of block copolymers with BR and PS building blocks has not been successful to the present day. Studies aiming at this target have been described by Jenkins [173].

2.3.5
Copolymerization of Butadiene with Ethylene or 1-Alkenes

By the use of special Nd-catalysts the copolymerization of BD and ethylene is achieved. Publications are available which describe the use of neodymocene catalyst systems for this type of copolymerization. In these studies the most active neodymocene compounds are the dimethylsilyl-bridged dicyclopentadienyl complex [Me$_2$Si(3-Me$_3$SiC$_5$H$_3$)$_2$]NdCl [306, 505] (Sect. 2.1.1.6, Scheme 10), the dimethylsilyl-bridged bisfluorenyl neodymium chloride complex [Me$_2$Si(C$_{13}$H$_8$)$_2$]NdCl [307] and the dimethylsilyl-bridged cyclopentadienyl fluorenyl Nd complex [Me$_2$Si(C$_5$H$_4$)-(C$_{13}$H$_8$)]NdCl [308, 309].

Copolymerization of BD and ethylene is achieved by the use of the catalyst system [Me$_2$Si(3-Me$_3$SiC$_5$H$_3$)$_2$]NdCl/BuLi/DIBAH at a molar ratio of 1/10/10. The catalyst system yields BD/ethylene copolymers with a BD-content up to 42 mol %. The copolymers are predominantly alternating and exhibit a high *trans*-1,4-content of the BD moieties. This system is also active in the homopolymerization of BD [505]. Further investigations in the activation of neodymocene catalysts confirm that the cocatalyst combination BuLi/DIBAH is superior over catalyst activation by MgR_2 [306]. These studies disclose that by the use of an optimized BuLi/DIBAH activation up to 62 mol % of BD is incorporated in the copolymer. The BD/ethylene copolymerization parameters are estimated (r_{BD} = 0.08 and $r_{Ethylene}$ = 0.25) and the observed limitation of molar masses is attributed to β-H-elimination and to irreversible chain transfer to the aluminum alkyl cocatalyst. Catalyst systems of this type were claimed in patents by Michelin and Atofina [506, 507].

By application of the dimethylsilyl-bridged bisfluorenyl neodymium chloride complex $[Me_2Si(C_{13}H_8)_2]NdCl$ (cocatalysts: BuLi, DIBAH, BuMgCl, BuMgOct) a complete new class of polymers is accessible by the copolymerization of ethylene and BD [307]. By reaction of 1 eq. BD and 2 eq. of ethylene 1,2-cyclohexane units are formed along the polymer chain. More than half of the inserted BD units (53–58 mol %) are involved in the ring closure reaction. The residual non-cyclicized BD units exhibit a nearly identical ratio of *trans*-1,4- and 1,2-units.

With the related cyclopentadienyl fluorenyl Nd complex $[Me_2Si(C_5H_4)-(C_{13}H_8)]NdCl$ which is activated by butyloctylmagnesium Thuillez et al. [308, 309] recently succeeded to establish another catalyst system which allows the alternating ethylene/BD polymerization. The catalytic activity is in the range of $100–300\,kg(polymer)\,mol^{-1}(Nd)\,h^{-1}$ and increases on addition of BD to the monomer feed. The diene units in the copolymer exhibit *trans*-1,4-configuration of up to 99.6% which is suggested to be a consequence of s-η^4-*trans*-coordination (Sect. 4.3) of the BD monomer to the catalyst center. With the same complex $[Me_2Si(C_5H_4)-(C_{13}H_8)]NdCl$ and appropriate cocatalysts copolymers of BD and other 1-alkenes such as propylene, hexene and octene were obtained [310]. The content of incorporated 1-alkenes is around 30 mol %. The work in the field of dimethylsilyl-bridged Nd sandwich complexes performed by Boisson et al. is reviewed in [310].

Copolymerizations of BD with 1-alkenes such as 1-octene and 1-dodecene aim at short chain branching of BR. Kaulbach et al. used the ternary catalyst system NdO/TIBA/EASC (n_{TIBA}/n_{Nd} = 25, n_{Cl}/n_{Nd} = 3) for the respective copolymerizations of BD/1-octene and BD/1-dodecene [508]. These authors showed that only small amounts of 1-alkenes are incorporated and that no neighboring 1-alkene moieties are present in the copolymer. The copolymerization parameters have been determined by the method of Kelen–Tüdös: r_{BD} = 25 and $r_{1\text{-octene}} \approx 0$; r_{BD} = 18 and $r_{1\text{-dodecene}}$ = 0.1. With increasing amounts of 1-alkene in the monomer feed catalyst activity decreases drastically. The *cis*-1,4-contents of the BD units in the copolymer were around 90% and were barely affected by increases of the 1-alkene content in the monomer feed.

3
Other Polymerization Technologies

In contrast to the polymerization of dienes with Ti-, Co- and Ni-based catalyst systems in Nd-catalysis the tendency of monomer dimerization, branching and gelling of the polymer is particularly low. Even at high monomer conversion these favorable features are maintained. Therefore Nd-catalysis can be applied for solvent-free diene polymerization. From an economic point of view this approach is advantageous as solvents have to be bought, stored, purified and recycled. In addition, the ecological impact of organic solvents can be avoided. There are two technologies which take advantage of these beneficial features of Nd-based catalyst systems. One application is the polymerization in mass or bulk. The other approach applies polymerization in the gas phase.

Another positive aspect of Nd catalysis is the high selectivity of Nd catalysts towards the polymerization of dienes. This feature is exploited in two ways. First, homopolymerization of BD in the presence of styrene as a solvent and second, selective homopolymerization of IP in the crude C_5 cracking fraction.

3.1
Polymerization in Bulk/Mass and in Suspension

Already in the early 1980s the potential of Nd-catalysis for the polymerization of BD and IP in mass was recognized by Enoxy/Enichem. By the application of this technique the use of solvent is avoided. The progress made in this area is documented in a series of patents. In the first patent (1983) it is demonstrated that bulk polymerization is feasible in the temperature range of 30 to 50 °C at a reaction time of 1 to 2 hours. Under these conditions no gel is formed, catalyst activities are high and residual Nd-contents are in the range of 68 to 75 ppm [226, 227]. Polymer concentrations from 25 to 70 wt. % are best handled in tubular reactors with plug flow characteristics or in appropriate self-cleaning single or twin screw extruders with residence times between 20 to 60 min. The polymerization is performed under partially adiabatic conditions with a monomer inlet temperature between 50 to 60 °C and an outlet temperature up to 130 °C. Additional temperature control is achieved by evaporation cooling [509, 510].

In a later patent (1986) it is disclosed that phase separation of *cis*-1,4-BR and BD occurs at 30 to 35 °C. Below 30 °C there is a single phase and above 35 °C there are two distinct phases of BR and BD. By the application of two polymerization steps the first of which is performed below and the second above the critical solution temperature molar mass is decreased and costs for aluminum alkyls which are used for molar mass control are reduced [511, 512]. Control of molar mass is further improved by the sequential addition of aluminum alkyls during the course of the polymerization. The reaction is shortstopped and residual monomer is removed in a separate degassing device. A wash step for the removal of catalyst residues is not required [389, 390]. By the addition of powdering agents the operation of the process is improved [513, 514]. Bulk polymerizations are best performed in a reactor with a special design [515, 516].

Bulk polymerization processes described so far are performed without the use of substantial amounts of solvents. Only small amounts of solvents (< 2 wt. %, and > 98 wt. % monomer) are required for preparation and addition of the catalyst solutions.

In the suspension (or slurry) polymerization of BD organic solvents (diluents) are used which dissolve the monomer BD but not the polymer. Preferred diluents comprise saturated hydrocarbons with a specified C_4-content (C_4-content > 80 wt. %) [517, 518]. As a consequence of the heterogeneous nature of the reaction mixture the viscosity of the slurry is low even at relatively high solid concentrations (25–40 wt. %). Because of the low slurry viscosity the heat of polymerization can be easily removed and reaction temperature can be well controlled even in an ordinary stirred tank reactor. In order to avoid particle agglomeration partitioning agents such as silica are added. The catalyst components are either added as solutions or impregnated on appropriate catalyst supports. In the slurry process polymerization activities are low and residual Nd concentrations in the polymer are high (180–250 ppm). Contrary to polymerizations which are performed in solution control of molar mass by 1,2-butadiene is possible under slurry conditions.

3.2
Polymerization in the Gas Phase

In addition to the polymerization in bulk or mass, polymerization in the gas phase provides another option to perform Nd-catalyzed polymerizations without the use of solvents. In gas-phase polymerization the reaction medium is a diene monomer in the gaseous state or a mixture of monomer with an inert gas such as nitrogen. In the gas-phase process polymer granules are dispersed in the gaseous medium. As the viscosity of the heterogeneous polymer/monomer mixture is low the production of BR with high molar mass is not a critical issue. On the other hand heat capacities of gases are rather low and the rate of the gas flow has to be carefully controlled in order to avoid blow out of polymer granules from the reactor. Therefore, the removal of the polymerization heat is a decisive factor in a gas-phase process as overheating of polymer granules has to be avoided.

Reactor
In Scheme 18 a gas-phase stirred bed reactor for BD polymerization is shown. A 0.25 L semi-batch reactor was used by Ni et al. for studies on the influence

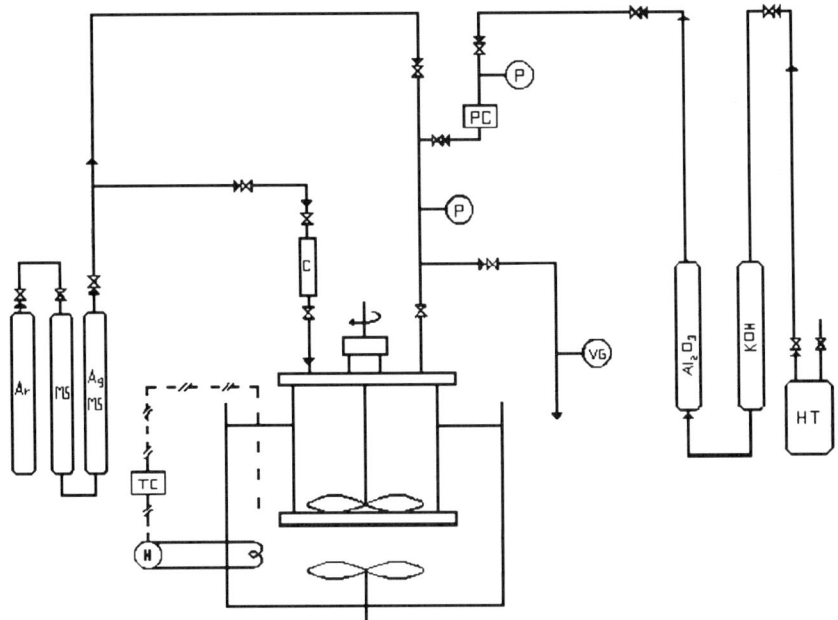

Scheme 18 Gas-phase stirred bed reactor for BD polymerization: TC = temperature control, PC = pressure control, HT = monomer tank, VG = vacuum gauge, C = catalyst injection column, MS = molecular sieves column, H = heat exchanger [388], reprinted with permission of John Wiley & Sons, Inc.

of the cocatalysts DIBAH and TIBA in the Nd-catalyzed gas-phase polymerization with the systems NdN/EASC/DIBAH and/or TIBA [388]. Before the start of the polymerization the reactor is evacuated at 95 °C for one hour and purged with argon several times. The polymerization is started in an argon atmosphere by the consecutive additions of supported catalyst and monomer.

Stickiness and Gelling

For the polymerization of ethylene and propylene large-scale gas-phase processes are well established. The implementation of gas-phase technology to the production of "sticky polymers" such as the ethylene/propylene-based rubbers EPM and EPDM was pioneered by UCC [519]. In a series of patents, UCC describes various approaches to overcome the inherent stickiness of rubber granules in the gas-phase polymerization. These approaches include the use of anti-agglomerants such as carbon black, silica, inorganic salts or appropriate catalyst supports and antistatic voltage etc. [520–535]. The addition of fluidization or anti-agglomeration aids is described by Zöllner et al., silica is used in particular [536, 537].

Contrary to the production of EP-rubbers, from a statistical point of view the formation of gel is much more pronounced in BD polymerization. This aspect is particularly important in a gas-phase process in which the monomer conversion in the individual polymer granules can be very high, especially when diffusion of monomer becomes the rate determining step. In addition, a local overshoot of temperature can easily occur especially at the start of the polymerization when the reactor granules are small and their respective heat capacities are still low.

Catalyst Supports and Catalyst Preparation

In 1992/1993 Bayer recognized the potential of Nd catalysis for the gas-phase polymerization of BD [538, 539]. The first patent in this field claims the use of supported Nd catalyst systems for the gas-phase polymerization of dienes. Soon after, gas-phase polymerization of BD is mentioned in a patent filed by UCC [540, 541]. Finally, Goodyear [542, 543] and Ube [544, 545] joined Bayer and UCC in their efforts to establish gas-phase technology for BR production.

In the first patent on gas-phase production of Nd-BR special attention is paid to the preparation of the supported catalyst [538, 539]. The components of a ternary Nd catalyst system which comprises a Nd precursor, an aluminum cocatalyst (including aluminum alkyls as well as alumoxanes) and a halide donor are pre-reacted in an inert solvent. By the subsequent removal of the solvent either at elevated temperature or at reduced pressure the catalyst system is impregnated on the support. The claimed catalyst support comprises a variety of particulate inorganic substances with a specific surface area $> 10\,m^2/g$ and a pore volume ranging from 0.3 to 15.0 cm^3/g [538, 539]. In a subsequent patent catalyst activity is increased by pre-reaction of the catalyst components in the presence of a small amount of diene monomer [546, 547]. Studies performed by Zhang et al. confirm the beneficial effect of dienes being present during the pre-reaction of the catalyst components [548, 549]. Sequential impregnation of the different catalyst components on the support further increases activity [550, 551]. The use of the catalyst support "Aktiv-Kohle" (activated carbon) results in a particularly high polymerization activity [552, 553]. In contrast to the before-mentioned inorganic catalyst supports, Steinhauser uses a support based on organic polymers such as microporous poly(propylene) and microporous poly(ethylene) [554, 555].

In the patents mentioned so far, the different catalyst components are either simultaneously or sequentially impregnated on the support. By these methods the supported catalyst is present in the active form and is sensitive to impurities and aging prior to use. Changes in activity which can occur during storage are avoided by the impregnation of the various catalyst components on at least two different supports. In doing so it is advantageous to separate the Nd-component and the aluminum alkyl component. The halide donor is added to either of the two supports [556–559].

Owing to the heat capacity of the catalyst support (this holds especially true for inorganic supports) overheating is avoided particularly in the early stages of the polymerization after the supported catalyst is introduced to the gas-phase reactor. Another approach is taken by UCC who describe the use of catalyst solutions rather than supported catalysts. In the UCC-approach catalyst solutions are injected into the gas-phase reactor and the polymerization heat is removed by evaporation of the solvent [560–563].

Control of Molar Mass

As in Nd-catalyzed solution processes in gas-phase polymerization of BD regulation of molar mass is a serious problem as there are no agents for the control of molar mass readily available. Vinyl chloride and toluene are no viable options. Vinyl chloride is ruled out due to ecological reasons and toluene is not applicable due to low transfer efficiencies and the required low concentrations if applied in a gas-phase process. For the control of molar mass and MMD in the polymerization of dienes a combination of different methods is recommended [457, 458]: (1) temperature of polymerization, (2) partial pressure of BD, (3) concentration of cocatalyst (or molar ratio of n_{Al}/n_{Nd}), (4) type of cocatalyst, (5) residence time of the rare earth catalyst in the polymerization reactor.

In a scientific paper on the control of molar mass in gas-phase polymerization, the importance of n_{Al}/n_{Nd} is also emphasized. In addition, a combination of the cocatalysts TIBA and DIBAH is recommended. By the use of two aluminum alkyl compounds the concentration ratio of two different active Nd-species is adjusted. As these two species produce different molar masses and MMDs the combination of TIBA and DIBAH allows for the control of these two parameters [229, 230].

Another option for the production of gas-phase-BR with a well-controlled molar mass comprises a two-step process. In the first step the production of (ungelled) BR with a high molar mass (ML 1 + 4/100 °C) ranging from 70–180 MU is performed. This rubber is well suited for the gas-phase process as it is less sticky than BR with low molar mass. In a subsequent process step the high molar mass BR is masticated in order to make the rubber processable in standard compounding equipment. In the second process step Mooney viscosities (ML 1 + 4/100 °C) are adjusted to the range 10–70 MU [564–567].

In addition to the patent literature available on the production of BR in the gas-phase there is some scientific literature which mainly refers to the modeling of reaction kinetics. Details on the experimental procedure for the determination of the macroscopic kinetics of the Nd-mediated gas-phase polymerization of BD in a stirred-tank reactor are reported [568, 569]. Special emphasis is given to video microscopy of individual supported catalyst particles, individual particle growth and particle size distribution (PSD). These studies reveal that individual particles differ in polymerization activity [536, 537, 570, 571]. Reactor performance and PSD are modeled on the basis of individual particle growth and dynamic population balances [572, 573]. A multigrain model is used for the simulation of reactor performance [574, 575] and a kinetic model is given which uses two kinds of active species that exhibit first- and second-order decay [576, 577].

Miscellaneous

In addition to the studies in which supported catalysts are exclusively used for gas-phase polymerizations one study is available in which the supported catalyst is optimized in a solution process prior to its application in the gas phase. Tris-allyl-neodymium [Nd(η^3–C$_3$H$_5$)·dioxane] which is a known catalyst in solution BD polymerization is heterogenized on various silica supports differing in specific surface area and pore volume. The catalyst is activated by MAO. In solution polymerization the best of the supported catalysts is 100 times more active (determined by the rate constant) than the respective unsupported catalyst [408]. In addition to the polymerization in solution, the supported allyl Nd catalyst is applied for the gas-phase polymerization of BD [578, 579] the performance of which is characterized by macroscopic consumption of gaseous BD and in-situ-analysis of BD insertion [580].

The application of Nd-mediated gas-phase polymerization on IP [479, 480] and the copolymerization of BD and IP [482, 483] confirms the features elaborated for the gas-phase polymerization of BD.

3.3
Homopolymerization of Dienes in the Presence of Other Monomers

Many Nd-catalysts are highly selective and allow for the homopolymerization of BD and IP in double-bond containing media. In this context the homopolymerization of BD in styrene (St) monomer and the homopolymerization of IP in the crude C$_5$ cracking fraction deserve particular attention.

Various patents on the homopolymerization of BD in the presence of styrene are available [581–590]. According to these patents, St is used as a solvent in which BD is selectively polymerized by the application of NdV/DIBAH/EASC. At the end of the polymerization a solution of BR in St is obtained. In subsequent reaction steps the unreacted styrene monomer is either polymerized radically, or acrylonitrile is added prior to radical initiation. During the subsequent radical polymerization styrene or styrene/acrylonitrile, respectively, are polymerized and *cis*-1,4-BR is grafted and partially crosslinked. In this way BR modified (or impact modified) thermoplast blends are obtained. In these blends BR particles are dispersed either in poly(styrene) (yielding HIPS = high impact poly(styrene) or in styrene-acrylonitrile-copolymers (yielding ABS = acrylonitrile/butadiene/styrene-terpolymers). In comparison with the classical bulk processes for HIPS and ABS, this new technology allows for considerable cost reductions as the overall complexity can be reduced. In the established HIPS- and bulk ABS-processes Li- and Co-BR-grades are used which are polymerized in solution. These BR-grades have to be isolated from the solution by steam stripping, water removal, drying and baling prior to the use in HIPS- or bulk ABS-production for which BR has to be dissolved in either styrene monomer or styrene/acrylonitrile-mixtures.

For the preparation of poly(isoprene), the monomer 2-methyl-1,3-butadiene (= isoprene = IP) is required as feedstock. This monomer can be obtained by various condensation methods that utilize four principles to create the C_5 skeleton. In the more modern process IP is obtained from the C_5 cracking fraction which contains various double-bond containing hydrocarbons with 5 C-atoms (e.g. among other C_5-compounds the fraction contains cyclopentadiene, various pentadienes and pentenes) [478]. The preparation of pure IP by either of these two routes is cost intensive. By the direct and selective polymerization of IP in the crude C_5 cracking fraction the cost intensive isolation of pure IP is avoided. Thereby production costs for IR are considerably reduced [264, 265]. The selective polymerization of IP in the crude C_5 cracking fraction is achieved by the application of a NdP-based catalyst system. The latest patent of Michelin claims a process in which dehydrogenation of the C_5 cut is applied prior to polymerization. In this way an IP-enriched C5-fraction is obtained which does not contain a high quantity of disubstituted alkynes, terminal alkynes and cyclopentadiene. The unpurified C5-fraction is used as the feedstock for polymerization [591, 592].

4
Kinetic and Mechanistic Aspects of Neodymium-Catalyzed Butadiene Polymerization

Until today the mechanism of Nd-catalyzed BD polymerization is not understood in full detail. This state of understanding is reflected by contradictory results between numerous studies. In spite of these differences there are reaction models which allow for a qualitative explanation of most of the macroscopic features observed. In this section the common points of understanding between the various studies as well as the contradictions are addressed. In the first step results on the polymerization kinetics are summarized.

4.1
Kinetic Aspects

The rate of BD polymerization in Nd-based Ziegler/Natta polymerization catalysis strongly depends on the amount of the neodymium compound, the Al-cocatalyst, the halide donor, and the solvent. Beside these "chemical" factors the temperature has a strong influence on the rate.

An overview of the available information until 1988 was compiled by Pross et al. In this summary the impact of the various catalyst components on the reaction rate is given [204]. In the following, the literature is analyzed and reviewed according to the formalism given by Pross et al.:

$$r_p/k_p = c_{Bd}^w \cdot c_{Nd}^y \cdot c_{Al}^z \cdot c_X^u$$

r_p	rate of polymerization
k_p	rate constant of polymerization
c_i	concentration of component i
w, y, z, u	reaction order.

Amongst the majority of publications there is unanimous agreement on the first-order dependence of BD consumption ($r_p/k_p \sim c_{Bd}^w$, w = 1), e.g. [87, 88, 93, 94, 111, 133, 134, 173, 188, 593–602].

Nd-allyl-based catalyst systems exhibit a different pattern of monomer consumption. For the two catalyst systems (1) $Nd(C_3H_4R)_3/2AlMe_2Cl/30AlMe_3$ and (2) $Nd(C_3H_4R)_3/2AlEt_2Cl/10AlEt_3$ BD consumption follows second-order kinetics ($r_p/k_p \sim c_{Bd}^w$, w = 2) [293]. Regarding the first-order dependence of r_p/k_p on the concentration of the Nd component ($r_p/k_p \sim c_{Nd}^y$, y = 1) there is agreement among the majority of authors, too, e.g. [88, 593–600]. The catalyst systems for which the dependence $r_p/k_p = c_{Nd}^1$ is well established are based on the Nd-components $Nd(CH_2Ph)_3$, $NdCl_3$, NdV, NdO and NdN. Exceptions with exponents y < 1 are also reported: $r_p/k_p \sim c_{Nd}^y$, y = 0.5 [425, 426] and $r_p/k_p \sim c_{Nd}^y$, y = 0.83 [188]. Reaction orders which differ from $r_p/k_p \sim c_{Nd}^1$ were also determined for Nd-alcoholate and for Nd-phosphate-based catalyst systems [272].

For the dependence of $r_p/k_p \sim c_{Al}^z$ different values for z are reported. For example, for the catalyst system $NdCl_3 \cdot 3^i PrOH/TEA$ Hu et al. report z = 0.5 [593, 594]. For the catalyst system $NdO/DIBAH/CH_2=CH-CH_2Cl$ Sun et al. determined z = 0.33 [599, 600].

For the two catalyst systems NdV/DIBAH/EASC and NdV/TIBA/EASC the ratios of n_{Al}/n_{Nd} were varied over a broad range from 5 to 50. It was shown that k_p depends on the nature of the cocatalyst (TIBA and DIBAH) and on the respective concentrations of these cocatalysts. As k_p is not constant the authors suggest that it is better considered as an apparent rate constant k_a. For the cocatalyst TIBA the dependence of k_a on n_{Al}/n_{Nd} results in a S-shaped curve with a maximum in $k_a = 311 \text{ L} \cdot \text{mol}^{-1} \cdot \text{min}^{-1}$ at $n_{TIBA}/n_{NdV} = 50$ while activation with DIBAH leads to an inverse u-shaped curve with a maximum $k_a \approx 220 \text{ L} \cdot \text{mol}^{-1} \cdot \text{min}^{-1}$ at $n_{DIBAH}/n_{NdV} = 35$ (Fig. 11 in Sect. 2.2.8) [179].

To the best of our knowledge, there is no publication which gives the dependence of $r_p/k_p = f(c_X)$ by an equation of the form $r_p/k_p \sim c_X^u$. The dependence of r_p on the concentration of the halide donor is quite complex and

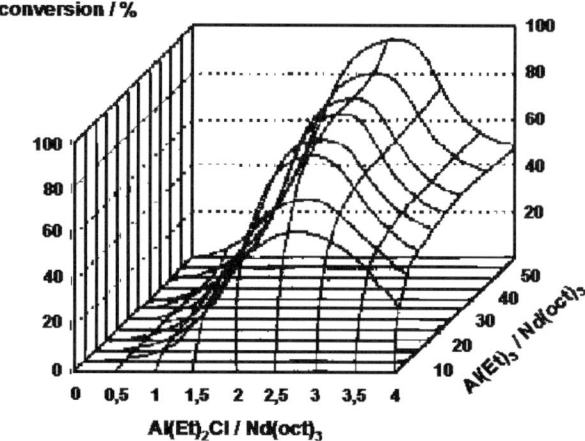

Fig. 12 Dependence of monomer conversion on the amount of halide donor and cocatalyst in BD polymerization with the catalyst system NdO/TEA/DEAC [168], reprinted with permission from Elsevier

cannot be described by such a simple equation. In the literature usually the dependence of r_p on the molar ratio of n_X/n_{Nd} is given by graphical representations. In these plots r_p passes through a maximum at $n_X/n_{Nd} \approx 2$ to 3 (X = Cl, Br, I) (Sect. 2.1.5). The precise location of the r_p-maximum depends on parameters such as the way of mixing the catalyst components. Premixing, sequence of consecutive dosing, presence of monomer during catalyst preformation etc. also influence r_p in this context.

For the two catalyst systems (1) NdO/TEA/DEAC [168] and (2) NdO/TIBA/EASC [163] Oehme et al. gave 3-dimensional plots for the dependence of monomer conversion (determined after a fixed period of time) on the concentrations of cocatalyst as well as halide donor (Fig. 12).

Aspects associated with the impact of n_{Cl}/n_{Nd} on r_p, cis-1,4-content, PDI etc. are discussed in detail in Sect. 2.1.5.

4.2
Active Species and its Formation

In scientific literature many speculations on the nature of the active Nd species can be found. Unfortunately, most of these speculations are not supported by experimental or analytical evidence. In the following, the discussions on this issue are summarized for catalyst systems which are based on different Nd-precursors. At the end of this subsection special attention is given to the state of discussions about the fraction of active Nd centers in the catalyst mixture.

NdX$_3$-Based Systems

The first suggestions on active species in rare earth halide-based catalyst systems date back to 1976 [603, 604]. In 1980 Shen et al. formulated the structural element of the active species in Ln-halide-based systems on the basis of these early suggestions (Scheme 19) [92].

Though NdX$_3$-based catalyst systems are among the first Nd Ziegler/Natta systems, studies directed at the elucidation of the active species became more numerous since 1986. Essentially there are the research groups of Hsieh, Iovu and Monakov who made major contributions to this topic.

Hsieh and co-workers focused on the catalyst system NdCl$_3$/EtOH/TEA. They formulated several Nd allyl species which are supposedly present during the polymerization of dienes [139].

The group of Iovu studied the catalyst system NdCl$_3 \cdot$ 2TBP/TIBA for which Nd dichloromono*iso*butyl is claimed as the first Nd intermediate. According to these authors Nd dichloromono*iso*butyl is converted into an allyl Nd compound after the first insertion of BD (Scheme 20) [131, 279].

Furthermore, on the basis of mathematic models Iovu et al. suggest heteronuclear Nd-Al species which are present in the catalyst system NdCl$_3 \cdot$ 2TBP/TIBA (Scheme 21) [279]. The results of these investigations have been summarized [605, 606].

Heteronuclear Nd-Al-species are also suggested by Monakov and co-workers [607, 608]. According to these authors four different intermediates are present which are formed by the association of two-Nd-species and by the association of one Nd-intermediate with one Al-organic compound (Scheme 22).

These intermediates contain Nd-Nd and Nd-Al species which are clustered by Cl/alkyl-bridges as well as by Cl/Cl-bridges. In various papers Monakov et al. emphasize the oligomeric and polymeric nature of these species [609–

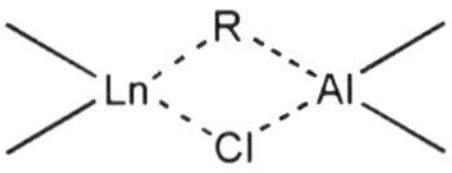

Scheme 19 Suggestion for the active structural element in LnX$_3$-based systems [92], reprinted with permission of John Wiley & Sons, Inc.

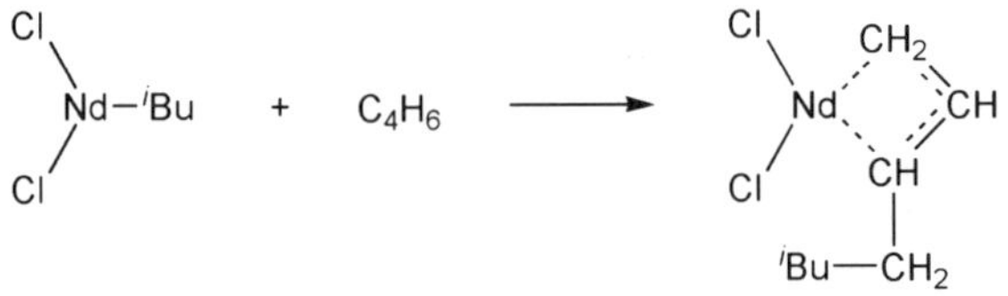

Scheme 20 Reaction mechanism of diene insertion with a catalyst system based on NdCl$_3 \cdot$ 2 TBP and TIBA [131], reprinted with permission from Elsevier

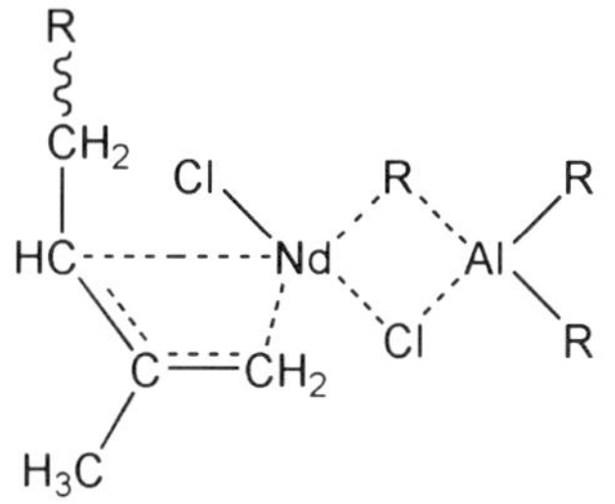

Scheme 21 Suggested active species in the catalyst system $NdCl_3 \cdot 2$ TBP/TIBA [279], reprinted with permission from Elsevier

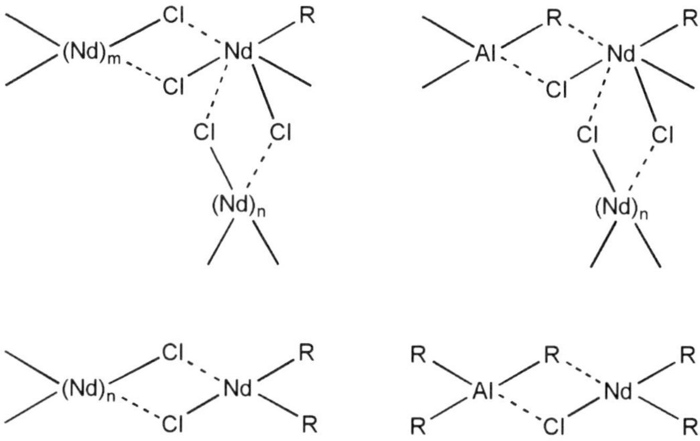

Scheme 22 Catalytically active species in $NdCl_3$-based catalyst systems according to Monakov et al. [607, 608], with kind permission of Springer Science and Buisiness Media

614]. Also mathematical models for the polymerization of dienes with catalyst systems of the type $NdX_3 \cdot$ TBP/TIBA were developed [615, 616].

Nd-Carboxylate-Based Systems

In spite of numerous studies on Nd-carboxylate-based catalyst systems a mechanistic understanding of the formation and the chemical structure of the active species is far off. By the application of new analytical methods some progress has recently been achieved. Kwag and co-workers applied synchrotron X-ray absorption in combination with UV-VIS-spectroscopy. These authors succeeded to characterize the reaction product obtained by the preformation of the catalyst components $NdV \cdot HV^9$/TIBA/DEAC. The following essential features are reported about the reaction product: (1) Nd – C bonds (bond length = 1.41 Å) with covalent and ionic character, (2) Nd – Cl bonds

---

[9] HV = versatic acid; $NdV \cdot HV$ is equivalent with $NdHV_4$; in literature given by Kwag et al. $NdV \cdot HV$ is also referred to as $NDV_4 = NdHV_4$

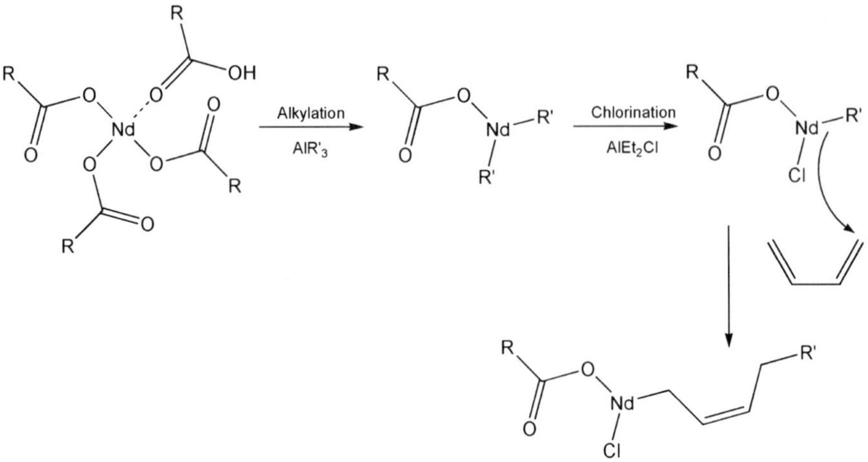

Scheme 23 Postulated reaction mechanism for BD polymerization with the system NdV·HV/TIBA/DEAC according to Kwag [156], reprinted with permission of the American Chemical Society

(bond length = 2.14 Å) and (3) Nd remains in the oxidation state +III [617]. The same group suggests an activation mechanism for the Nd-carboxylate-based compound NdV·HV. Supposedly, the reaction product is monomeric (Scheme 23) [156].

According to this mechanism the Nd precursor is first alkylated and subsequently chlorinated. The first and subsequent insertions of BD occur into a Nd–C bond. The proposed active species exhibits the same structural features as the active species suggested earlier for NdX_3-based catalyst systems. Furthermore, Kwag emphasizes the importance of one excess molecule of versatic acid in the NdV precursor (NdV·HV). By the excess of versatic acid NdV clusters are broken into monomeric species and every Nd atom becomes accessible for alkylation and subsequently participates in polyinsertion catalysis [156, 618]. In the most recent publication Kwag et al. give a suggestion for the active catalytic species in a NdV·HV-based system. This species consists of a Nd center which is coordinated by a monodentate versatate, a chloride and a poly(butadien)yl chain which is bound in the η^3-mode. In this species the penultimate double bond of the poly(butadien)yl chain stabilizes the active Nd center [367].

Evans et al. sequentially reacted the Nd carboxylate precursor $\{Nd[O_2CC(CH_3)_2CH_2CH_3]_3\}_x$ first with DEAC and then with TIBA. By this reaction catalytically active systems are obtained which polymerize IP. In the first reaction step in which Nd carboxylate is reacted with DEAC mixed ligand complexes are formed which contain neodymium and aluminum as well as halide and ethyl groups. Upon crystallization $NdCl_3$-based compounds are obtained in which solvent is coordinated. These compounds exhibit a more complex structure than that of pure NdCl$_3$. In the preformation reaction carboxylate groups attached to Nd are replaced by halide ligands which are transferred from DEAC. From this study it becomes evident that the halogenation of NdV does not result in the straightforward formation of NdCl$_3$ [366].

Fischbach et al. isolated and characterized products which are obtained from the reaction of various highly substituted neodymium carboxylates with trimethyl aluminum. Several reaction schemes are presented which account for the whole variety of reactions and isolated Nd species [185, 186].

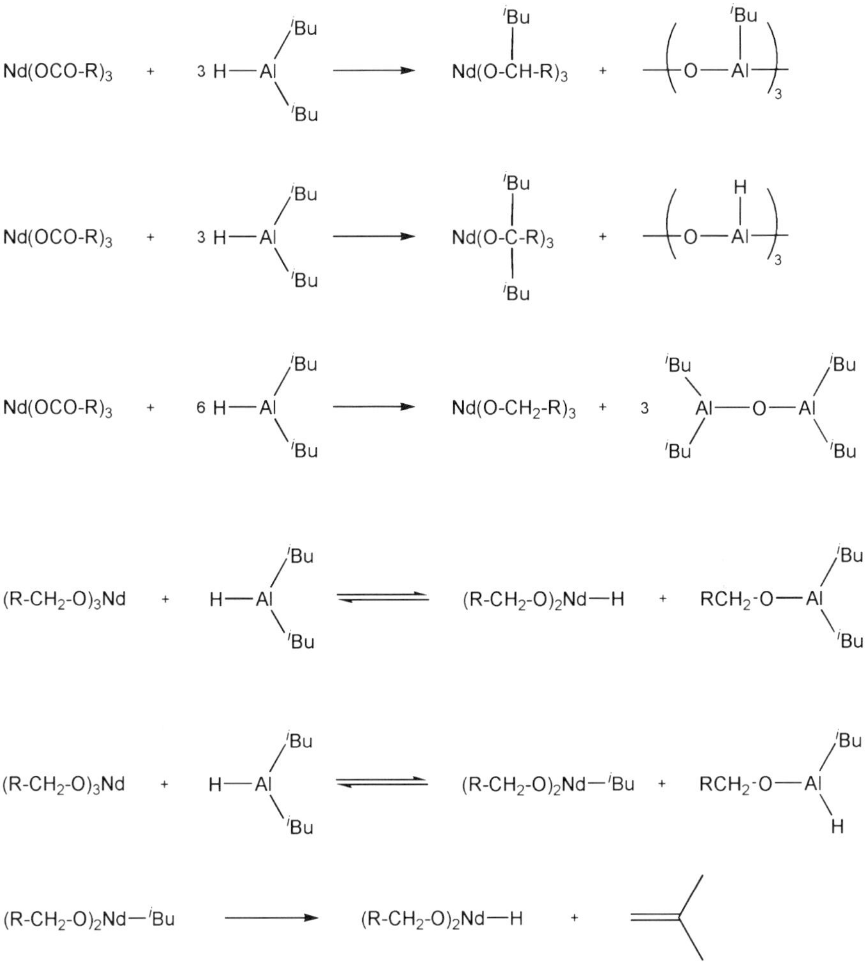

Nd(OCO-R)$_3$ + 3 H—Al(^{i}Bu)$_2$ ⟶ Nd(O-CH-R)$_3$ + ($-$O—Al(^{i}Bu)$-$)$_3$

Nd(OCO-R)$_3$ + 3 H—Al(^{i}Bu)$_2$ ⟶ Nd(O-C(^{i}Bu)-R)$_3$ + ($-$O—Al(H)$-$)$_3$

Nd(OCO-R)$_3$ + 6 H—Al(^{i}Bu)$_2$ ⟶ Nd(O-CH$_2$-R)$_3$ + 3 (^{i}Bu)$_2$Al—O—Al(^{i}Bu)$_2$

(R-CH$_2$-O)$_3$Nd + H—Al(^{i}Bu)$_2$ ⇌ (R-CH$_2$-O)$_2$Nd—H + RCH$_2$·O—Al(^{i}Bu)$_2$

(R-CH$_2$-O)$_3$Nd + H—Al(^{i}Bu)$_2$ ⇌ (R-CH$_2$-O)$_2$Nd—^{i}Bu + RCH$_2$·O—Al(^{i}Bu)(H)

(R-CH$_2$-O)$_2$Nd—^{i}Bu ⟶ (R-CH$_2$-O)$_2$Nd—H + CH$_2$=C(CH$_3$)$_2$

Scheme 24 Postulated sequence of consecutive reactions for NdV + DIBAH (formation of Nd alcoholate by the reduction of NdV and subsequent formation of a Nd hydrido species either by hydride transfer or by β-hydride elimination of an isobutyl group) [178], reprinted with permission of Wiley-VCH Verlag GmbH & Co. KGaA

A sequence of chemical reactions is postulated to account for the features observed in the kinetics of BD polymerization catalyzed by NdV/DIBAH/EASC [178]. From the reaction scheme which is hypothetically put forward 3 to 7 eq. of DIBAH are required for the generation of one active Nd species starting from one equivalent of NdV (Scheme 24). In the first reaction step the reduction of Nd carboxylate to Nd alcoholate is assumed to occur. Experimental evidence in favor of the reduction of Nd-carboxylate comes from NMR studies in which the reduction of the respective lanthanum carboxylate by DIBAH was observed (halide donor EASC) (Windisch, 2006, personal communication). This reduction does not occur when Nd carboxylates are reacted with TMA (Anwander and Fischbach, 2006, personal communication). DIBAH might be capable of reducing carboxylic acids to the respective al-

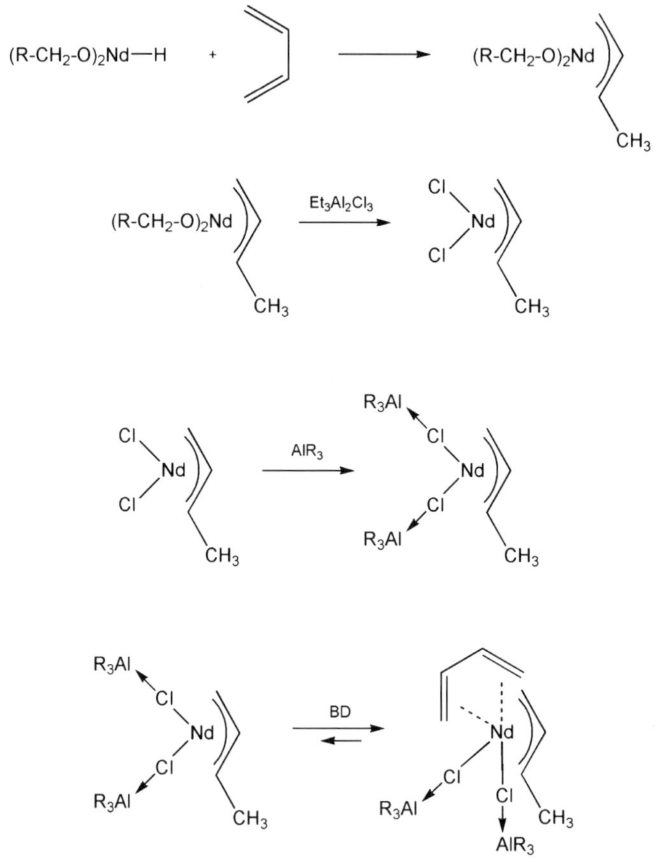

Scheme 25 Reaction sequence for the formation of a Nd allyl species by the reaction of Nd hydride with BD, subsequent chlorination of the reaction product, interaction of the chlorinated Nd allyl with Lewis acids and coordination of BD [178], reprinted with permission of Wiley-VCH Verlag GmbH & Co. KGaA

cohols [178]. Carboxylates in general, however, are considered to be inert to a reduction by DIBAH [619]. From this evidence the reduction of Nd carboxylate to Nd alcoholate by DIBAH is doubtable. As a consequence of these conflicting findings on the reduction of rare earth carboxylates by DIBAH and by TMA further work is needed to clarify the relevance of the conceivable reduction of Nd carboxylates to Nd alcoholates.

According to the reaction scheme put forward the reduction of NdV requires between 3 to 6 eq. of DIBAH. As a result of this reduction Nd alcoholate is obtained. The Nd-hydrido species is considered to be an important intermediate which can be formed from the alcoholate by two different reaction routes. If the Nd hydrido species is formed by hydride transfer from DIBAH to Nd alcoholate another equivalent of DIBAH is needed. If the Nd hydride species is formed by β-hydride elimination no additional equivalent of DIBAH is required. As a consequence from these considerations in total 3–7 eq. of DIBAH are required for the generation of one Nd-hydrido derivative, which is considered to be the first active species. The total sequence of consecutive reactions which result in the neodymium-catalyzed polyinsertion of BD is summarized in Scheme 25.

It is important to note that the above schemes of consecutive reactions are partially hypothetical and bound to further clarification.

Nd-Alcoholate-Based Systems

A ternary catalyst system based on Nd alcoholate provides one of the first examples in which the active species could be isolated and analytically characterized [620]. The preformation of Nd(O^{i}Pr)$_3$/TEA/DEAC (molar ratio = 1/10/1.5) results in a crystalline precipitate. Further characterization of single crystals gives evidence for a polynuclear Nd – Al bimetallic complex in which Nd atoms are coordinated by Cl, alcoholate and alkyl ligands.

In 2004 Carpentier and co-workers were the first who isolated and characterized the active species from a Nd alcoholate/MgR$_2$ mixture. The reaction of either a trinuclear or a monomeric Nd alcoholate precursor with Mg(CH$_2$SiMe$_3$)$_2 \cdot$(Et$_2$O) yields an alkyl neodymium complex and a new heterobimetallic Nd-Mg complex. The latter complex is considered to be the active species for the investigated catalyst system [621].

Anwander et al. isolated and characterized the products which are obtained from the reaction of various highly substituted neodymium aryloxides with trimethyl aluminum. A reaction scheme is put forward which accounts for the whole variety of species isolated [185, 234].

Nd-Phosphate-Based Systems

To our knowledge there is no study available on the formation and the nature of the active species present in Nd-phosphate-based catalyst systems. Such a study should be designed to elucidate the question why extraordinary small amounts of Al alkyls ($n_{Al}/n_{NdP} \leq 4$) are sufficient for catalyst activation.

Neodymium Amide Systems

Monteil et al. suggest mechanisms which account for the activation and the polymerization mediated by $Nd(N(SiMe_3)_2)_3/[HNMe_2Ph]^+ [B(C_6F_5)_4]^-/AlR_3$ [320]. The reaction model put forward by these authors comprises the formation of a cationic heterometallic Nd/Al complex that is capable of coordinating diene monomers followed by monomer insertion.

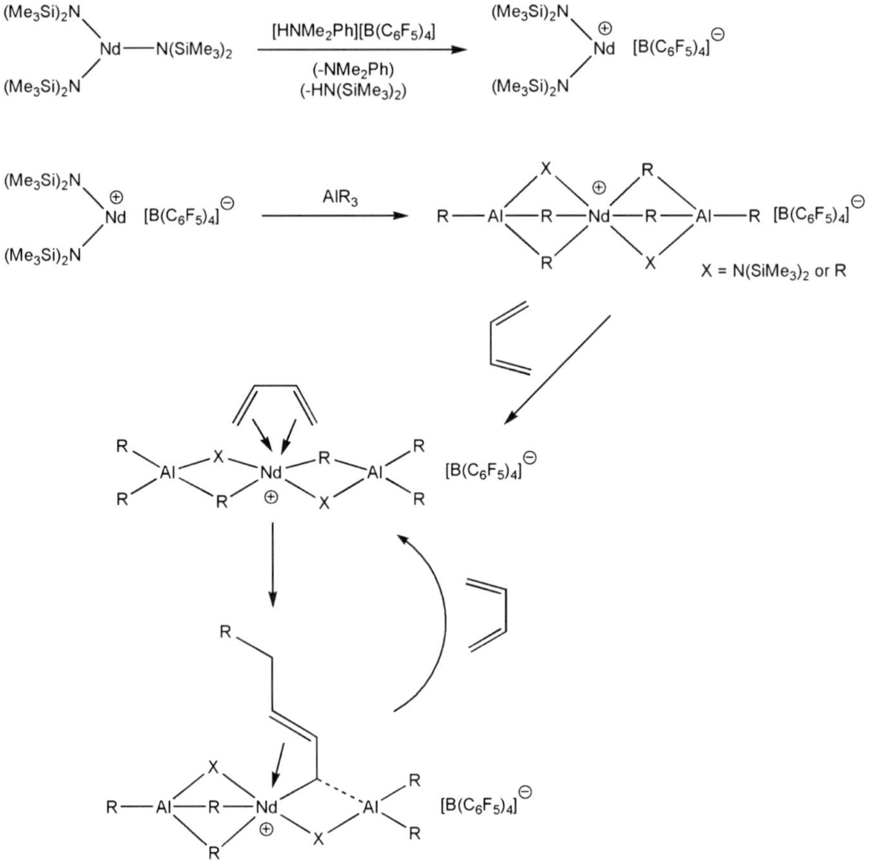

Scheme 26 Proposed mechanism for the polymerization of dienes by the catalyst system $Nd(N(SiMe_3)_2)_3/[HNMe_2Ph]^+ [B(C_6F_5)_4]^-/AlR_3$ [320], copyright Soc Chem Ind. Reproduced with permission. Permission is granted by John Wiley& Sons Ltd. on behalf of the SCI

Neodymium Allyl Systems (Sect. 2.1.1.5)

In their most-recent publication Taube et al. suggest a sequence of reactions which account for the importance of the halogenation of the Nd center and the abstraction of the halide in order to create a vacant coordination site on the Nd center [293]. This scheme emphasizes the fact that allyl-based systems

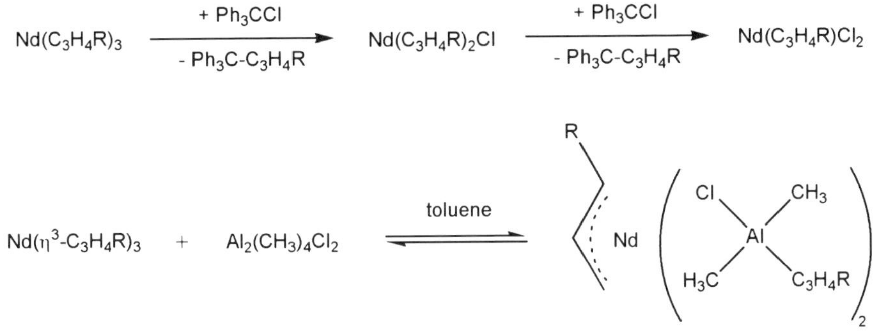

Scheme 27 showing reactions:

Nd(C$_3$H$_4$R)$_3$ →[+ Ph$_3$CCl / − Ph$_3$C-C$_3$H$_4$R] Nd(C$_3$H$_4$R)$_2$Cl →[+ Ph$_3$CCl / − Ph$_3$C-C$_3$H$_4$R] Nd(C$_3$H$_4$R)Cl$_2$

Nd(η^3-C$_3$H$_4$R)$_3$ + Al$_2$(CH$_3$)$_4$Cl$_2$ ⇌[toluene] [Nd(C$_3$H$_4$R)·(Al(Cl)(CH$_3$)(H$_3$C)(C$_3$H$_4$R))$_2$]

Scheme 27 Two potential reaction routes for the formation of active catalyst species from tris allyl neodymium according to Taube et al. (electrical charges are omitted for clarity) [293], reprinted with permission of Wiley-VCH Verlag GmbH & Co. KGaA

without the application of further catalyst components (halide donors) will not exhibit *cis*-1,4-stereospecificity. In Scheme 27 two potential pathways for the interaction of halide donors with the Nd-site are given. Either a halide donor (Ph$_3$CCl) transfers a halide atom to Nd in exchange with an allyl anion or a halide containing Lewis acid (Al$_2$(CH$_3$)$_4$Cl$_2$) abstracts an allyl group from Nd(allyl)$_3$ which leads to the formation of a cationic Nd allyl species.

Neodymium Boranate Systems

Bonnet et al. describe the activation of Nd boranate complexes by dialkyl magnesium reagents [349]. According to the authors a Nd(μ-BH$_4$)Mg containing moiety is formed for which strong indication was provided by NMR spectroscopy (Scheme 28).

Fraction of Active Neodymium

Various papers address the aspect that only a small fraction of the total amount of neodymium is active in polymerization. All figures published on the percentage of active Nd are below 10%. Monakov et al. determine an efficiency factor of 7% [93, 94]. According to Pan et al. the efficiency factor is between 0.4 to 8% of the total amount of Nd present [595–598]. Sun et al. give the percentage of active neodymium in the range of 0.4 to 0.5% [599, 600]. Yu

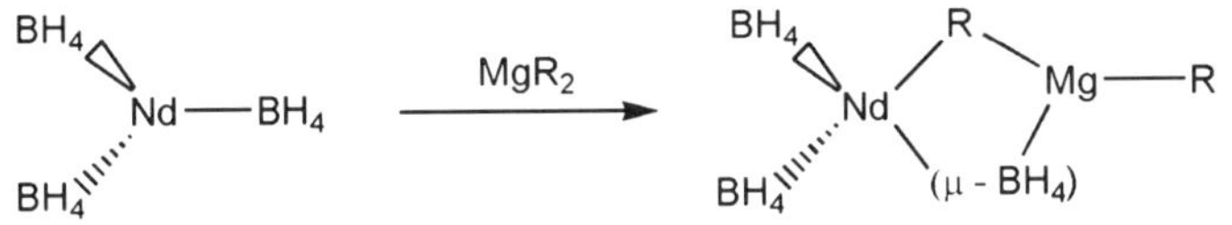

Scheme 28:

(BH$_4$)(BH$_4$)Nd—BH$_4$ →[MgR$_2$] (BH$_4$)(BH$_4$)Nd(μ-BH$_4$)Mg—R with R

Scheme 28 Formation of the active bimetallic species Nd(μ-BH$_4$)Mg upon addition of dialkylmagnesium to Nd(BH$_4$)$_3$(THF)$_3$ (solvent molecules are omitted for clarity) [349], reprinted with permission of the American Chemical Society

et al. determined 3% [601, 602], Porri et al. report on 6% [162] and Gehrke et al. determined 0.7% of active Nd for $Nd^iO/TEA/AlBr_3$ (1/25/0.8) [159] and 2.2% for $Nd^iO/TIBA/DIBAC$ (1/15/2.5) [160].

In the late 1980s two explanations for the observed low concentrations of active catalyst were given by Porri et al.: (1) low degree of alkylation of Nd centers and (2) low stability of the resulting Nd – C bonds [162, 165, 166]. Since then, additional explanations were put forward in the literature. According to these publications the low efficiency factors can also be attributed to partial insolubility of the Nd precursors in organic solvents, the formation of heterogeneities during the preformation of catalyst components and the agglomeration or clustering of reactive intermediates.

Clustering of active Nd-species or the presence of oligomeric or polymeric Nd structures is emphasized by various authors. For example declustering is achieved by small changes in the composition of a Nd-versatate-based catalyst. By the coordination of one molecule of versatic acid to NdV during catalyst preparation $NdH(neodecanoate)_4$ (= NdV · versatic acid = NDV_4 = $NdV \cdot HV$ = $NdHV_4$) is formed [156, 417]. $NdH(neodecanoate)_4$ exhibits a significant increase in polymerization activity over common NdV. This advantage in activity is attributed to the break-up of NdV-clusters. By this mechanism a greater portion of the Nd becomes accessible for alkylation and the subsequent reactions. A detailed kinetic study in which quantitative aspects of the declustering are addressed is missing up to today but there is strong evidence that the more qualitative information put forward for the explanation of the increase of catalyst activity will be confirmed in further investigations.

Declustering seems to be also possible by the application of appropriate catalyst supports. For tris-allyl-neodymium $Nd(\eta^3-C_3H_5)\cdot$ dioxane the rate constants for monomer consumption are determined for supported as well as for unsupported catalysts. Under comparable experimental conditions the respective rate constants differ by a factor of 100. This observation can be explained by differences in Nd efficiency or by differences in the concentration of active Nd sites [408].

In spite of the presence of Nd-clusters, partial alkylation and micro heterogeneities the number of active Nd-species seems to be fairly constant during the course of a polymerization. Otherwise neither consistent polymerization kinetics (particularly 1st-order monomer consumption up to high monomer conversion) nor linear increases of molar mass during the whole course of the polymerization would be observed in so many studies. It therefore can be concluded that the fraction of active Nd as well as the number of active catalyst species are fixed either at an early stage of the polymerization or even prior to initiation of the polymerization. It can be speculated whether the fixation of the number of active species occurs during catalyst preformation/activation or even during the preparation of the Nd compound. In contrast to this consideration Jun et al. report on the decay of active centers during the course of the polymerization. These authors use the catalyst system NdN/DIBAH/TIBA/DIBAC [622].

In summary, the majority of results can be interpreted on the basis of a constant concentration of active Nd-centers during the major part of the polymerization. This conclusion, however, is not applicable to all catalyst systems.

Conclusion

Presently, the importance of Nd allyl compounds as intermediates in Nd carboxylate- and other Nd-based catalyst systems is widely accepted. As various Nd allyl compounds have been synthesized, characterized and successfully tested as polymerization catalysts this view is supported by solid experimental evidence (Sect. 2.1.1.5 and the references therein). Selected Nd allyl compounds exhibit significant polymerization activities without the addition of cocatalysts. In these cases the active species is neutral. But also cationic active Nd species are taken into consideration (Sect. 2.1.1.5) [288, 291]. Cationic species also prevail in the presence of non-coordinating anions.

The active portion of the Nd precursor was determined in various studies. There is common agreement that the fraction of active Nd is very low. Some recent studies are available, however, which provide evidence that Nd-efficiency can be considerably increased. The use of supported catalysts and the modification of the readily available NdV catalyst allow for significant increases of overall catalyst activities. These increases can be explained by increases of the active portion of the Nd precursor used.

4.3
Polyinsertion Reaction and Control of Microstructure

The polyinsertion reaction of conjugated dienes can proceed in three modes which yield three different isomers: 1,2-polymers, 3,4-polymers and 1,4-polymers [297]. The situation is more complicated as the 1,4-isomers either exhibit *cis*-1,4 or *trans*-1,4 configuration. In addition, the 1,2-/3,4-polymers can have an atactic, isotactic or syndiotactic structure (Scheme 1 in Sect. 1.2) [623]. The various moieties are either randomly distributed along the polymer chain or are aligned block wise.

Most reaction models which describe the mechanism of diene polymerization by Nd catalysts have been adopted from models developed for the polymerization of ethylene and propylene by the use of Ti- and Ni-based catalysts systems. A monometallic insertion mechanism which accounts for many features of the polymerization of α-olefins has been put forward by Cossee and Arlman in 1964 [624–626]. Respective bimetallic mechanisms date back to Patat, Sinn, Natta and Mazzanti [627, 628]. The most important and generally accepted mechanisms for the polymerization of dienes by Nd-based catalysts are discussed in the following.

The adoption of reaction models available for the polymerization of conjugated dienes by Ni- and Ti-catalysts to the polymerization of BD by Nd catalysis is justified by the similarities of the respective metal carbon bonds. In each of these mechanistic models the last inserted monomer is bound to the metal in a η^3-allyl mode. The existence of Ni-η^3-allyl-moieties was demonstrated by the reaction of the deuterated nickel complex [(η^3– C_4D_6H)NiI]$_2$ with deuterated BD (deuterated in the 1- and 4-position). After each monomer insertion a new η^3-allyl-bond is formed [629]. As π-allyl-complexes are known for Ti and Ni this knowledge has been adopted for Nd-based polymerization catalysts [288, 289, 293, 308, 309, 630–636, 638–645].

As there are two reactive sites on C_1 and C_3 of the allyl group a π-allyl group present at the end of the polymer chain allows for two insertion modes: 1,2- and 1,4-insertion. If BD is exclusively polymerized in either of these modes the two poly(butadiene) isomers 1,2-poly(butadiene) and 1,4-poly(butadiene) are obtained (Scheme 29) [297].

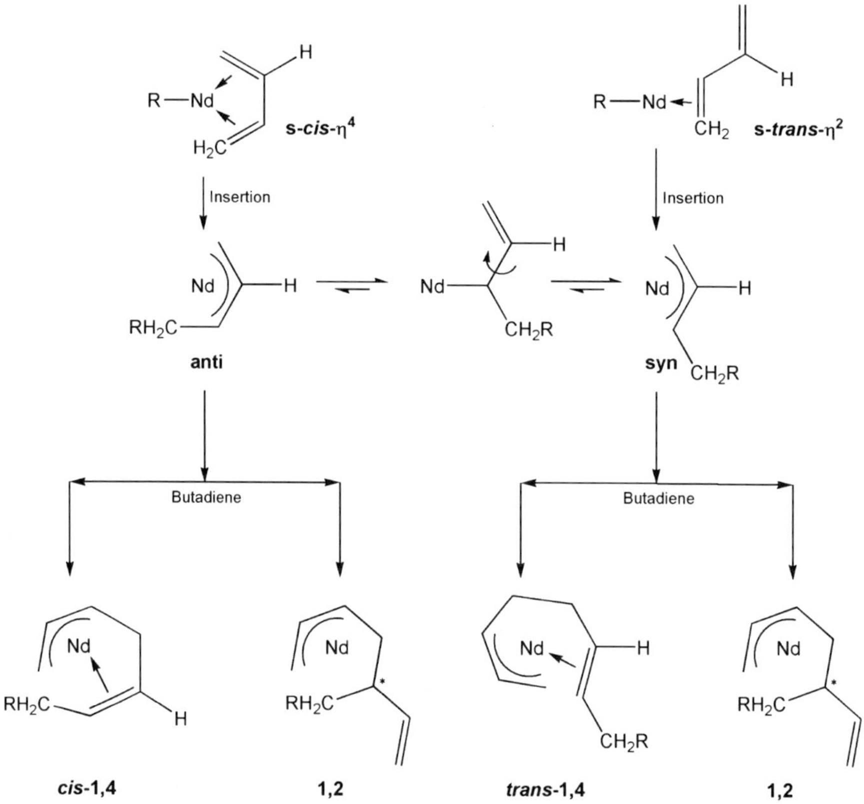

Scheme 29 Polymerization of butadiene by 1,2- and 1,4-insertion with neodymium-based Ziegler/Natta-catalysts (charge and ligands of neodymium are omitted for clarity)

In addition, the terminal allyl unit of the polymer chain can coordinate to the metal site in either anti- or syn-conformation. The conformation is assigned by the relative positions of the hydrogen atom bound to C_2 and the $-CH_2-R$ group attached to C_3 (Scheme 29). Hydrogen and $-CH_2-R$ are ecliptic in the syn-conformation and staggered in the anti-conformation. The anti-π-allyl polymer chain originates from a single-*cis*-η^4-coordinated BD monomer and the syn-π-allyl polymer chain results from a single-*trans*-η^2-coordinated monomer. Therefore, syn- and anti- coordination of monomer to the metal site are decisive factors for the control of the microstructure of the resulting poly(butadiene). The anti-isomer results in *cis*-1,4-polymerization and the syn-isomer yields *trans*-1,4-polymer. According to Kobayashi et al. the newly generated double bond in the polymer chain stabilizes the catalyst by means of π-electron back donation [176].

The mechanism is complicated by the possibility of anti-syn-isomerization and by $\pi - \sigma$-rearrangements ($\pi - \eta^3$-allyl $\Delta\sigma - \eta^1$-allyl). In the case of C_2-unsubstituted dienes such as BD the syn-form is thermodynamically favored [646, 647] whereas the anti-isomer is kinetically favored [648]. If monomer insertion is faster than the anti-syn-rearrangement the formation of the *cis*-1,4-polymer is favored. A higher *trans*-1,4-content is obtained if monomer insertion is slow compared to anti-syn-isomerization. Thus, the microstructure of the polymer (*cis*-1,4- and *trans*-1,4-structures) is a result of the ratio of the relative rates of monomer insertion and anti-syn-isomerization. As a consequence of these considerations an influence of monomer concentration on *cis/trans*-content of BR can be predicted as demonstrated by Sabirov et al. [649]. A reduction of monomer concentration results in a lower rate of monomer insertion and yields a higher *trans*-1,4-content. On the other hand the *cis*-1,4-content increases with increasing monomer concentration. These theoretical considerations were experimentally verified by Dolgoplosk et al. and Iovu et al. [133, 650, 651]. Furthermore, an increase of the polymerization temperature favors the formation of the kinetically controlled product and results in a higher *cis*-1,4-content [486]. 1,2-poly(butadiene) can be formed from the anti- as well as from the syn-isomer. In both cases 2,1-insertion occurs [486]. By the addition of electron donors the number of vacant coordination sites at the metal center is reduced. The reduction of coordination sites for BD results in the formation of the 1,2-polymer. In summary, the microstructure of poly(diene) depends on steric factors on the metal site, monomer concentration and temperature.

Another feature of Nd-catalyzed BD polymerization which is reported by various authors is the reduction of *cis*-1,4-content by the addition of donor compounds. The respective increase of *trans*-1,4-moieties and decrease of *cis*-1,4-moieties is also observed when the amount of the alkyl aluminum cocatalyst is increased. This feature does not depend on the nature of the Nd component and has for example been comparatively studied for three different catalyst systems (1) NdV/DIBAH/EASC, (2) neodymium neopen-

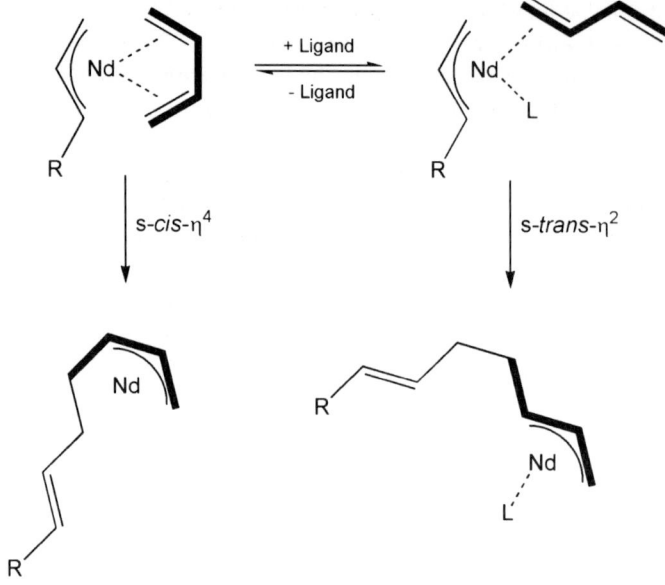

Scheme 30 Influence of ligands L (e.g. electron donors such as triphenylphosphine and metal-alkyls such as DIBAH, TIBA and ZnEt$_2$) on the microstructure of BR (charge and ligands of neodymium are omitted for clarity)

tanolate/DIBAH/EASC and (3) NdP/DIBAH/EASC [272]. It can be speculated that cocatalysts such as DIBAH are coordinated to vacant neodymium sites. As a result of higher coordinative saturation of the Nd site the extent of η^4-coordination of BD is reduced and the coordination of BD in the η^2-mode is preferred. As a result *trans*-1,4-insertion is favored (Scheme 30, DIBAH = ligand) [178]. As a consequence, the *cis*-1,4-content decreases with increasing amounts of Al cocatalyst. By means of this mechanism the decrease of *cis*-1,4-content can also be explained for other cocatalyst components such as TIBA, ZnEt$_2$ etc. [179, 180].

A similar reaction scheme which accounts for the coordination mode of diene monomers was already suggested by Witte in 1981 [49]. Though DIBAH or TIBA are known as Lewis acids, in an anhydrous hydrocarbon environment aluminum alkyls can act as electron donors which are capable of coordinating to vacant Nd sites. This is evidenced as DIBAH forms oligomers [461]. Thus, DIBAH acts as an electron donor as well as an electron acceptor. In addition, complexes of aluminum alkyls with various Nd precursors were isolated and characterized, e.g. [184–186, 234].

Scheme 30 is also valid for "real" electron-donating ligands. By the addition of electron donors such as PPh$_3$ the *cis*-1,4-content decreases drastically and the content of *trans*-1,4- and 1,2-moieties increases accordingly [364]. The same impact is observed when electron donors such as diethoxydiethylsilane are added to Nd-based catalyst systems [114]. The addition of CS_2 and PPh_3 has a similar effect in Ni- and Co-based Ziegler/Natta catalysts [364].

For the coordination of aromatic solvents such as toluene to the Nd center a similar mechanistic concept was published [318].

A result which is contradictory to the commonly observed influence of electron donors is reported for the catalyst system NdV · HV/TIBA/DIBAH/ DEAC. By the addition of different triphenylphosphine derivatives the *cis*-1,4-content increases from 97.1% to 99.7%. The increase in *cis*-1,4-content goes along with a decrease of catalyst activity and an increase of molar mass. The authors explain the impact of phosphines on *cis*-1,4-content by the preference of anti-conformation and by the acceleration of 1,4-insertion relative to anti-syn-isomerization [367]. The effect of phosphines reported by Kwag et al. is quite surprising as ligands in general and electron donors in particular enhance *trans*-1,4-polymerization (Scheme 30). It is equally possible that in this catalyst system triphenylphosphines selectively deactivate one or more catalytic species which produce BR with a high *trans*-1,4-content. By selective catalyst poisoning the influence of triphenylphosphines on the reduction of polymerization activity, on the increase of molar mass and on the increase of the overall *cis*-1,4-content can be easily explained.

A detailed review on the multiple aspects of the reaction mechanisms of diene polymerization with transition-metal catalysts was published by Porri, Giarrusso and Ricci in 1991 [486].

4.4
Living Polymerization

To the present day, numerous studies on various features of Nd-catalyzed diene polymerizations have been performed and nearly every aspect of the multidimensional system has been addressed. In spite of the large number of studies, to this day, no reaction model is available that allows for the simultaneous prediction of monomer conversion, molar mass and MMD as a function of polymerization time, polymerization temperature, Nd-concentration and the ratio of catalyst components. It therefore has to be concluded that our knowledge on Nd-catalyzed polymerizations is still rather limited and that important aspects are not comprehensively understood. A good example for the ongoing discussions about Nd-catalyzed diene polymerizations is about the living character of the polymerization. Since living polymerizations allow for various synthetic options as for example the preparation of block copolymers, end-group functionalization and the precise control of molar mass and MMD the aspect of living polymerization was in the focus of many studies. In summary, a lot of studies are available that provide evidence for the living character of Nd-catalyzed polymerization. This evidence is not strictly conclusive and the terms "partially living", "quasi-living", "pseudo-living", "living/controlled" etc. prevail for the description of the nature of the polymerization. Subsequently, some of the studies that provide evidence in favor and against the living character of Nd-catalyzed polymerizations are reviewed.

An essential feature of a strictly living polymerization is the absence of transfer reactions [652]. This requirement was found to be valid for the polymerization of IP catalyzed by $NdCl_3 \cdot TBP/TIBA$ in hexane. The lack of chain termination reactions, the second requirement for a living polymerization [652], was confirmed by the application of a mathematical model to the experimental data [279]. Bruzzone et al. realized that transfer reactions occur in the polymerization of BD. In spite of this observation the "pseudo-living" character of the polymerization was assigned to the superposition of chain growth and chain transfer both of which exhibit a different dependence on monomer conversion [87].

Another indicator in favor of a living polymerization is a linear increase of number average molar masses (M_n) on monomer conversion. Hsieh et al. reported on linear plots for the dependence of inherent viscosities on monomer conversion. These plots were highly linear and Hsieh et al. assigned the term "quasi-living" to these polymerizations [134, 139]. Kwag et al. attributed the living character of Nd-catalyzed diene polymerizations to the ionic character of the Nd allyl bond and the stable oxidation state of Nd. In addition, theoretical frontier orbital analysis confirmed these results [653].

A serious contradiction with the requirements of a strictly living polymerization are broad or even bimodal MMDs which are in the focus of many studies, e.g. [87, 178, 620]. This observation of broad and at least bimodal MMDs is the result of the presence of at least two active catalyst species which show different activities. This feature is in contradiction with a strictly living polymerization. Wilson attributed the polymer fraction with a high molar mass to "insoluble" catalyst species which are invisible to the naked eye whereas the low molar mass fraction of the polymer is supposedly produced by "soluble" sites which operate in a "quasi-living" manner [89]. In his study Wilson used catalyst systems of the type $Nd(carboxylate)_3/DIBAH/^tBuCl$.

Most of the non-theoretical studies on the living character of Nd-catalyzed diene polymerization only address selected aspects of a living polymerization whereas other studies only provide semi-quantitative information in favor of the living nature. To this day, stringent proof for a completely living nature of Nd-catalyzed diene polymerizations is lacking. This is particularly true as not every criterion put forward for a truly living polymerization fully applies to Nd-catalyzed polymerizations of dienes [652, 654–656]. According to the definitions given in these references a living polymerization has to simultaneously comply with all of the following requirements:

1. First order kinetics with respect to monomer conversion (no irreversible chain termination);
2. Linear increase of M_n with monomer conversion (no irreversible chain transfer);

3. Linear dependence of M_n on monomer/catalyst-ratio at constant monomer conversion;
4. Increase of molar mass on sequential monomer addition;
5. Feasibility of block copolymer formation;
6. Feasibility of end-group functionalization.

In the following sections the available literature on Nd-catalyzed polymerizations of dienes is discussed regarding the compliance with each of the criteria put forward for a strictly living polymerization.

1. First-order kinetics with respect to monomer conversion
(no irreversible chain termination)
In the following section some of the reports quoted earlier are selected and discussed in more detail regarding the kinetics of monomer consumption.

For binary $NdCl_3$-based catalyst systems such as $NdCl_3 \cdot nL/AlR_3$ a few publications are available in which linear first-order plots regarding monomer conversion are given. In this context the study by Hsieh et al. deserves special attention as the first-order plots with respect to monomer consumption were reported to be ideally linear at different polymerization temperatures (30, 50 and 70 °C) [134].

For carboxylate-based ternary catalyst systems first-order reaction kinetics with respect to BD conversion were reported by many authors. Pross et al. used the catalyst system NdV/TIBA/EASC and found first-order kinetics at different monomer concentrations [204]. This catalyst system was also used by Friebe et al. who varied the molar ratios n_{TIBA}/n_{NdV}. These authors confirmed the earlier results by Pross et al. [179]. Also for NdV/DIBAH/EASC the kinetics of BD polymerization was studied [178]. In this study n_{DIBAH}/n_{NdV}-ratios were varied and linear first-order plots for BD conversion were obtained irrespective of n_{DIBAH}/n_{NdV}-ratios. Also for the catalyst system NdO/TIBA/DEAC linear first-order plots were obtained at different n_{TIBA}/n_{NdO}-ratios up to high monomer conversions [188]. Wilson studied the kinetics of various systems of the type $Nd(OCOR)_3/DIBAH/^tBuCl$ with regard to the structure of the Nd carboxylate (isodecanoate and neodecanoate) [183]. First-order kinetics were particularly pronounced for Nd-isodecanoate-based catalyst systems. Pan et al. found for a series of catalyst systems with the overall composition Nd compound/$AlR_3/^iBu_2AlCl$ that on one hand chain propagation is first order with respect to monomer conversion and the concentration of active catalyst. On the other hand these authors realized that chain termination occurred simultaneously with chain propagation. Chain termination was reported to be second order with respect to the concentration of active Nd [597, 598]. In catalyst systems with the overall composition Nd-precursor/DIBAH/EASC (NdV, Nd neopentanolate and NdP) only NdV yielded first-order monomer consumption. For the Nd-alcoholate and the Nd-phosphate-based catalyst systems different patterns of monomer consumption were observed. Apparently these Nd-alcoholate and Nd-phosphate-based catalyst systems belong to the few exceptions for which requirement 1 for a living polymerization does not apply [272].

In a detailed study Maiwald et al. used the Nd allyl complex $Nd(\eta^3-C_3H_5)_3$ for the polymerization of BD in toluene. A linear relationship between $-\ln(1-x)$ and polymerization time was obtained. An intercept at 10 to 12 min was attributed to an induction period [291]. Similar results were obtained for a variety of catalyst mixtures based on Nd allyl compounds [293].

Various studies by Jenkins focused on the use of magnesium alkyl cocatalysts. In the first of these studies Jenkins investigated the kinetics of the *trans*-1,4-specific catalyst system didymium versatate/$MgBu_2$ [173]. A linear first-order plot with regard to monomer consumption was obtained with a deviation for the early phase of the polymerization. According to Jenkins the formation of the active catalyst was sluggish and caused an induction period during the first 50 min of the reaction. A careful study of BD polymerization at 20 °C by NdV/$MgBu_2$/DEAC ($n_{MgBu2}/n_{Nd} = 10$, $n_{Cl}/n_{Nd} = 15$) revealed the existence of two linear first-order plots. The first section up to 60 min had a lower slope than the second section. Again, Jenkins assumed that the active catalyst was formed during the first sluggish period of the polymerization [157].

For the polymerization of BD in hexane Gromada et al. also used a catalyst system which contained a Mg-cocatalyst (Nd(O-2,6-*t*-Bu_2-4-Me-Ph)$_3$(THF)/Mg(^{n}Hex)$_2$). A linear first-order plot for the applied polymerization conditions was obtained "which suggests a constant number of active species throughout the polymerization course" [235].

The majority of catalyst systems yield polymerization kinetics which comply with the requirement of "first-order kinetics with respect to monomer conversion". Some of the investigated Nd-alcoholates and Nd-phosphates exhibit a deviating pattern of monomer consumption and belong to the few exceptions. Also for some of the Nd-carboxylate-based catalyst systems deviations from the first-order dependence of monomer consumption are reported for the first stage of the polymerization ("induction periods"). Induction periods are usually observed when the active catalyst is prepared in-situ (Sect. 2.1.6). In these cases the formation of the active Nd-species is slow in relation to the rate of polymerization.

2. Linear Increase of M_n with Monomer Conversion
(no Irreversible Chain Transfer)

Linear increases of intrinsic viscosity, inherent viscosity, dilute solution viscosity (DSV), M_w, M_n or M_v on monomer conversion are reported in many studies. For example, Wilson emphasized the unique character of Nd-based catalysts in comparison to conventional transition-metal systems in the following terms: "the typically linear relationship between molecular weight and conversion of the Nd-based polymerization of BD indicates a quasi-living system" [89]. Reports on linear relationships between intrinsic viscosity and monomer conversion date back to 1980 when Shen et al. comprehensively reported on "The Characteristics of Lanthanide Coordination Catalysts and the *cis*-Polybutadienes Prepared Therewith". Shen et al. report on linear dependencies of the viscosity average molar mass (M_v) on monomer conversion. According to Shen et al. this indicates "that under the given conditions the polymerization proceeds by a mechanism in which termination of the kinetic chain and transfer of the growing polymer chain are almost entirely absent" [92]. Another study in which it was shown that inherent viscosities increase linearly with monomer conversion was published by Hsieh et al. in 1985. In this study catalyst systems of the type $NdCl_3 \cdot nROH/AlR_3$ were used and the polymerization temperature was varied (30, 50 and 70 °C). From this study it is not quite clear, however, how inherent viscosities evolve at the start of the polymerization (conversion < 20%). According to Hsieh et al. the "quasi-living" character is especially pronounced when a non-hydride containing aluminum cocatalyst was used and when the polymerizations were performed at low n_{Al}/n_{Nd}-ratios and at low polymerization temperatures [134].

Contrary to viscosity measurements GPC provides number average molar mass data (M_n). For a few studies on the polymerization of BD with Nd-carboxylate-based catalyst systems GPC was systematically applied for the monitoring of M_n as a function of monomer conversion. In these studies three catalyst systems were used: (1) NdV/DIBAH/EASC [178], (2) NdV/TIBA/EASC [179] and (3) NdO/TIBA/DEAC [188]. In the first two studies linear M_n-conversion plots were obtained at various molar ratios n_{Al}/n_{NdV}. In these studies, however, molar mass data at low monomer conversions (<20%) are lacking and positive intercepts on the M_n-axis were found. For the ternary catalyst system NdO/TIBA/DEAC used in the third of these studies the concentrations of Nd and TIBA were varied. A linear increase of M_w and M_n on monomer conversion was found. Deviations from linearity were also observed for low monomer conversions.

Jenkins published two studies in which he used *trans*-1,4-specific catalyst systems for the polymerization of BD. For didymium versatate/"$MgBu_2$" molar mass increased linearly up to 50% conversion. As a consequence Jenkins suggests "that this system contains a high proportion of living chain ends" [173]. For the catalyst system $NdV/MgBu_2/DEAC$ ($n_{MgBu2}/n_{Nd} = 10$, $n_{Cl}/n_{Nd} = 15$) molar masses increased linearly to over 70% conversion [157]. Extrapolations of the straight lines, however, did not pass through the origin and MMDs were rather broad (PDI > 5). Gromada et al. also used a *trans*-1,4-specific catalyst system $Nd(O-2,6-t-Bu_2-4-Me-Ph)_3(THF)/Mg(^nHex)_2$. According to Gromada et al. this catalyst system allowed the control of both molar mass as well as PDI. Therefore, the authors considered this polymerization as "living/controlled". At monomer conversions > 20%, considerable deviations between the calculated and the experimentally determined molar masses were observed. PDI also increased with increasing monomer conversion. On the basis of these observations Gromada et al. concluded that transfer reactions were "insignificant for low molar masses but cannot be neglected at higher conversions" [235].

In this context it is particularly interesting to note that straight-line dependencies of M_n on monomer conversion were obtained with Nd-alcoholate and Nd-phosphate-based catalyst systems even though these two catalyst systems did not comply with requirement No. 1 for a living polymerization "linear first-order plots" [205].

In summary, it has to be mentioned that in many studies intrinsic viscosity, inherent viscosity and dilute solution viscosity (DSV) were used in order to monitor the increase of molar mass on monomer conversion. Unfortunately, only a few studies use GPC rather than viscosity measurements. For a few Nd-carboxylate-based catalyst systems linear dependencies of M_n on monomer conversion were established and proof in favor of requirement No. 2 "Linear increase of M_n with monomer conversion (no irreversible chain transfer)" was provided. A more detailed analysis of the data, however, reveals deviations from linearity particularly at low monomer conversions (< 20%). These deviations are particularly pronounced for polymerizations with induction periods. Also the extrapolation of the straight lines to zero monomer conversion reveals intercepts on the M_n-axis.

If requirement No. 2 "linear increase of M_n with monomer conversion (no irreversible chain transfer)" is substituted by the more stringent requirement "linear dependence of M_n on monomer conversion which passes through the origin" it has to be concluded that the compliance of Nd-catalyzed polymerizations with requirement No. 2 is rather an exception than a general rule.

3. Linear Dependence of M_n on Monomer/Catalyst-Ratio at Constant Monomer Conversion

The dependence of molar mass on the ratio of monomer to catalyst (n_M/n_{Nd}) was investigated in only a few studies. In his pioneering study on the use of ternary catalyst systems Throckmorton investigated the influence of n_M/n_{Ce} on dilute solution viscosities (DSV) [34]. Quite surprisingly, for two catalyst systems (1) Ce octanoate/TIBA/EtAlCl$_2$ and (2) Ce octanoate/DIBAH/HBr DSV decreased with increasing ratios of n_M/n_{Ce}. This observation is not at all understood and is in contradiction with the requirements for a living system.

At constant Nd-concentrations the polymerization of IP with the binary catalyst system NdCl$_3$ · 3TBP/TIBA resulted in the requested linear correlations between monomer concentration and M_n as well as with M_w [133]. Linear increases of the viscosity average molar mass with n_M/n_{Nd} were also established by Gehrke et al. in a study on the polymerization of BD with Nd^{i}O/TEA/AlBr$_3$ [159]. The same group also reported on a linear increase of the viscosity average molar mass on n_M/n_{Nd}-ratios for BD polymerization with the catalyst system Nd^{i}O/TIBA/DIBAC [160]. Apart from these studies there are no further reports on Nd-carboxylate-based catalyst systems in which the dependence of M_n on n_M/n_{Nd} was determined. Therefore, it has to be concluded that for Nd-carboxylate-based catalyst systems stringent proof for the compliance with requirement No. 3 is missing.

Compliance with requirement No. 3 could be fully established for various well-defined allyl Nd compounds. The combination of the two Nd-precursors $Nd(\eta^3-C_3H_5)_2Cl \cdot 1.5$ THF and $Nd(\eta^3-C_3H_5)Cl_2 \cdot 2$ THF with the cocatalysts HIBAO and MAO allowed for the adjustment of molar masses by the ratio n_M/n_{Nd} [288]. Also the use of $Nd(\eta^3-C_3H_5)_3$ provided the controlled build-up of M_n by the adjustment of $((c_{BD,0}/c_{Nd}) \cdot$ conversion) [291]. Confirmations also comes from a variety of catalyst systems based on other Nd allyl compounds [293].

In conclusion, the number of reports on the impact of the monomer/catalyst-ratio (n_M/n_{Nd}) on M_n are rather limited. Studies in which M_n was determined are only available for binary catalyst systems of the type $NdCl_3 \cdot 3TBP/TIBA$ and for allyl Nd compounds. The data reported on these catalyst systems are in full compliance with requirement No. 3 for a living polymerization: "Linear dependence of M_n on monomer/catalyst-ratio at constant monomer conversion". Comparable studies with Nd carboxylates and other Nd precursors are missing.

4. Increase of Molar Mass on Sequential Monomer Addition

In an ideally living polymerization, molar mass increases on the sequential addition of two or more monomer batches. In detail, M_n should proportionally increase with increasing ratios of n_M/n_{Nd} as pointed out for criterion No. 3 "Linear dependence of M_n on monomer/catalyst-ratio at constant monomer conversion". Usually the storage temperature and the length of shelf life prior to the addition of the second monomer batch are not further defined.

Shen et al. reported on the first experiment with sequential polymerization of two batches of IP monomer with a Ln-based catalyst system. They demonstrated that the monomer added in the second batch was completely polymerized and that solution viscosity increased by a factor of about 2 [92]. Jun et al. studied sequential monomer addition with the catalyst system NdN/TIBA/DIBAH/EASC. These authors found that a second batch of diene monomer is polymerized even 10 days after the polymerization of the first batch has gone to full completion [657]. In a similar experimental set-up two charges of BD were sequentially polymerized with the catalyst system NdV/DIBAH/EASC (Fig. 13) [205]. It was found that M_n increased linearly upon addition of the two monomer batches. The slopes of the polymerizations of the two batches are not the same due to different monomer/catalyst ratios in the two stages of the experiment.

The results obtained from these experiments with sequential monomer addition performed with the two Nd-carboxylate-based catalyst systems (1)

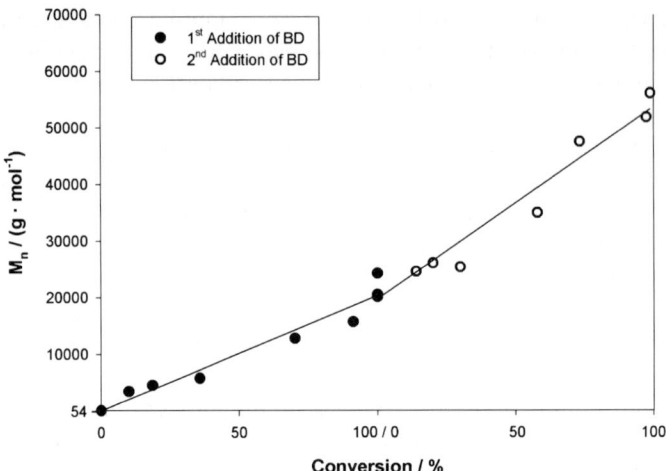

Fig. 13 Evolution of M_n with conversion for BD polymerization with the system NdV/DIBAH/EASC after sequential addition of BD [205]

NdN/TIBA/DIBAH/EASC and (2) NdV/DIBAH/EASC are in compliance with requirement No. 4: "Increase of molar mass on sequential monomer addition". These results provide definitive proof in favor of a living polymerization.

5. Formation of Block Copolymers

A truly living polymerization allows for the preparation of block copolymers without the formation of the respective homopolymers. Reports are available on experiments in which Nd catalysis was applied to synthesize both diene/diene as well as diene/non-diene block copolymers.

The first report is available from Shen et al. who studied the preparation of BR/IR block copolymers by sequential polymerization of BD and IP [92]. Shen et al. found that the polymerization of the second monomer batch resulted in an increase of solution viscosity by 100%. The viscosity increase was considered as strong evidence in favor of block copolymer formation. Further evidence came from stress strain measurements in which the respective BD/IP block copolymers were compared with blends of BR and IR (at the same molar masses). It was found that the block copolymer exhibited higher elongation at break and higher tensile strength. Unfortunately, M_n data were not provided. Therefore, these results are not fully relevant regarding requirement No. 5 for a living polymerization.

Efforts to apply lanthanide metal catalysts for diene/non-diene block copolymerization by sequential monomer addition were reported for samarium catalysts (comparable to Nd) [658] and Nd catalysts [659–661]. In both cases the efforts towards the preparation of block copolymers comprising a poly(diene) building block (BD or IP) and a poly-ε-caprolactone building block were reported. Barbier-Baudry et al. applied a Sm catalyst system for the block copolymerization of IP and CL. They did not find any side reactions which led to the formation of the respective homopolymers. The authors assumed that the formation of the block copolymers occurred in a quantitative way and that no homopolymers (neither unreacted nor newly formed) were present after the addition of the second monomer batch (CL). Proof for the build-up of the block copolymers comes from GPC, DSC, AFM and NMR. The end product was not fractionated and M_n data on the first building block are lacking. Friebe et al. applied the catalyst system NdV/DIBAH/EASC on the preparation of a BR-*block*-poly(ε-caprolactone) block copolymer. The polymer mixture obtained after the addition of the second monomer batch was fractionated with the temperature-dependent demixing solvent pair N,N-dimethyl formamide (DMF)/methyl cyclohexane (MCH). In this way it was found that the yield of the block copolymer was only 11 wt. %. On the basis of these results it was concluded that the Nd-based catalyst system NdV/DIBAH/EASC did not allow for the quantitative formation of block copolymers. Therefore, the Nd-mediated copolymerization of dienes and ε-caprolactone—in contrast to the described Sm system—does not comply with the requirement "Formation of block copolymers". It is not clear whether this negative conclusion has to be confined to the copolymerization of BD/CL and to the catalyst system NdV/DIBAH/EASC. In summary, it has to be concluded that there are no results available which unambiguously comply with requirement No. 5.

6. End-Group Functionalization

Regarding end-group functionalization of Nd-BR there is a vast number of patents in which end-group functionalization is claimed on the assumption that the polymerization is living. To our knowledge, there is no scientific publication in which quantitative information is given on the number of active end groups and on the percentage of living chain ends which can be actually functionalized. In one paper the results of various attempts to couple living end groups were described with the following words: "the two groups of experiments do not rule out the possibility of using living chain ends for further reaction" [173]. It can be speculated that the effectiveness of end-group functionalization will depend on the catalyst system applied and on the nature of the living chain ends (attached to either Nd or Al). In addition, there will be an influence of the modification agent on the efficiency of end-group modification.

In summary, the various hints scattered in the literature do not provide stringent proof in favor of requirement No. 6 "end-group functionalization (which occurs with 100% efficiency)". It therefore has to be concluded that Nd-catalyzed polymerizations of dienes do not comply with requirement No. 6 "end-group functionalization".

Conclusion

From the review of the available literature it is clear that Nd-catalyzed polymerizations of dienes do not simultaneously comply with all criteria for a "living polymerization" as discussed herein [652, 655, 656, 662, 663]. It therefore has to be concluded that Nd-mediated polymerizations of dienes are not living in the strict sense. Only some of the features which are considered to be essentials for a strictly living polymerization apply for selected Nd-based-catalyst systems. Therefore, the terms "pseudo-living", "quasi-living" or "partially living" seem to be adequate for the description of some of the features, though the use of these expressions is discouraged [652].

4.5
Molar Mass Regulation

As discussed in Sects. 2.1 and 2.2.8 control of molar mass is an important aspect in the large-scale polymerization of dienes. In Nd-catalyzed polymerizations the control of molar mass is unique amongst Ziegler/Natta catalyst systems as standard molar mass control agents such as hydrogen, 1,2-butadiene and cyclooctadiene which are well established for Ni- and Co-systems do not work with Nd catalysts [82, 206, 207]. The only known additives which allow for the regulation of molar mass without catalyst deactivation are aluminum alkyls, magnesium alkyls, and dialkyl zinc.

As early as 1980 Shen et al. demonstrated that aluminum compounds such as DIBAH are active chain transfer agents [92]. This result was subsequently confirmed by several authors, e.g. [87, 134, 141, 165, 166, 178, 460]. It was also recognized that cocatalysts of the type $HAlR_2$ are more active chain transfer agents than the respective hydrogen free AlR_3 cocatalysts [134, 179].

Until 1999 the transfer of living poly(butadien)yl chains from Nd onto Al was considered to be irreversible [162, 165, 166, 169, 188, 279]. In 1995 a reaction scheme which accounts for the reduction of molar mass by irreversible transfer of living poly(butadien)yl chains was put forward by Nickaf et al. (Scheme 31).

According to Nickaf et al. there is an irreversible exchange of poly(butadien)yl chains attached to active Nd sites and isobutyl groups attached to Al. The reaction product from poly(butadien)yl anions with TIBA is considered to be inactive (first reaction in Scheme 31). The assumption about irreversible termination of growing polymer chains is supported by the broadening of MMDs with increasing monomer conversion. The second reaction in Scheme 31 shows the insertion of BD into the newly formed Nd isobutyl species. This reaction results in the formation of a new living polymer chain. As active Nd sites are not deactivated by the irreversible transfer of polymer chains the number of active Nd sites remains constant during the whole course of the polymerization. From the reaction scheme

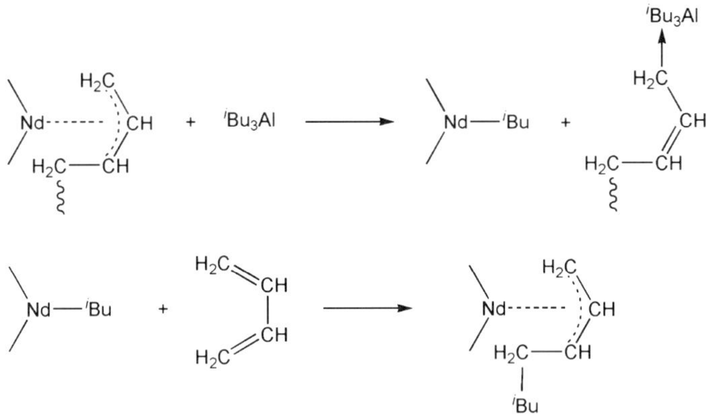

Scheme 31 TIBA-mediated molar mass control in the polymerization of BD by NdO/TIBA/DEAC; depicted as given in [188], reprinted with permission of John Wiley & Sons, Inc.

put forward by Nickaf et al. it can be derived that the formal number of poly(butadiene) chains (i.e., number of chains formed per Nd atom) increases with BD conversion. Oehme et al. agreed on the irreversible transfer of living poly(butadien)yl chains [169]. Also for $NdCl_3$-based systems an irreversible chain exchange reaction was put forward by Iovu et al. [279].

In contrast to these observations, no influence on molar mass was found when the amount of aluminum alkyl cocatalyst was varied for two allyl Nd-based catalyst systems: (1) $Nd(\eta^3-C_3H_5)_2Cl \cdot 1.5$ THF and (2) $Nd(\eta^3-C_3H_5)Cl_2 \cdot 2$ THF [292].

Recently, the impact of the metal alkyls TIBA, DIBAH and $ZnEt_2$ on molar mass was comparatively studied. In these studies ternary NdV-based catalyst systems were used [178–182]. The first two of these studies focus on molar mass control by DIBAH and by TIBA. Linear dependencies of M_n on monomer conversion were obtained. In addition, PDIs decreased with increasing monomer conversion. On the basis of these observations it was concluded that chain transfer of living poly(butadien)yl chains between Nd and Al is fully reversible. A reaction mechanism which accounts for these features is outlined in Scheme 32.

Molar mass reduction with DIBAH was significantly more effective than with TIBA. The higher activity of DIBAH was assigned to the hydride group. The differences in the relative activities of hydride and isobutyl moieties towards substitution by living polybutadienyl chains were qualitatively described by different equilibrium positions (Scheme 33) [179, 180, 205].

The lengths of the arrows in Scheme 33 give an estimation of the equilibrium positions in order to account for the different molar mass control activities of DIBAH and TIBA. In these studies it was also demonstrated that β-hydride-elimination from TIBA which resulted in the formation of DIBAH

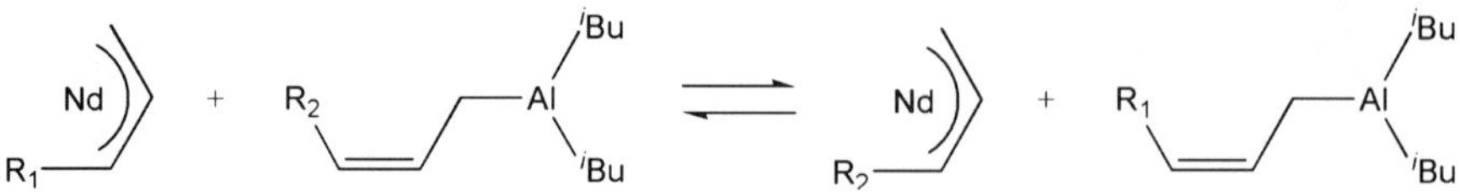

Scheme 32 Reversible exchange of poly(butadien)yl chains (R_1 and R_2) between Nd and Al in the ternary catalyst systems NdV/DIBAH/EASC and NdV/TIBA/EASC (charge and ligands of Nd are omitted for clarity)

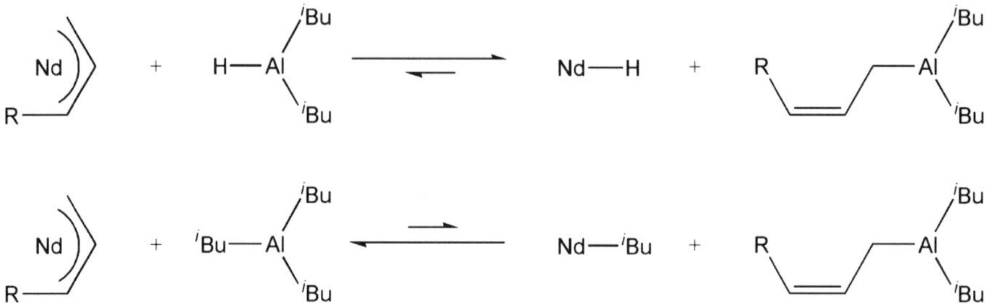

Scheme 33 Reversible exchange of active polymer chains from Nd to Al in the catalyst systems NdV/DIBAH/EASC (*top*) and NdV/TIBA/EASC (*bottom*) [179, 180, 205], reproduced by permission of Taylor & Francis Group, LLC., http://www.taylorandfrancis.com

and isobutylene does not occur at the applied polymerization temperature (60 °C) [205].

The addition of $ZnEt_2$ to the catalyst system NdV/DIBAH/EASC had the same effect on the decrease of molar mass as aluminum alkyls. Therefore, a reversible exchange of polybutadienyl chains between Nd and Zn was also assumed to apply for molar mass control by $ZnEt_2$ (Schemes 34 and 35) [180].

A comparison of the molar mass control efficiencies of $ZnEt_2$, DIBAH and TIBA allowed for the establishment of the ranking DIBAH > $ZnEt_2$ > TIBA [180, 205]. The reaction mechanism which accounts for the control of molar mass by reversible transfer of (poly)butadienyl chains between Nd and Al on one hand and between Nd and Zn on the other hand was established for the catalyst system NdV/DIBAH/EASC. In addition, it was concluded that in this system the (poly)butadienyl chains are only active during the period in which they are attached to Nd. The (poly)butadienyl chains are dormant in the period during which they are attached to Al or Zn. In the context of these results it is not clear whether the irreversible transfer of polybutadienyl

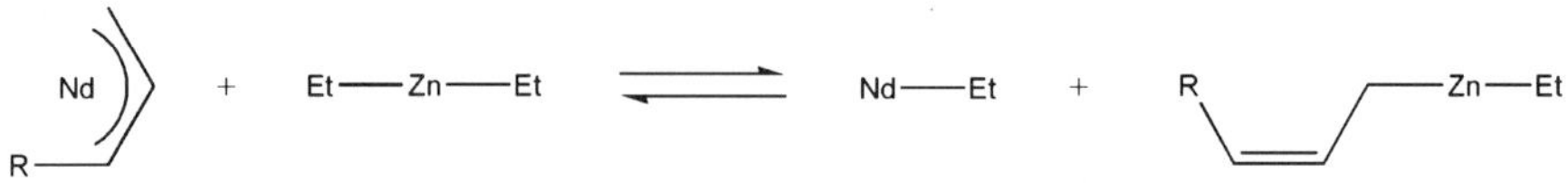

Scheme 34 Molar mass control by $ZnEt_2$ in the system NdV/DIBAH/EASC. Reversible exchange of poly(butadien)yl chains from Nd to Zn by poly(butadien)yl-ethyl interchange (charge and ligands of Nd are omitted for clarity), reproduced by permission of Taylor & Francis Group, LLC., http://www.taylorandfrancis.com

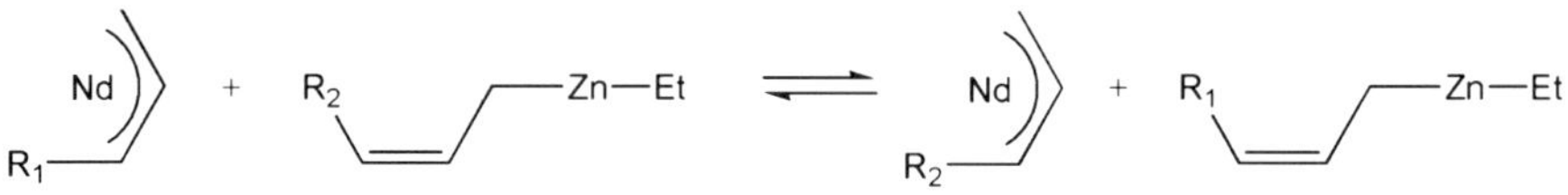

Scheme 35 Reversible exchange of polymer chains (R_1 and R_2) between Nd and Zn in the catalyst system NdV/DIBAH/EASC/ZnEt$_2$ (charge and ligands of Nd are omitted for clarity)

chains from Nd to Al reported in the earlier studies has to be attributed to the catalyst system used or whether the earlier results have to be revised in the light of the later results.

5
Open Questions

In this section some unresolved questions concerning Nd-based diene polymerization are addressed. By definition, this collection is highly subjective as it reflects the specific background and knowledge of the authors. Such a list cannot be complete. It may also contain questions which some readers may consider to be "wrong", "based on the wrong assumptions", "not necessary" or "solved a long time ago" etc. Our intention is to emphasize that in the field of Nd-initiated diene polymerization not everything is understood at least not in full detail and that there are observations which cannot be explained. By these questions we want to show the present frontier of knowledge. In our opinion the first step to an answer very often is the precise formulation of the respective question. By this list of questions we want to raise interest in the topic. We also want to stimulate further work in the field of Nd-catalyzed polymerizations.

The questions are clustered under two headlines ("*Catalyst Systems*" and "*Kinetic and Mechanistic Aspects*") and are compiled in an arbitrary order.

Catalyst Systems

"Why do Nd-phosphates require less aluminum alkyl cocatalyst for activation?"
It is not clear how many equivalents of aluminum alkyls are required in order to generate one active species from Nd-phosphates. Clarification is needed why Nd phosphates require less aluminum alkyl for activation than for example Nd carboxylates do.

"Is there a reduction of Nd-carboxylates to Nd-alcoholates
by the aluminum alkyl cocatalyst?"
It is not clear whether Nd-carboxylates are reduced by aluminum alkyls to the respective alcohols prior to the formation of the active catalyst. It has to

be clarified whether the cocatalysts TMA, TEA, TIBA and DIBAH perform differently in this respect.

"What is the minimum amount of Al-cocatalyst to activate Nd-halide based precursors?"

In the activation of halide-based Nd-precursors the halide is not reduced by aluminum alkyls. Therefore, the generation of the active Nd-species should be possible with a very small or even the molar equivalent of an aluminum alkyl. The consumption of aluminum alkyl will somewhat depend on the donor present in the $NdCl_3$ system. To our opinion systematic studies lack in this respect.

"How can the size of $NdCl_3$-clusters be minimized and what is the impact of cluster size on polymerization activities and MMD?"

First results demonstrate that polymerization activities increase reciprocally with the size of $NdCl_3$ nanoparticles. It was also discussed that the size of $NdCl_3$-based heterogeneities have an influence on the MMD. In this context the impact of very small $NdCl_3$ clusters on polymerization activities and on MMD deserves interest.

"Is catalyst heterogeneity associated with gel formation?"

There is evidence that heterogeneous or micro-heterogeneous Nd systems lead to higher levels of gel than homogeneous systems. Most certainly, there are other factors which contribute to gel formation. In this context it has to be clarified which factors trigger gel formation and what is the quantitative contribution of catalyst heterogeneity.

"What is the size and the size distribution of heterogeneities in heterogeneous catalyst systems?"

"Prereaction" of catalyst components results in so called "preformed catalysts", which are sometimes aged prior to use. Preformed catalysts are often heterogeneous to the naked eye whereas "unpreformed catalyst systems" seem to be homogenous to the naked eye. Often it can be argued whether there are micro-heterogeneities present even in "homogeneous catalysts". In this respect quantitative measurements and correlations with other features of the polymerization are required (gel formation, polymerization kinetics, product properties etc.).

"Are there additives which allow for selective poisoning of active species?"

Many Nd-based catalyst systems comprise more than two or three components. Quite often various donors are added the role of which is not really understood. It may be speculated whether some of these additives poison selected catalyst sites or catalyst species.

Kinetic Aspects and Mechanistic Aspects

"What are the characteristics of the various active Nd species?"
With many Nd catalyst systems multimodal or at least bimodal MMDs are obtained. From this observation it is concluded that there are several active species. A particularly highly active species which causes high molar masses at low monomer conversions is present at the start of the polymerization. Once the initial "induction period" is over and consistent kinetics are observed, often this species is no longer active. The fate of the species is not clear. Does this species die off after the initial high activity or is it transformed to a species which exhibits a significantly lower activity? Is the "early" species heterogeneous and the "normal" species homogeneous?

"Are there metal-free chain modification agents for Nd-mediated polymerizations?"
In Co- and Ni-mediated polymerizations of dienes the molar mass is commonly regulated by metal-free agents such as hydrogen, cyclooctadiene, 1,2-butadiene etc. In Nd-catalysis these additives do not result in a reduction of molar mass. What is the reason for this difference and are there metal-free chain modification agents which work with Nd-catalysts?

"What is the rate of chain transfer reactions?"
In various studies on molar mass regulation by variation of the amount of cocatalyst different interpretations concerning the rate of chain transfer between Mg, Al and Zn on one hand and Nd on the other hand are given. Detailed kinetic studies on the complex molar mass regulation processes between Nd centers and the cocatalyst are still lacking. What are the relative rates of chain transfer and chain propagation?

"How many polymer chains grow per Nd center?"
In the systems in which the overall number of polymer chains formed per Nd-atom is > 1 there is the theoretical possibility that the functionality of a Nd atom is also > 1. So far, there are not even vague ideas about the functionality of a Nd-atom. In particular, it will be difficult to determine the functionality of a Nd atom without interference from the cocatalyst. In this context allyl Nd-based catalysts deserve particular interest.

"What is the reason for inconsistent apparent rate constants?"
In ternary Nd-based catalyst systems the variation of the cocatalyst component (e.g. TIBA or DIBAH) results in a range of apparent rate constants k_a which differ by magnitudes. It is not clear whether these differences correlate with the number of active sites per Nd or with the total portion of active Nd.

"Why are we so ignorant about the magnitude of propagation rate constants?"
Apparent rate constants k_a are dependent on the total Nd-concentration (c_{Nd}), the fraction of active Nd (eff_{Nd}), the functionality of Nd (f_{Nd}) and the propagation rate constant k_p:

$$k_a = \mathit{eff}_{Nd} \cdot c_{nd} \cdot f_{Nd} \cdot k_p\,.$$

As pointed out in the previous questions many question marks are contained in this equation. Because of many assumptions we are so ignorant about the real value of the propagation rate constant. To the present day not even the magnitude of k_p is clear. For the solution of this question kinetic studies with catalyst systems are required for which $\mathit{eff}_{Nd} = 100\%$.

"Is it possible to increase the portion of active Nd to 100%?"
So far, we are only able to qualitatively assign variations of catalyst efficiencies. A quantification of this effect is required. Can the portion of active Nd be increased to 100%?

"Are there further strategies to increase the active portion of Nd?" and
"Are there other strategies by which Nd-efficiency can be increased?"
So far, four major principles are known for the increase of the active portion of Nd:

- Choice of aluminum alkyl cocatalyst and quantity;
- Synthetic route for the preparation of the Nd precursor;
- Support of catalysts;
- Use of $NdCl_3$ nanoparticles.

"Why does catalyst activation by Al- and Mg-based cocatalysts result in completely different BR-microstructures?"
In ternary Nd Ziegler/Natta catalyst systems of the type Nd compound/cocatalyst/halide donor Al alkyl cocatalysts yield high *cis*-1,4-BR whereas Mg alkyl cocatalysts yield *trans*-1,4-poly(butadiene). To date, it is not clear why the use of these two classes of cocatalysts leads to completely different polymer microstructures.

"Is it possible to obtain copolymers of BD and alkenes with cis-1,4-poly(butadiene) units?"
So far, the successful copolymerization of BD and alkenes goes along with an increase of 1,2- and *trans*-1,4-configuration. Sometimes even 1,2-cyclohexane units are formed. For various applications copolymers with a high *cis*-content are of high interest. It is not clear neither from experimental nor from a theoretical point of view whether copolymers with a *cis*-1,4-content are feasible and how the respective catalysts have to be modified in order to meet this requirement.

"Nature of chain ends and impact on modification reactions?"
There is an equilibrium between "living" chain ends which are attached to Nd and "dormant" chain ends which can be attached to Al, Mg and Zn. Very likely these chain ends exhibit differences in reactivity towards modification agents and polar monomers. This assumption will possibly result in the interpretation of inconsistent results observed in end-group functionalization and block copolymer formation, e.g. with polar comonomers such as ε-caprolactone.

6
Evaluation of Nd-BR-Technology

To date for the large-scale production of high *cis*-1,4-BR (*cis*-1,4-contents > 90%) solution processes are applied in which catalyst systems on the basis of Co, Ni, Ti and Nd are used. In order to elucidate the specific benefits and disadvantages of Nd-technology it is compared with the three other available technologies. This comparison is performed in two respects:

1. Features of the existing solution processes.
2. Features of the available products.

In Table 29 some basic features of the available solution processes are compared:

The comparison of the existing BR-solution processes reveals a series of specific advantages in favor of the Nd-based solution process. These advantages include:

- Low tendency to gel formation;
- Monomer conversion up to 100%;
- No necessity for external cooling/full adiabatic process with a maximum polymerization temperature up to 120 °C;
- Low rate of vinyl cyclohexene formation.

The first three benefits are a direct consequence from the extremely low tendency of the Nd-catalyst to form branches and gel. Because of this remarkable feature, Nd-catalysts allow monomer conversions up to 100%. Therefore, the polymerization reaction does not have to be shortstopped below a critical monomer conversion in order to avoid gel. In addition, polymerization temperature does not have to be controlled within a well-defined temperature range. As the maximum polymerization temperature (at complete monomer conversion) can be as high as 120 °C the polymerization process can be performed in a fully adiabatic manner. In this case energy costs for cooling and for the removal of low molar mass residuals can be very low. Another benefit of the Nd-catalyst is the low tendency to catalyze the Diels–Alder dimerization of BD to vinyl cyclohexene.

Table 29 Technological features of the existing solution processes for the production of high cis-1,4-BR

Technology	Co	Ni	Ti	Nd
Solvent	benzene, cyclohexane	aliphates, toluene, benzene	benzene, toluene	hexane, cyclohexane
Total solids[a]/wt.-%	14–22	15–16	11–12	18–22
Monomer conversion/%	55–80	< 85	< 95	~ 100
Tendency to gel formation	medium	medium	high	very low
Formation of vinyl cyclohexene (VCH)	low	high	high	very low
Max. polymerization temperature/°C	80	80	50	120
Necessity for external or evaporation cooling	yes/only partially adiabatic	yes/only partially adiabatic	yes/only partially adiabatic	no/fully adiabatic
Transition metal content in rubber (without efficient removal of metal residues)/ppm	10–50	50–100	200–250	100–200
Molar ratio of $n_{Al}/n_{transition\ metal}$	70–80/1	40/1	5/1	(5) 10–15/1
Availability of metal-free agents for molar mass control	yes	yes	no	no

[a] Non-volatile residue after evaporation of solvent

The major disadvantages of Nd catalysts are directly related to comparatively low catalytic activities. As a consequence of this, high amounts of the relatively expensive Nd-catalyst have to be used and high contents of Nd residues are left in the final products if no efficient residue removal technique is applied. At fixed molar ratios n_{Al}/n_{Nd} = 10–15 high Nd-loadings go along with high amounts of Al-cocatalyst. Both factors contribute to the relatively high overall catalyst costs for Nd-based BR. Another disadvantage of Nd-technology is the lack of availability of metal-free molar mass control agents. Contrary to Co- and Ni-based catalyst systems in Nd-technology molar mass is controlled by the molar ratio of n_{Al}/n_{Nd} (Sects. 2.1.4 and 2.2.8). For the production of Nd-BR with a low molar mass particularly high amounts of the expensive Al-cocatalyst have to be applied. As a direct consequence, the overall catalyst costs are extremely high for low-Mooney Nd-BR-grades.

The second aspect of evaluation are product features such as product quality and the range of available products. Important aspects are summarized in Table 30.

Table 30 Product features of high cis-1,4-BR obtained with Ziegler/Natta catalyst systems

Technology	Co	Ni	Ti	Nd
cis-1,4-Content/%	96	97	93	98
$T_g/°C$	– 106	– 107	– 103	– 109
Linearity of products	adjustable	branched	linear	highly linear
Distribution of molar mass	medium	broad	narrow	broad
Cold flow	adjustable	low	high	high
Gel content	variable, can be extremely low (dependent on technology)	medium	medium	extremely low
Color	colorless	(colorless)	colored	colorless
Availability of tire grades	yes	yes	yes	yes
Availability of HIPS-grades	yes	no	no	yes
Availability of ABS-grades	yes	no	no	no

As can be seen in Table 30 Co-, Ni-, Ti- and Nd-based catalyst systems yield BR with cis-1,4-contents >90%. A direct consequence of high cis-1,4-contents is strain-induced crystallization of raw rubbers as well as of the respective vulcanizates. As the cis-1,4-content is extraordinary high in Nd-BR, strain-induced crystallization is particularly pronounced for this BR-grade. As strain-induced crystallization beneficially influences the tack of raw rubbers as well as tensile strength and resistance of vulcanizates to abrasion and fatigue the high cis-1,4-content makes Nd-BR particularly useful for tire applications.

The use of BR for the modification of ABS and HIPS is limited by strict requirements regarding gel content, color and solution viscosity. Amongst the commercially available high cis-BR rubbers only special Co-based grades fully meet the demands for both application areas (HIPS and ABS). Li-BR grades also do the job but are not discussed here. Regarding gel content and color Nd-BR completely meets the requirements for the use in HIPS and ABS. Concerning solution viscosity only special newly developed Nd-grades meet the requested specifications for the application in HIPS. Even lower solution viscosities are required for the use in ABS. To the present day, there is no Nd-BR grade available which meets this criterion.

The disadvantages of Nd-based BR have been known for a long time and a lot of work has been performed to overcome these deficiencies. Literature on Nd-based BD polymerization catalysis has been reviewed in a more comprehensive manner in the previous sections. In the following concluding section specific approaches are selected that are considered to deserve particular attention in the context of overcoming existing deficiencies in the production process as well as in product performance.

7
Remarks on Present Developments in Nd-Technology and Speculations about Future Trends

As shown in the previous section, Nd-based technology exhibits significant technological advantages over the other technologies for the production of high cis-1,4-BR. On the other hand there are several drawbacks the elimination of which have been in the focus of past and ongoing development work. From the vast amount of available literature the achievements which are relevant in the context of the major goals of development are briefly highlighted:

- Reduction of process costs related to energy consumption.
- Overall reduction of catalyst costs.
- Reduction of Nd-costs by the increase of Nd-activity (this goes along with reduction of residual Nd in the product).
- Reduction of cocatalyst-cost by the application of catalyst systems that require less Al-based cocatalyst (this goes along with reduction of residual Al in the product).
- Application of metal-free molar mass control agents.
- Further reduction of solution viscosities for the modification of ABS (narrowing of MMD and higher degree of branching).

One interesting option towards the reduction of energy consumption comprises the reduction of the amount of solvent to be recycled in the process. In this context the increase of total solids from $\sim$ 20 wt. % commonly used the solution process up to 25–40 wt. % by the application of a slurry technology is particularly advantageous (Sect. 3.1) [517, 518]. According to the authors' opinion slurry technology avoids the risks associated with the implementation of more advanced technologies such as bulk (Sect. 3.1) or gas-phase polymerization (Sect. 3.2).

Increases in catalyst activity can theoretically be achieved by an increase of the active portion of the total amount of Nd applied as for commonly used Nd-based catalyst systems the percentage of active Nd is only in the range of 0.4% to 8% (Sect. 4.2) [93, 94, 159, 162, 595–600]. An increase of up to 100% efficiency of total Nd used would result in an increase of catalyst activity by a factor of 12.5 to 250. On the basis of this consideration there is an enormous theoretical potential for the reduction of the amount of catalyst and associated costs.

Three independent routes by means of which catalyst efficiency can be increased are reported in the literature. In the first of these routes supported

Nd-catalysts are used in a solution process (Sect. 2.1.7) [400–404]. A factor of up to 100 is reported for the increase of catalyst activity by the application of appropriate catalyst supports [408].

The second route towards increasing Nd catalyst activity focuses on the fragmentation of Nd carboxylate clusters which are present in non-polar solvents. Fragmentation is achieved by simple changes in the preparation of the Nd-carboxylate. This change is reported to result in exceptionally high catalyst activities (Sects. 2.1.1.2 and 4.2) [156, 367, 417–419].

The third route to achieve higher catalytic activities is related to $NdCl_3$-based catalysts and concerns the use of $NdCl_3$ nanoparticles (Sect. 2.1.1.1). The nanoparticles are prepared by a sequence of preparative steps [132]. It is demonstrated that the size of the $NdCl_3$ particles correlates with catalytic activity.

In residual Nd-concentration a critical threshold is 100 ppm, as pointed out by Quirk [187]. Ideally, this Nd level should be achieved without a separate process step for catalyst removal. The available strategies to increase catalyst activity provide grounds for the achievement of this target.

The reduction of the molar ratio of n_{Al}/n_{Nd} constitutes another key to the reduction of overall catalyst costs. Especially in NdV-based catalyst systems a considerable portion of the expensive Al-alkyl is consumed in the activation of Nd-carboxylates and other reactions prior to the formation of the active Nd-species (Sect. 4.2) [178]. The formation of the active Nd-species from Nd-phosphate precursors requires a comparably low amount of Al-alkyls [272] as Nd-phosphate-based catalyst systems are active at very low molar ratios of $n_{Al}/n_{Nd} \leq 5$ (Sects. 2.1.1.4 and 2.1.4) [268, 269, 272].

From a theoretical point of view the application of readily available metal free molar mass control agents should allow the reduction of overall catalyst costs for BR-grades with low molar masses. Unfortunately, to the present day such additives are not known for Nd-catalyzed solution processes. The only exception seems to be slurry technology in which control of molar mass by the use of 1,2-butadiene is possible [517, 518].

In order to meet the requirements for ABS-modification it is essential to further reduce solution viscosities of the existing Nd-BR grades. Without compromising cold flow this target will be achieved by the introduction of branches. In this context two approaches deserve particular attention [395, 396, 431, 432] (Sect. 2.2.6).

Regarding the improvement of overall economy of ABS and HIPS-production the polymerization of BD in the presence of styrene monomer is particularly worth mentioning (Sect. 3.3) [581–590]. By the implementation of this technology the total number of process steps and overall process complexity would be considerably reduced. In particular, the isolation of BR from the polymer solution and the drying and baling of BR at the end of the finishing process could be avoided. In addition, the preparation of BR solutions in styrene (for HIPS-production) or in styrene/acrylonitrile-mixtures (for ABS-production) would no longer be necessary.

Many important technology developments are only described in patent literature. On the basis of this information it is sometimes difficult to recognize the full potential of a technology described. Even though, according to the authors' point of view, there is sufficient evidence provided that Nd-technology has a great deal of additional potential for further improvements in the manufacturing process as well as in the extension of the product range to meet the specific requirements for the modification of ABS.

So far, the potential of Nd-technology has been highlighted regarding the production of high *cis*-1,4-BR. As pointed out in the Sects. 2.3.1 and 2.3.2 the potential of Nd-technology is even greater for the production of high *cis*-1,4-IR as Nd-technology allows the production of *cis*-1,4-contents up to 99.0–99.6% [266, 267]. This *cis*-1,4-content is significantly higher than *cis*-1,4-contents achieved by the established Ti- and Li-based technologies and seems to be even higher than that of natural rubber. Depending on the availability of IP monomer, Nd-technology will be the technology of choice for the production of high *cis*-1,4-IR in the future. In terms of cost reduction the selective polymerization of IP in the crude C5 cracking fraction by Nd-technology deserves particular attention (Sect. 3.3) [264, 265, 591, 592].

In summary, the application of Nd-technology to BR and IR production has many benefits over the other readily available technologies. Therefore, it is quite surprising that Nd-technology is the technology which is the least applied in the production of high *cis*-1,4-BR and high *cis*-1,4-IR. The most simple and most straightforward explanations of this striking fact are: "Historical reasons and business environment". Nd-technology became commercially established as the last of the high *cis*-technologies. As the last of the high *cis*-technologies it had to compete not only with the other established high *cis*-technologies but also with NR. In the past, NR-consumption had been subject to massive changes in availability and in price as the markets for BR and even more that of IR had been and still are characterized by drastic changes in the annual volumes consumed. On the other hand crude oil is the feedstock for the monomers BD and IP which also has been subject to substantial changes in production volumes and even more in price. As a consequence of these factors and as a result of the normal fluctuations of economy the impact on the utilization of BR- and IR-capacities had been and still is enormous. Quite naturally these factors had and still have a strong influence on the respective profit margins. It is easy to understand that in this business environment changes in production technology such as the introduction of Nd-technology occurred in a sluggish manner especially when financial commitments of any sort were necessary.

Presently, 20 years after Nd-technology had been introduced into large-scale production, the superiority of Nd-technology is widely acknowledged in industry. The availability of knowledge on further process and product improvements renders Nd-technology and Nd-catalysis based products even more attractive. Strong indication in favor of this view comes from the increasing number of patents filed by companies which so far had no stake in Nd-technology. These patent activities provide strong evidence that the shares of high *cis*-1,4-BR and high *cis*-1,4-IR manufactured by Nd-technology will increase in future. The expiration of the basic patents on Nd-technology will certainly favor the proliferation of this technology.

Acknowledgements The authors thank Prof. Dr. Wieder, Dr. U. Wolf, Dr. H. Kloppenburg and Dr. S. Mücke for valuable advice and suggestions and J. C. Sauceda and D. E. Herbert for checking the manuscript. L. F. thanks Bayer AG for financial support during his diploma and Ph.D. thesis. L. F. also gratefully acknowledges a Postdoctoral Fellowship granted by the German Academic Exchange Service (Deutscher Akademischer Austauschdienst DAAD). L. F. thanks Prof. I. Manners for excellent working conditions in his group.

References

1. Ziegler K, Breil H, Holzkamp E, Martin H (1955) Angew Chem 67:541
2. Wilke G (2003) Angew Chem 115:5150
3. Wilke G (2003) Angew Chem Int Ed 42:5000
4. Ziegler K, Breil H, Holzkamp E, Martin H (1953) Ger Patent 973626 (1960) Chem Abstr 54:14794
5. Natta G (1955) J Polym Sci 16:143
6. Natta G (1956) Angew Chem 68:393
7. Natta G, Pino P, Corradini P, Danusso F, Mantica E, Mazzanti G, Moragli G (1955) J Am Chem Soc 77:1708
8. http://www.nobel.se/chemistry/laureates/1963/press.html (accessed on Jan/19/2006)
9. Horne SE (1960) US Patent 3114743
10. Horne SE (1960) Chem Abstr 54:100169
11. Horne SE, Kiehl JP, Shipman JJ, Folt VL, Gibbs CF, Willson EA, Newton EB, Reinhart MA (1956) Ind Eng Chem 48:784
12. Horne SE, Kiehl JP, Shipman JJ, Folt VL, Gibbs CF, Willson EA, Newton EB, Reinhart MA (1956) Chem Abstr 50:62619
13. Carlson CJ, Horne SE (1973) US Patent 3728325
14. Carlson CJ, Horne SE (1973) Chem Abstr 79:6532
15. Phillips Petroleum Co (1960) GB Patent 920244
16. Phillips Petroleum Co (1963) Chem Abstr 58:82802
17. Tucker H (1962) Ger Patent 1128143
18. Tucker H (1963) Chem Abstr 58:9470
19. Ueda K, Onishi A, Yoshimoto T, Hosono J, Maeda K (1965) US Patent 3170905
20. Ueda K, Onishi A, Yoshimoto T, Hosono J, Maeda K (1965) Chem Abstr 62:91881
21. Natta G, Porri L, Corradini P, Morero D (1958) Chim Ind (Milan) 40:362
22. Natta G, Porri L, Corradini P, Morero D (1959) Chem Abstr 53:1614
23. Natta G, Porri L, Mozzanti G (1955) Ital Patent 536631
24. Natta G, Porri L, Mozzanti G (1959) Chem Abstr 53:20463
25. Natta G, Porri L, Zanini G, Palvarini A (1959) Chim Ind (Milan) 41:1165
26. Natta G, Porri L, Zanini G, Palvarini A (1961) Chem Abstr 55:111532

27. Natta G, Porri L, Zanini G, Fiore L (1959) Chim Ind (Milan) 41:526
28. Natta G, Porri L, Zanini G, Fiore L (1960) Chem Abstr 54:6376
29. Natta G, Porri L (1955) Ital Patent 538453
30. Natta G, Porri L (1958) Chem Abstr 52:27895
31. Shen T, Gung CY, Chung ZG, Ou Y (1964) J Sci Sin (Engl Transl) 13:1339
32. Von Dohlen WC, Wilson TP, Caflisch EG (1964) Belg Patent 644291
33. Von Dohlen WC, Wilson TP, Caflisch EG (1965) Chem Abstr 63:32659
34. Throckmorton MC (1969) Kautsch Gummi Kunstst 22:293
35. Yang JH, Tsutsui M, Chen Z, Bergbreiter DE (1982) Macromolecules 15:230
36. Bruzzone M (1982) ACS Symp Ser 193:33
37. Carbonaro A, Ferraro D, Bruzzone M (1983) Eur Patent 76535
38. Carbonaro A, Ferraro D, Bruzzone M (1983) Chem Abstr 99:88736
39. Sylvester G, Witte J, Marwede G (1980) Ger Patent 2830080
40. Sylvester G, Witte J, Marwede G (1980) Chem Abstr 92:130367
41. Lugli G, Mazzei A, Modini G (1973) Ger Patent 2257786
42. Lugli G, Mazzei A, Modini G (1973) Chem Abstr 79:54072
43. Lugli G, Mazzei A, Poggio S (1974) Makromol Chem 175:2021
44. De Chirico A, Lanzani PC, Raggi E, Bruzzone M (1974) Makromol Chem 175:2029
45. Bruzzone M, Mazzei A, Giuliani G (1974) Rubber Chem Technol 47:1175
46. Gargani L, Giuliani GP, Mistrali F, Bruzzone M (1976) Angew Makromol Chem 50:101
47. Sylvester G, Witte J, Marwede G (1977) Ger Patent 2625390
48. Sylvester G, Witte J, Marwede G (1978) Chem Abstr 88:75165
49. Witte J (1981) Angew Makromol Chem 94:119
50. Porri L, Giarrusso A (1989) In: Eastmond GC, Ledwith A, Russo S, Sigwalt P (eds) Comprehensive Polymer Science, vol 4, part II. Pergamon Press, Oxford, p 53
51. http://www.iisrp.com (accessed on Jan/19/2006); IISRP (2002) Worldwide Rubber Statistics.
52. Kakiuchi S, Saito T, Tomita S (1987) Jap Patent 6289750
53. Kakiuchi S, Saito T, Tomita S (1987) Chem Abstr 107:97986
54. Yokota M (2004) Jap Patent 2004337231
55. Yokota M (2004) Chem Abstr 142:24284
56. Higuchi H, Kataoka N, Namna M, Sasaki H (2004) Jap Patent 2004180733
57. Higuchi H, Kataoka N, Namna M, Sasaki H (2004) Chem Abstr 141:72837
58. Higuchi H, Kataoka N, Namna M, Sasaki H (2004) Jap Patent 2004180716
59. Higuchi H, Kataoka N, Namna M, Sasaki H (2004) Chem Abstr 141:72836
60. Burkhart CW, Kerns ML, Castner KF, Rachita MJ, Mallamaci MP (2003) US Patent 2003013556
61. Burkhart CW, Kerns ML, Castner KF, Rachita MJ, Mallamaci MP (2003) Chem Abstr 138:91263
62. Nesbitt RD (2003) US Patent 2003100384
63. Nesbitt RD (2003) Chem Abstr 138:402976
64. Higuchi H, Nanba M (2002) Jap Patent 2002355339
65. Higuchi H, Nanba M (2002) Chem Abstr 137:385870
66. Bissonnette LC, Halko RD, Bulpett DA, Mallamaci MP (2002) US Patent 2002119837
67. Bissonnette LC, Halko RD, Bulpett DA, Mallamaci MP (2002) Chem Abstr 137:186883
68. Shindo J (2001) Jap Patent 2001149508
69. Shindo J (2001) Chem Abstr 135:6755
70. Binette ML, Kennedy TJ, Nesbitt RD, Sullivan MJ (2001) Int Patent 2001052947
71. Binette ML, Kennedy TJ, Nesbitt RD, Sullivan MJ (2001) Chem Abstr 135:108432

72. Kennedy TJ, Nesbitt RD, Binette ML, Neill JT, Sullivan MJ (2001) Int Patent 2001052945
73. Kennedy TJ, Nesbitt RD, Binette ML, Neill JT, Sullivan MJ (2001) Chem Abstr 135:123589
74. Kennedy TJ, Nesbitt RD, Binette ML, Neill JT, Sullivan MJ (2001) Int Patent 2001052943
75. Kennedy TJ, Nesbitt RD, Binette ML, Neill JT, Sullivan MJ (2001) Chem Abstr 135:123587
76. Sone T, Hattori I (1999) Eur Patent 920886
77. Sone T, Hattori I (1999) Chem Abstr 131:20139
78. Sasaki H, Higuchi H, Kataoka N, Nanba M (2004) Jap Patent 2004180729
79. Sasaki H, Higuchi H, Kataoka N, Nanba M (2004) Chem Abstr 141:72830
80. Baranwal K, Ohm R, Fell RR, Rodgers B (1997) Rubber, Natural. In: Kirk-Othmer: Encyclopedia of Chemical Technology, vol 21, 4th ed. Wiley, Weinheim, Germany, pp 562–591
81. Taube R, Sylvester G (2002) In: Cornils B, Herrmann WA (eds) Applied Homogeneous Catalysis with Organometallic Compounds, vol 1. VCH, Weinheim, Germany, p 285
82. Ullmann's Encyclopedia of Industrial Chemistry Electronic Release (2004) 7th ed. VCH, Weinheim, Germany, chapter "Rubber"
83. Hamed GR (2001) Materials and Compounds. In: Gent AN (ed) Engineering with Rubber, 2nd ed. Carl Hanser Verlag, München, Germany, p 15
84. Thorn-Csanyi E, Luginsland HD (1997) Rubber Chem Technol 70:222
85. Eisele U (1990) Introduction to Polymer Physics. Springer, Berlin Heidelberg New York, Germany
86. Customer Brochure by Bayer AG Rubber Business Group, Order No.: KA 34287. Edition 10.98
87. Bruzzone M, Gordini S (1991) Lanthanide catalysts for diene polymerization: Some peculiarities and unresolved questions. Meeting of the ACS Rubber Division, Detroit, MI, USA, Oct 8–11
88. Sylvester G, Stollfuß B (1988) Paper No. 32, Meeting of the ACS Rubber Division, Dallas, TX, USA, Apr 19–22
89. Wilson DJ (1993) Makromol Chem Macromol Symp 66:273
90. Wu X, Wu Z, Liu H, Yin J, Mo K, Qiao S (2002) J Polym Mater 19:329
91. Fukahori Y (2004) Int Polym Sci Technol 31:11
92. Shen Z, Ouyang J, Wang F, Hu Z, Fu Y, Qian B (1980) J Polym Sci Polym Chem Ed 18:3345
93. Monakov YB, Marina NG, Savel'eva IG, Zhiber LE, Kozlov VG, Rafikov SR (1982) Dokl Akad Nauk SSSR 265:1431
94. Monakov YB, Marina NG, Savel'eva IG, Zhiber LE, Kozlov VG, Rafikov SR (1983) Chem Abstr 98:54523
95. Tadaki T (2003) Int Polym Sci Technol 31:5
96. Korea Kumho Petrochemical Co., Ltd., Annual Report 2003
97. Anon (2004) Vestnik Khimicheskoi Promyshlennosty 32:43
98. Evans WJ, Giarikos DG, Allen NT (2003) Macromolecules 36:4256
99. Sanyagin AA, Kormer VA (1985) Dokl Akad Nauk SSSR 283:1209
100. Sanyagin AA, Kormer VA (1986) Chem Abstr 104:51150
101. Yang J, Pang S, Ouyang J (1984) Gaofenzi Tongxun 73
102. Yang J, Pang S, Ouyang J (1984) Chem Abstr 100:210508
103. Qi S, Liu X, Chang Y, Chen W (1986) Yingyong Huaxue 3:73

104. Qi S, Liu X, Chang Y, Chen W (1986) Chem Abstr 105:209457
105. Qi S, Gao X, Xiao S, Chen W (1986) Yingyong Huaxue 3:63
106. Qi S, Gao X, Xiao S, Chen W (1986) Chem Abstr 104:186909
107. Yeh GHC, Martin JL, Hsieh HL (1985) US Patent 4544718
108. Yeh GHC, Martin JL, Hsieh HL (1986) Chem Abstr 104:51252
109. Yang J, Hu J, Pan S, Xie D, Zhong C, Ouyang J (1980) Sci Sin 23:734
110. Yang J, Hu J, Pan S, Xie D, Zhong C, Ouyang J (1980) Chem Abstr 93:115054
111. Shen Z (1987) Inorg Chim Acta 140:7
112. Zavadovskaya EN, Yakovlev VA, Tinyakova EI, Dolgoplosk BA (1992) Vysokomol Soedin Ser A 34:54
113. Zavadovskaya EN, Yakovlev VA, Tinyakova EI, Dolgoplosk BA (1992) Chem Abstr 117:192403
114. Rao GSS, Upadhyay VK, Jain RC (1997) Angew Makromol Chem 251:193
115. Rao GSS, Upadhyay VK, Jain RC (1999) J Appl Polym Sci 71:595
116. Shamaeva ZG, Marina NG, Monakov YB, Rafikov SR (1982) Izv Akad Nauk SSSR Ser Khim 846
117. Shamaeva ZG, Marina NG, Monakov YB, Rafikov SR (1982) Chem Abstr 97:12631
118. Monakov YB, Marina NG, Tolstikov GA (1988) Chem Stosow 32:547
119. Monakov YB, Marina NG, Tolstikov GA (1990) Chem Abstr 112:199162
120. Zavadovskaya EN, Yakovlev VA, Tinyakova EI, Dolgoplosk BA (1992) Vysokomol Soedin Ser A 34:54
121. Zavadovskaya EN, Yakovlev VA, Tinyakova EI, Dolgoplosk BA (1992) Chem Abstr 117:192403
122. Qu Y, Yu W, Li Y, Yu G (1998) Yingyong Huaxue 15:21
123. Qu Y, Yu W, Li Y, Yu G (1998) Chem Abstr 130:25385
124. Iovu H, Hubca C, Dimonie M, Simionescu E, Badea EG (1997) Mater Plast 34:5
125. Iovu H, Hubca C, Dimonie M, Simionescu E, Badea EG (1997) Chem Abstr 127:51038
126. Gallazzi MC, Bianchi F, Depero L, Zocchi M (1988) Polymer 29:1516
127. Asahi Chemical Industry Co. (1984) Jap Patent 59113003
128. Asahi Chemical Industry Co. (1985) Chem Abstr 102:7256
129. Marina NG, Monakov YB, Rafikov SR, Gadeleva KK (1984) Vysokomol Soedin Ser A 26:1123
130. Marina NG, Monakov YB, Rafikov SR, Gadeleva KK (1984) Chem Abstr 101:55542
131. Iovu H, Hubca G, Simionescu E, Badea E, Hurst JS (1997) Eur Polym J 33:811
132. Kwag G, Kim D, Lee S, Bae C (2005) J Appl Polym Sci 97:1279
133. Iovu H, Hubca G, Simionescu E, Badea EG, Dimonie M (1997) Angew Makromol Chem 249:59
134. Hsieh HL, Yeh HC (1985) Rubber Chem Technol 58:117
135. Duvakina NV, Monakov YB (2002) Dokl Chem 384:129
136. Duvakina NV, Monakov YB (2002) Chem Abstr 137:232959
137. Porri L, Shubin N (2001) Int Patent 2001087991
138. Porri L, Shubin N (2001) Chem Abstr 136:7504
139. Hsieh HL, Yeh GHC (1986) Ind Eng Chem Prod Res Dev 25:456
140. Iovu H, Hubea G, Hurst SJ (1996) Acta Polym 47:450
141. Porri L, Ricci G, Giarrusso A, Shubin N, Lu Z (2000) ACS Symp Ser 749:15
142. Balducci A, Porri L, Choubine N (1999) Eur Patent 919573
143. Balducci A, Porri L, Choubine N (1999) Chem Abstr 131:5700
144. Taylor MD, Cartes CP (1962) J Inorg Nucl Chem 24:387
145. Kwag GH, Kim DH, Jang YC (2000) Eur Patent 994131
146. Kwag GH, Kim DH, Jang YC (2000) Chem Abstr 132:280329

147. Nasyrov IS, Bazhenov YP, Abdullin AN, Bokin AI, Iskakov BA, Petrunia AV (2004) Russ Patent 2220909
148. Nasyrov IS, Bazhenov YP, Abdullin AN, Bokin AI, Iskakov BA, Petrunia AV (2004) Chem Abstr 140:377438
149. Bulgakov RG, Kuleshov SP, Zuzlov AN, Mullagaleev IR, Khalilov LM, Dzhemilev UM (2001) J Organomet Chem 636:56
150. Stere C, Obrecht W, Müller HG, Szczesniak K, Sylvester G (2002) Ger Patent 10110241
151. Stere C, Obrecht W, Müller HG, Szczesniak K, Sylvester G (2002) Chem Abstr 137:225917
152. Monakov YB, Marina NG, Duvakina NV, Rafikov SR (1977) Dokl Akad Nauk SSSR 236:617
153. Monakov YB, Marina NG, Duvakina NV, Rafikov SR (1977) Chem Abstr 87:185045
154. Sylvester G, Witte J, Marwede G (1980) Eur Patent 0007027
155. Sylvester G, Witte J, Marwede G (1980) Chem Abstr 92:130367
156. Kwag G (2002) Macromolecules 35:4875
157. Jenkins DK (1985) Polymer 26:152
158. Wilson DJ, Jenkins DK (1992) Polym Bull 27:407
159. Gehrke K, Kohlheim J, Gebauer U (1994) Plaste Kautsch 41:213
160. Gehrke K, Krueger G, Gebauer U, Lechner MD (1996) Kautsch Gummi Kunstst 49:760
161. Ricci G, Boffa G, Porri L (1986) Makromol Chem Rapid Commun 7:355
162. Ricci G, Italia S, Cabassi F, Porri L (1987) Polym Commun 28:223
163. Oehme A, Gebauer U, Gehrke K, Beyer P, Hartmann B, Lechner MD (1994) Macromol Chem Phys 195:3773
164. Wilson DJ, Jenkins DK (1995) Polym Bull 34:257
165. Cabassi F, Italia S, Ricci G, Porri L (1988) Transition Met Catal Polym, Proc Int Symp, 2nd:655
166. Cabassi F, Italia S, Ricci G, Porri L (1989) Chem Abstr 111:115796
167. Porri L, Ricci G, Shubin N (1998) Macromol Symp 128:53
168. Oehme A, Gebauer U, Gehrke K (1993) J Mol Catal 82:83
169. Oehme A, Gebauer U, Gehrke K, Lechner MD (1996) Angew Makromol Chem 235:121
170. Oehme A, Gebauer U, Gehrke K, Lechner MD (1997) Kautsch Gummi Kunstst 50:82
171. Gebauer U, Schmidt S, Gehrke K (1992) Plaste Kautsch 39:297
172. Gehrke K, Boldt D, Gebauer U, Lechner MD (1996) Kautsch Gummi Kunstst 49:510
173. Jenkins DK (1985) Polymer 26:147
174. Wilson DJ (1995) J Polym Sci Part A Polym Chem 33:2505
175. Wilson DJ (1996) Polym Int 39:235
176. Kobayashi E, Hayashi N, Aoshima S, Furukawa J (1998) J Polym Sci Part A: Polym Chem 36:1707
177. Kobayashi E, Hayashi N, Aoshima S, Furukawa J (1998) J Polym Sci Part A: Polym Chem 36:241
178. Friebe L, Nuyken O, Windisch H, Obrecht W (2002) Macromol Chem Phys 203:1055
179. Friebe L, Nuyken O, Windisch H, Obrecht W (2004) J Macromol Sci Pure Appl Chem 41:245
180. Friebe L, Müller JM, Nuyken O, Obrecht W (2006) J Macromol Sci Pure Appl Chem 43:11
181. Obrecht W, Sylvester G, Müller JM, Friebe L, Nuyken O (2005) Ger Patent 102004008921
182. Obrecht W, Sylvester G, Müller JM, Friebe L, Nuyken O (2005) Chem Abstr 143:230399

183. Wilson DJ (1993) Polymer 34:3504
184. Fischbach A, Perdih F, Sirsch P, Scherer W, Anwander R (2002) Organometallics 21:4569
185. Fischbach A (2004) In: Lanthanoid-basierte Ziegler-Mischkatalysatoren. Verlag Dr Hut, München, Germany
186. Fischbach A, Perdih F, Herdtweck E, Anwander R (2006) Organometallics 25:1626
187. Quirk RP, Kells AM (2000) Polym Int 49:751
188. Nickaf JB, Burford RP, Chaplin RP (1995) J Polym Sci Part A Polym Chem 33:1125
189. Dong W, Yang J, Shan C, Pang S, Huang B (1997) Cuihua Xuebao 18:234
190. Dong W, Yang J, Shan C, Pang S, Huang B (1997) Chem Abstr 127:66261
191. Quirk RP, Kells AM, Yunlu K, Cuif JP (2000) Polymer 41:5903
192. Zhao C, Xu L, Zhong C, Tang X (1994) Huagong Xueyuan Xuebao 15:226
193. Zhao C, Xu L, Zhong C, Tang X (1994) Chem Abstr 123:112882
194. Sinn H, Kaminsky W, Vollmer HJ, Woldt R (1980) Angew Chem 92:396
195. Sinn H, Kaminsky W, Vollmer HJ, Woldt R (1980) Angew Chem Int Ed Engl 19:390
196. Sinn H, Kaminsky W (1980) Adv Organomet Chem 18:99
197. Dong W, Yang J, Shan C, Pang S, Huang B (1998) Yingyong Huaxue 15:1
198. Dong W, Yang J, Shan C, Pang S, Huang B (1998) Chem Abstr 129:276406
199. Nekhaeva LA, Frolov VM, Konovalenko NA, Vyshinskaya LI, Tikhomirova IN, Khodzhaeva VL, Shklyaruk BF, Antipov EM (2003) Vysokomol Soedin Ser A Ser B 45:540
200. Nekhaeva LA, Frolov VM, Konovalenko NA, Vyshinskaya LI, Tikhomirova IN, Khodzhaeva VL, Shklyaruk BF, Antipov EM (2003) Chem Abstr 139:197798
201. Jenkins DK (1992) Polymer 33:156
202. Jenkins DK (1984) GB Patent 2140435
203. Jenkins DK (1985) Chem Abstr 102:167341
204. Pross A, Marquardt P, Reichert KH, Nentwig W, Knauf T (1993) Angew Makromol Chem 211:89
205. Friebe L (2004) In: Butadienpolymerisation mit Neodym-ZIEGLER-Katalysatoren. Verlag Dr Hut, München, Germany
206. Throckmorton MC (1987) US Patent 4663405
207. Throckmorton MC (1987) Chem Abstr 107:176668
208. Jiang L, Zhang S (1987) Hecheng Xiangjiao Gongye 10:189
209. Jiang L, Zhang S (1987) Chem Abstr 107:24555
210. Roberts JE (1961) J Am Chem Soc 83:1087
211. Anwander R (1996) Top Curr Chem 179:33
212. Anwander R (1999) Top Organomet Chem 2:1
213. Nicolini LF, De Albuquerque Campos CR, De Lira CH, De Andrade Coutinho PL (2000) Eur Patent 1055659
214. Nicolini LF, De Albuquerque Campos CR, De Lira CH, De Andrade Coutinho PL (2000) Chem Abstr 134:17861
215. Tocchetto Piraes NM, Nicolini LF, De Albuquerque Campos CR, De Andrade Coutinho PL (2002) US Patent 6482906
216. Tocchetto Piraes NM, Nicolini LF, De Albuquerque Campos CR, De Andrade Coutinho PL (2000) Chem Abstr 134:17861
217. Huang CP, Chu CN (2000) Eur Patent 968992
218. Huang CP, Chu CN (2000) Chem Abstr 132:64653
219. Knauf T, Obrecht W (1992) Eur Patent 512346
220. Knauf T, Obrecht W (1993) Chem Abstr 119:96395
221. Alas M, Yunlu K (1998) Int Patent 9839283

222. Alas M, Yunlu K (1998) Chem Abstr 129:231157
223. Misra SN, Misra TN, Mehrotra RC (1963) J Inorg Nucl Chem 25:195
224. Pedretti U, Lugli G, Poggio S, Mazzei A (1979) Ger Patent 2833721
225. Pedretti U, Lugli G, Poggio S, Mazzei A (1979) Chem Abstr 90:169316
226. Carbonaro A, Ferraro D (1983) Eur Patent 92270
227. Carbonaro A, Ferraro D (1984) Chem Abstr 100:23431
228. Andreussi P, Lauretti E, Miani B (1994) Europrene Neocis: Improved cis-1,4-Polybutadiene by Neodymium Catalysts, Stepol, Milano, Italy, Jun 6–10
229. Biagini P, Lugli G, Garbassi F, Andreussi P (1995) Eur Patent 667357
230. Biagini P, Lugli G, Garbassi F, Andreussi P (1995) Chem Abstr 123:341326
231. Dong W, Masuda T (2003) Polymer 44:1561
232. Dong W, Endo K, Masuda T (2003) Macromol Chem Phys 204:104
233. Taniguchi Y, Dong W, Katsumata T, Shiotsuki M, Masuda T (2005) Polym Bull 54:173
234. Fischbach A, Herdtweck E, Anwander R, Eickerling G, Scherer W (2003) Organometallics 22:499
235. Gromada J, Le Pichon L, Mortreux A, Leising F, Carpentier JF (2003) J Organomet Chem 683:44
236. Garbassi F, Biagini P, Andreussi P, Lugli G (1995) Eur Patent 638598
237. Garbassi F, Biagini P, Andreussi P, Lugli G (1995) Chem Abstr 123:201711
238. Garbassi F, Biagini P, Andreussi P, Lugli G (1996) US Patent 5484897
239. Garbassi F, Biagini P, Andreussi P, Lugli G (1995) Chem Abstr 123:201711
240. Zheng YS, Shen ZQ (1999) Chin Chem Lett 10:597
241. Zheng YS, Shen ZQ (1999) Chem Abstr 132:122973
242. Ni XF, Zhang YF, Shen ZQ, Tao J (2001) Zhejiang Daxue Xuebao Gongxueban 35:132
243. Ni XF, Zhang YF, Shen ZQ, Tao J (2001) Chem Abstr 136:20271
244. Ni XF, Li WS, Zhang YF, Shen ZQ, Zhang TX (2000) Gaodeng Xuexiao Huaxue Xuebao 21:1936
245. Ni XF, Li WS, Zhang YF, Shen ZQ, Zhang TX (2001) Chem Abstr 134:311477
246. Dong W, Masuda T (2002) J Polym Sci Part A: Polym Chem 40:1838
247. Shen Y, Shen Z, Zhang Y, Yao K (1996) Macromolecules 29:8289
248. Ling J, Shen Z, Huang Q (2001) Macromolecules 34:7613
249. Berg AA, Monakov YB, Budtov VP, Rafikov SR (1978) Vysokomol Soedin Ser B 20:295
250. Berg AA, Monakov YB, Budtov VP, Rafikov SR (1978) Chem Abstr 89:7349
251. Anon (1985) Jap Patent 60023406
252. Anon (1985) Chem Abstr 103:23637
253. Yu L, Yang M, Shen Z (1985) Cuihua Xuebao 6:260
254. Yu L, Yang M, Shen Z (1987) Chem Abstr 106:19082
255. Sun J, Wu L, Feng L, Shen Z (1988) Zhejiang Daxue Xuebao Ziran Kexueban 22:203
256. Sun J, Wu L, Feng L, Shen Z (1989) Chem Abstr 110:173825
257. Chen X, Zhang Y, Shen Z (1994) Chin J Polym Sci 12:28
258. Zhang Y, Zheng R, Shen Z (1995) Chin Sci Bull 40:1182
259. Jiang L, Zhang Y, Shen Z (1997) Eur Polym J 33:577
260. Jiang L, Shen Z, Zhang Y (2000) Eur Polym J 36:2513
261. Wu LB, Li BG, Cao K, Li BF (2001) Eur Polym J 37:2105
262. Xu X, Ni X, Shen Z (2005) Polym Bull 53:81
263. Liu L, Gong Z, Zheng Y, Jing X, Zhao H (1999) Macromol Chem Phys 200:763
264. Laubry P (2002) Int Patent 2002048218
265. Laubry P (2002) Chem Abstr 137:34344
266. Laubry P (2002) Int Patent 2002038635

267. Laubry P (2002) Chem Abstr 136:370934
268. Laubry P (2002) Int Patent 2002038636
269. Laubry P (2002) Chem Abstr 136:370935
270. Laubry P (2003) Int Patent 2003048221
271. Laubry P (2003) Chem Abstr 139:23140
272. Friebe L, Nuyken O, Obrecht W (2005) J Macromol Sci Pure Appl Chem 42:839
273. Porri L, Natta G, Gallazzi MC (1964) Chim Ind 46:428
274. Porri L, Natta G, Gallazzi MC (1964) Chem Abstr 61:11619
275. Babitskii BD, Dolgoplosk BA, Kormer VA, Lobach MI, Tinyakova EI, Yakovlev VA (1965) Isv Akad Nauk SSSR Ser Chem 1507
276. Babitskii BD, Dolgoplosk BA, Kormer VA, Lobach MI, Tinyakova EI, Yakovlev VA (1965) Chem Abstr 63:99630
277. Wilke G, Bogdanovic B, Hardt P, Heimbach P, Keim W, Kröner M, Oberkirch W, Tanaka K, Steinrucke E, Walter D, Zimmermann H (1966) Angew Chem 78:157
278. Wilke G, Bogdanovic B, Hardt P, Heimbach P, Keim W, Kröner M, Oberkirch W, Tanaka K, Steinrucke E, Walter D, Zimmermann H (1966) Angew Chem Int Ed Engl 5:151
279. Iovu H, Hubca G, Racoti D, Hurst JS (1999) Eur Polym J 35:335
280. Sabirov ZM, Minchenkova NK, Monakov YB (1990) Vysokomol Soedin Ser B 32:803
281. Sabirov ZM, Minchenkova NK, Monakov YB (1991) Chem Abstr 114:123133
282. Monakov YB, Marina NG, Savel'eva IG, Zhiber LE, Duvakina NV, Rafikov SR (1984) Dokl Akad Nauk SSSR 278:1182
283. Monakov YB, Marina NG, Savel'eva IG, Zhiber LE, Duvakina NV, Rafikov SR (1985) Chem Abstr 102:79372
284. Mazzei A (1981) Macromol Chem Phys Suppl 4:61
285. Taube R, Windisch H, Maiwald S (1995) Macromol Symp 89:393
286. Taube R, Maiwald S, Sieler J (1996) J Organomet Chem 513:37
287. Taube R, Schmidt U, Gehrke JP, Boehme P, Langlotz J, Wache S (1993) Makromol Chem Macromol Symp 66:245
288. Taube R, Windisch H, Maiwald S, Hemling H, Schumann H (1996) J Organomet Chem 513:49
289. Maiwald S, Weissenborn H, Windisch H, Sommer C, Müller G, Taube R (1997) Macromol Chem Phys 198:3305
290. Maiwald S, Taube R, Hemling H, Schumann H (1998) J Organomet Chem 552:195
291. Maiwald S, Weissenborn H, Sommer C, Müller G, Taube R (2001) J Organomet Chem 640:1
292. Maiwald S, Sommer C, Müller G, Taube R (2001) Macromol Chem Phys 202:1446
293. Maiwald S, Sommer C, Müller G, Taube R (2002) Macromol Chem Phys 203:1029
294. Porri L, Giarrusso A, Shubin N, Lu Z, Ricci G (1998) Polym Prepr (Am Chem Soc Div Polym Chem) 39:214
295. Wu W, Chen M, Zhou P (1991) Organometallics 10:98
296. Lorenz V, Goerls H, Thiele SKH, Scholz J (2005) Organometallics 24:797
297. Thiele SKH, Wilson DR (2003) J Macromol Sci Polym Rev 43:581
298. Yasuda H (1999) Top Organomet Chem 2:255
299. Yasuda H, Ihara E (1997) Adv Polym Sci 133:53
300. Yu G, Chen W, Wang Y (1984) Kexue Tongbao 29:412
301. Yu G, Chen W, Wang Y (1984) Chem Abstr 101:91491
302. Chen W, Xiao S, Wang Y, Yu G (1984) Kexue Tongbao 29:892
303. Chen W, Xiao S, Wang Y, Yu G (1985) Chem Abstr 102:7981
304. Cui LQ, Ba XW, Teng HX, Ying LQ, Li KC, Jin YT (1998) Polym Bull 40:729
305. Taube R, Maiwald S, Sieler J (2001) J Organomet Chem 621:327

306. Boisson C, Monteil V, Ribour D, Spitz R, Barbotin F (2003) Macromol Chem Phys 204:1747
307. Monteil V, Spitz R, Barbotin F, Boisson C (2004) Macromol Chem Phys 205:737
308. Thuilliez J, Monteil V, Spitz R, Boisson C (2005) Angew Chem 117:2649
309. Thuilliez J, Monteil V, Spitz R, Boisson C (2005) Angew Chem Int Ed 44:2593
310. Boisson C, Monteil V, Thuilliez J, Spitz R, Monnet C, LLauro MF, Barbotin F, Robert P (2005) Macromol Symp 226:17
311. Kaita S, Hou Z, Wakatsuki Y (2000) Int Patent 2000052062
312. Kaita S, Hou Z, Wakatsuki Y (2000) Chem Abstr 133:223182
313. Kaita S, Hou Z, Wakatsuki Y (2003) Int Patent 2003082932
314. Kaita S, Hou Z, Wakatsuki Y (2003) Chem Abstr 139:292644
315. Kaita S, Hou Z, Nishiura M, Doi Y, Kurazumi J, Horiuchi AC (2003) Macromol Rapid Commun 24:179
316. Windisch H (1999) Ger Patent 19746266
317. Windisch H (1999) Chem Abstr 130:312237
318. Boisson C, Barbotin F, Spitz R (1999) Macromol Chem Phys 200:1163
319. Andersen RA, Templeton DH, Zalkin A (1978) Inorg Chem 17:2317
320. Monteil V, Spitz R, Boisson C (2004) Polym Int 53:576
321. Thiele SKH, Monroy VM, Stoye H, Wilson DR (2003) Int Patent 2003033545
322. Thiele SKH, Monroy VM, Stoye H, Wilson DR (2003) Chem Abstr 138:321739
323. Rafikov SR, Monakov YB, Bieshev YK, Valitova IF, Murinov YI, Tolstikov GA, Nikitin YE (1976) Dokl Akad Nauk SSSR 229:1174
324. Rafikov SR, Monakov YB, Bieshev YK, Valitova IF, Murinov YI, Tolstikov GA, Nikitin YE (1976) Chem Abstr 85:160632
325. Anon (1983) Netherland Patent 8202603
326. Anon (1983) Chem Abstr 99:54873
327. Anon (1984) Jap Patent 59136306
328. Anon (1985) Chem Abstr 102:80067
329. Anon (1984) Jap Patent 59166508
330. Anon (1985) Chem Abstr 102:46429
331. Anon (1984) Jap Patent 59131609
332. Anon (1984) Chem Abstr 101:231771
333. Hattori I, Sakakibara M, Tsutsumi F, Yoshizawa M (1988) Eur Patent 267675
334. Hattori I, Sakakibara M, Tsutsumi F, Yoshizawa M (1988) Chem Abstr 109:191952
335. Hattori I, Tsutsumi F, Sakakibara M, Makino K (1990) Stud Surf Sci Catal 56:313
336. Sone T, Takashima A, Nonaka K, Hattori I (1998) Jap Patent 10306115
337. Sone T, Takashima A, Nonaka K, Hattori I (1998) Chem Abstr 130:39664
338. Sone T, Nonaka K, Hattori I, Takashima A (1998) Eur Patent 863165
339. Sone T, Nonaka K, Hattori I, Takashima A (1998) Chem Abstr 129:231842
340. Sone T, Takashima A, Nonaka K, Hattori I (1998) Eur Patent 846707
341. Sone T, Takashima A, Nonaka K, Hattori I (1998) Chem Abstr 129:55331
342. Sone T, Hattori I (1999) Eur Patent 920886
343. Sone T, Hattori I (1999) Chem Abstr 131:20139
344. Zhang Q, Ni X, Zhang Y, Shen Z (2001) Macromol Rapid Commun 22:1493
345. Thiele SKH, Wilson DR (2004) Int Patent 2004074333
346. Thiele SKH, Wilson DR (2004) Chem Abstr 141:244788
347. Sugiyama H, Gambarotta S, Yap GPA, Wilson DR, Thiele SKH (2004) Organometallics 23:5054
348. Bonnet F, Visseaux M, Pereira A, Bouyer F, Barbier-Baudry D (2004) Macromol Rapid Commun 25:873

349. Bonnet F, Visseaux M, Pereira A, Barbier-Baudry D (2005) Macromolecules 38: 3162
350. Bonnet F, Visseaux M, Barbier-Baudry D, Vigier E, Kubicki MM (2004) Chem Eur J 10:2428
351. Hsieh HL, Yeh GHC (1987) US Patent 4699962
352. Hsieh HL, Yeh GHC (1987) Chem Abstr 107:238396
353. Monakov YV, Marina NG, Duvakina NV, Zolotarev VL (1997) Vysokomol Soedin Ser A Ser B 39:787
354. Monakov YV, Marina NG, Duvakina NV, Zolotarev VL (1997) Chem Abstr 127:307683
355. Jenkins DK (1983) Eur Patent 91287
356. Jenkins DK (1983) Chem Abstr 99:213061
357. Jenkins DK (1994) Polymer 35:2897
358. Becke S, Denninger U, Kahlert S, Obrecht W, Schmid C, Windisch H (2002) Int Patent 0220629
359. Becke S, Denninger U, Kahlert S, Obrecht W, Schmid C, Windisch H (2002) Chem Abstr 136:248051
360. Gallazzi MC, Bianchi F, Giarrusso A, Porri L (1983) lnorg Chim Acta 94:108
361. Marina NG, Monakov YB, Rafikov SR, Gadeleva KK (1984) Polym Sci USSR 26:1251
362. Chigir NN, Sharaev OK, Tinyakova EI, Dolgoplosk BA (1983) Vysokomol Soedin Ser B 25:47
363. Chigir NN, Sharaev OK, Tinyakova EI, Dolgoplosk BA (1983) Chem Abstr 98:126695
364. Watanabe H, Masuda T (1997) In: Kobayashi S (ed) Catalysis in Precision Polymerization, part 1, Diene Polymerization. Wiley, Hoboken, NJ, USA, p 55
365. Evans WJ, Giarikos DG (2004) Macromolecules 37:5130
366. Evans WJ, Giarikos DG, Ziller JW (2001) Organometallics 20:5751
367. Kwag G, Kim P, Han S, Choi H (2005) Polymer 46:3782
368. Jang YC, Kwag GH, Kim AJ, Lee SH (2000) Eur Patent 1031583
369. Jang YC, Kwag GH, Kim AJ, Lee SH (2000) Chem Abstr 133:194429
370. Jang Y, Kwag G, Lee H (2000) Polymer J 32:456
371. Lee DH, Lee DH, Ahn TO (1988) Polymer J 29:713
372. Kormer VA, Bubnova SV, Drozdov BT, Shelokhneva LF, Bodrova VS, Kovalev NF (2003) Russ Patent 2206578
373. Kormer VA, Bubnova SV, Drozdov BT, Shelokhneva LF, Bodrova VS, Kovalev NF (2003) Chem Abstr 140:219206
374. Kormer VA, Bubnova SV, Drozdov BT, Shelokhneva LF, Kovalev NF (2003) Russ Patent 2206577
375. Kormer VA, Bubnova SV, Drozdov BT, Shelokhneva LF, Kovalev NF (2003) Chem Abstr 140:219205
376. Zhang Q, Ni X, Shen Z (2004) J Macromol Sci Pure Appl Chem 41:39
377. Zhang Q, Ni X, Shen Z (2002) Polym Int 51:208
378. Knauf T, Braubach W (2001) Eur Patent 1095952
379. Knauf T, Braubach W (2001) Chem Abstr 134:312297
380. Jang YC, Kwag KH, Kim PS (2000) Korean Patent 2000032230
381. Jang YC, Kwag KH, Kim PS (2002) Chem Abstr 136:341170
382. Sone T, Hattori I, Yamazaki D, Nonaka K (2001) Eur Patent 1099710
383. Sone T, Hattori I, Yamazaki D, Nonaka K (2001) Chem Abstr 134:354405
384. Ishino Y, Nakayama S, Mori Y, Hattori I (2002) US Patent 6391990
385. Ishino Y, Nakayama S, Mori Y, Hattori I (2000) Chem Abstr 133:151837
386. Fukahori T, Inamura K (1995) Int Patent 9504090

387. Fukahori T, Inamura K (1995) Chem Abstr 123:58498
388. Ni X, Li J, Zhang Y, Shen Z (2004) J Appl Polym Sci 92:1945
389. Gordini S, Carbonaro A, Spina S (1987) Eur Patent 207558
390. Gordini S, Carbonaro A, Spina S (1987) Chem 106:68561
391. Jenkins D, Ansell PJ (1991) US Patent 5017539
392. Jenkins D, Ansell PJ (1990) Chem Abstr 113:154135
393. Ansell P J, Williams HD (1997) US Patent 5686371
394. Ansell P J, Williams HD (1993) Chem Abstr 119:119251
395. Steinhauser N, Obrecht W (2002) Eur Patent 1176157
396. Steinhauser N, Obrecht W (2002) Chem Abstr 136:136085
397. Balducci A, Righi S (2004) Eur Patent 1431318
398. Balducci A, Righi S (2004) Chem Abstr 141:54809
399. Bergbreiter DE, Chen LB, Chandran R (1985) Macromolecules 18:1055
400. Barbotin F, Spitz R, Boisson C (2001) Macromol Rapid Commun 22:1411
401. Barbotin F, Boisson C, Spitz R (2001) Eur Patent 1092733
402. Barbotin F, Boisson C, Spitz R (2001) Chem 134:297042
403. Robert P, Spitz R (1993) Eur Patent 0599096
404. Robert P, Spitz R (1995) Chem Abstr 122:82304
405. Yu G, Li Y, Qu Y, Li X (1993) Macromolecules 26:6702
406. Fischbach A, Klimpel MG, Widenmeyer M, Herdtweck E, Scherer W, Anwander R (2004) Angew Chem 116:2284
407. Fischbach A, Klimpel MG, Widenmeyer M, Herdtweck E, Scherer W, Anwander R (2004) Angew Chem Int Ed 43:2234
408. Ruehmer T, Giesemann J, Schwieger W, Schmutzler K (1999) Kautsch Gummi Kunstst 52:420
409. Takeuchi Y, Sakakibara M, Shibata T (1984) US Patent 4461883
410. Takeuchi Y, Sakakibara M, Shibata T (1983) Chem Abstr 99:54873
411. Porri L, Di Corato A, Natta G (1970) Eur Polym J 6:751
412. Pross A, Marquardt P, Reichert KH, Nentwig W, Knauf T (1992) Angew Makromol Chem 211:147
413. Huang CP, Chu DN (2000) Eur Patent 968992
414. Huang CP, Chu DN (2000) Chem Abstr 132:64653
415. Carbonaro A, Ferraro D, Bruzzone M (1988) US Patent 4736001
416. Carbonaro A, Ferraro D, Bruzzone M (1987) Chem Abstr 106:121167
417. Kwag G, Kim A, Lee S, Jang Y, Kim P, Baik H, Yoon DI, Jeong H, Lee JG, Lee H (2002) Rubber Chem Technol 75:907
418. Kwag GH, Lee SH, Jang YC, Kim AJ (2001) Eur Patent 1134233
419. Kwag GH, Lee SH, Jang YC, Kim AJ (2001) Chem Abstr 135:242675
420. Yunlu K, He M, Cuif JP, Alas M (1999) Int Patent 9954335
421. Yunlu K, He M, Cuif JP, Alas M (1999) Chem Abstr 131:310944
422. Friebe L, Müller JM, Nuyken O, Obrecht W (2006) J Macromol Sci Pure Appl Chem 43:841
423. Mello IL, Coutinho FMB, Nunes DSS, Soares BG, Costa MAS, de Santa Maria LC (2004) Eur Polym J 40:635
424. Biagini P, Lugli G, Abis L, Millini R (1995) New J Chem 19:713
425. Chigir NN, Guzman IS, Sharaev OK, Tinyakova EI, Dolgoplosk BA (1982) Dokl Akad Nauk SSSR 263:375
426. Chigir NN, Guzman IS, Sharaev OK, Tinyakova EI, Dolgoplosk BA (1982) Chem Abstr 97:56281
427. Wieder W, Kuhlmann D, Nentwig W (1996) Ger Patent 4436059

428. Wieder W, Kuhlmann D, Nentwig W (1996) Chem Abstr 124:319448
429. Keim W (1990) Angew Chem 102:251
430. Keim W (1990) Angew Chem Int Ed 29:223
431. Andreussi P, Bianchi R, Bruzzone M (1990) Eur Patent 386808
432. Andreussi P, Bianchi R, Bruzzone M (1990) Chem Abstr 113:233166
433. Sone T, Hattori I, Nakayama A, Tanaka R, Tani K, Akema H (2001) Eur Patent 1099711
434. Sone T, Hattori I, Nakayama A, Tanaka R, Tani K, Akema H (2001) Chem Abstr 134:354406
435. Sone T, Nonaka K, Takashima A, Hattori I (2001) US Patent 6255416
436. Sone T, Nonaka K, Takashima A, Hattori I (1999) Chem Abstr 131:352466
437. Hattori I, Sakakibara M, Tsutsumi F, Yoshizawa M (1991) Eur Patent 0406920
438. Hattori I, Sakakibara M, Tsutsumi F, Yoshizawa M (1988) Chem Abstr 109:191952
439. Sone T, Yamazaki D, Nonaka K (2001) Eur Patent 1099710
440. Sone T, Yamazaki D, Nonaka K (2001) Chem Abstr 134:354405
441. Jang YC, Kim PS, Han S, Kwag GH, Lee SH (2004) US Patent 2004102589
442. Jang YC, Kim PS, Han S, Kwag GH, Lee SH (2004) Chem Abstr 141:8475
443. Sakakibara M, Ikeyama M, Aoki T (1985) Jap Patent 60223804
444. Sakakibara M, Ikeyama M, Aoki T (1986) Chem Abstr 104:177750
445. Wang C (2005) Mater Chem Phys 89:116
446. Dong WM, Pang SF, Hu JY, Yang JH (2000) Yingyong Huaxue 17:272
447. Dong WM, Pang SF, Hu JY, Yang JH (2000) Chem Abstr 133:194384
448. Jenkins DK, Ansell PJ (1991) US Patent 5017539
449. Jenkins DK, Ansell PJ (1990) Chem Abstr 113:154135
450. Knauf TF, Osman A (1994) Eur Patent 652240
451. Knauf TF, Osman A (1995) Chem Abstr 123:201712
452. Wieder W, Kuhlmann D, Nentwig W (1996) US Patent 5567784
453. Wieder W, Kuhlmann D, Nentwig W (1996) Chem Abstr 124:319448
454. Sylvester G, Wieder W (1982) ACS Symp Ser 193:57
455. Pires NMT, Coutinho FMB, Costa MAS (2004) Eur Polym J 40:2599
456. Odian G (1991) In: Principles of Polymerization, 3rd ed. Wiley, New York, p 650
457. Muruganandam N, Cann KJ, Apecetche MA, Moorhouse JH (2000) Eur Patent 1010710
458. Muruganandam N, Cann KJ, Apecetche MA, Moorhouse JH (2000) Chem Abstr 133:44735
459. Skuratov KD, Lobach MI, Shibaeva AN, Churlyaeva LA, Erokhina TV, Osetrova LV, Kormer VA (1992) Polymer 33:5197
460. Skuratov KD, Lobach MI, Shibaeva AN, Churlyaeva LA, Erokhina TV, Osetrova LV, Kormer VA (1992) Polymer 33:5202
461. Hoffmann EG (1960) Liebigs Ann Chem 629:104
462. Miller HJ, Hamada T, Ozawa Y, Pakdel P (2000) Int Patent 2000069928
463. Miller HJ, Hamada T, Ozawa Y, Pakdel P (2000) Chem Abstr 134:5826
464. Ren S, Gao X, Jiang L (1982) Gaofenzi Tongxum 435
465. Ren S, Gao X, Jiang L (1983) Chem Abstr 99:88604
466. Lynch TJ (1999) Eur Patent 964008
467. Lynch TJ (1999) Chem Abstr 132:36869
468. Zinck P, Barbier-Baudry D, Loupy A (2005) Macromol Rapid Commun 26:46
469. Schoenberg E, Marsh SJ, Walters SJ, Saltman WM (1979) Rubber Chem Technol 52:526
470. Saltman WM (1977) In: The Stereo Rubbers. Wiley, New York

471. Rachita MJ, Xu ZG, Stockdale MK (2001) Fall Technical Program, Rubber Expo 2001, 160th, Cleveland, Ohio, USA, Oct 16–20, p 2364
472. Carbonaro A, Ripani L (1983) Eur Patent 92271
473. Carbonaro A, Ripani L (1984) Chem Abstr 100:7405
474. Carbonaro A, Ferraro D (1986) Eur Patent 204373
475. Carbonaro A, Ferraro D (1987) Chem Abstr 106:139631
476. Sylvester G, Witte J, Marwede G (1980) Eur Patent 11184
477. Sylvester G, Witte J, Marwede G (1980) Chem Abstr 93:96555
478. Ullmann's Encyclopedia of Industrial Chemistry (2003) "Isoprene", 6th ed, vol 18. Wiley, Weinheim, Germany, pp 629–648
479. Zhang X, Li J, Ni X, Zhang Y (2003) Hecheng Xiangjiao Gongye 26:313
480. Zhang X, Li J, Ni X, Zhang Y (2003) Chem Abstr 140:95363
481. Gehrke K, Krüger G, Gebauer U, Lechner MD (1997) Kautsch Gummi Kunstst 50:266
482. Zhang X, Ni X, Li J, Zhang Y (2004) Hecheng Xiangjiao Gongye 27:107
483. Zhang X, Ni X, Li J, Zhang Y (2004) Chem Abstr 141:449901
484. Stere C, Obrecht W, Sondermann U, Sylvester G (2005) Eur Patent 1593694
485. Stere C, Obrecht W, Sondermann U, Sylvester G (2005) Chem Abstr 143:441779
486. Porri L, Giarrusso A, Ricci G (1991) Prog Polym Sci 16:405
487. Xie DM, Zhong CG, Yuan NA, Sun YF, Xiao SX, Ouyang J (1979) Gaofenzi Tongxun 233
488. Xie DM, Zhong CG, Yuan NA, Sun YF, Xiao SX, Ouyang J (1980) Chem Abstr 92:77008
489. Yeh HC, Hsieh HL (1984) Polym Prepr (Am Chem Soc Div Polym Chem) 25:52
490. Panasenko AA, Odinokov VN, Monakov YB, Kahlilov LM, Bezgina AS, Ignatynk VK, Rafikov SR (1977) Vysokomol Soedin B 19:656
491. Panasenko AA, Odinokov VN, Monakov YB, Kahlilov LM, Bezgina AS, Ignatynk VK, Rafikov SR (1977) Chem Abstr 87:185152
492. Bolognesi A, Destri S, Porri L, Wang F (1982) Makromolekul Chem Rapid Commun 3:187
493. Jin Y, Sun Y, Liu X, Li X, Ouyang J (1984) Fenzi Kexue Yu Huaxue Yanjiu 4:247
494. Jin Y, Sun Y, Liu X, Li X, Ouyang J (1984) Chem Abstr 101:131149
495. Wang F, Bolognesi A (1982) Gaofenzi Tongxun 238
496. Wang F, Bolognesi A (1982) Chem Abstr 97:182930
497. Ricci G, Zetta L, Porri L, Meille SV (1995) Macromol Chem Phys 196:2785
498. Windisch H, Trimbach J, Schertl P, Giebeler E, Engehausen R (2001) Eur Patent 1078939
499. Windisch H, Trimbach J, Schertl P, Giebeler E, Engehausen R (2001) Chem Abstr 134:194444
500. Windisch H, Sylvester G, Taube R, Maiwald S, Giesemann J, Rosenstock T (2001) Int Patent 0185814
501. Windisch H, Sylvester G, Taube R, Maiwald S, Giesemann J, Rosenstock T (2001) Chem Abstr 135:372913
502. Zhang Q, Li W, Shen Z (2002) Eur Polym J 38:869
503. Kaita S, Kobayashi E, Sakakibara S, Aoshima S, Furukawa J (1996) J Polym Sci Part A Polym Chem 34:3431
504. Oehme A, Gebauer U, Gehrke K (1995) Macromol Rapid Commun 16:563
505. Barbotin F, Monteil V, Llauro MF, Boisson C, Spitz R (2000) Macromolecules 33:8521
506. Barbotin F (2001) Eur Patent 1092731
507. Barbotin F (2001) Chem Abstr 134:296245

508. Kaulbach R, Gebauer U, Gehrke K, Lechner MD, Hummel K (1995) Angew Makromol Chem 226:101
509. Carbonaro A, Gordini S, Cucinella S, Roggero S (1984) Eur Patent 127236
510. Carbonaro A, Gordini S, Cucinella S, Roggero S (1985) Chem Abstr 102:114922
511. Carbonaro A, Gordini S, Cucinella S (1986) Eur Patent 201962
512. Carbonaro A, Gordini S, Cucinella S (1987) Chem Abstr 106:139630
513. Carbonaro A, Gordini S, Cucinella S (1986) Eur Patent 201979
514. Carbonaro A, Gordini S, Cucinella S (1987) Chem Abstr 106:121167
515. Gordini S (1987) Eur Patent 207540
516. Gordini S (1987) Chem Abstr 106:85299
517. Knauf T, Sylvester G, Schmid C, Osman A (2000) Eur Patent 1048675
518. Knauf T, Sylvester G, Schmid C, Osman A (2000) Chem Abstr 133:336398
519. Stakem FG, Paeglis AU, Collins JD (1992) Paper: Union Carbide Gas Phase EPM and EPDM Rubber, 142nd Technical Meeting of the ACS Rubber Division, Nashville, TE, USA, Nov 3-6
520. Baker EC, Scarola LS, Lee KH, Rhee SJ, Edwards DN, Karol FJ, Moorhouse JH (1991) Eur Patent 0422452
521. Baker EC, Scarola LS, Lee KH, Rhee SJ, Edwards DN, Karol FJ, Moorhouse JH (1991) Chem Abstr 114:166107
522. Rhee ASJ, Gregory JG (1993) Eur Patent 0530709
523. Rhee ASJ, Gregory JG (1993) Chem Abstr 119:74393
524. Boysen RL, Mure CR, Scarola LS, Rhee ASJ (1993) Eur Patent 0570966
525. Boysen RL, Mure CR, Scarola LS, Rhee ASJ (1994) Chem Abstr 120:324909
526. Hussein FD, Lee KH (1994) Eur Patent 584574
527. Hussein FD, Lee KH (1993) Chem Abstr 119:119252
528. Eisinger RS, Hussein FD, Edward DN, Lee KH (1994) Eur Patent 0605001
529. Eisinger RS, Hussein FD, Edward DN, Lee KH (1994) Chem Abstr 120:193833
530. Baker EC, Cevallos-Candau JF, Hussein FD, Lee KH, Noshay A (1994) Eur Patent 0614917
531. Baker EC, Cevallos-Candau JF, Hussein FD, Lee KH, Noshay A (1995) Chem Abstr 122:135795
532. Baker EC, Cevallos-Candau JF, Lucas EA, Victor JG, Noshay A (1996) Eur Patent 704464
533. Baker EC, Cevallos-Candau JF, Lucas EA, Victor JG, Noshay A (1996) Chem Abstr 125:13083
534. Eisinger RS, Hunniset CS, Hussein FD, Lee KH, Cann KJ (1996) Eur Patent 735058
535. Eisinger RS, Hunniset CS, Hussein FD, Lee KH, Cann KJ (1996) Chem Abstr 125:302976
536. Zöllner K, Reichert KH (2000) Chem Ing Tech 72:396
537. Zöllner K, Reichert KH (2000) Chem Abstr 133:17895
538. Sylvester G, Vernaleken H (1995) Eur Patent 647657
539. Sylvester G, Vernaleken H (1995) Chem Abstr 123:229280
540. Bernier RJN, Boysen RL, Brown RC, Scarola LS, Williams GH (1996) Eur Patent 697421
541. Bernier RJN, Boysen RL, Brown RC, Scarola LS, Williams GH (1995) Chem Abstr 123:341350
542. Nissim CJ, Joel JM, Castner HA, Floyd K (1999) US Patent 5859156
543. Nissim CJ, Joel JM, Castner HA, Floyd K (1999) Chem Abstr 130:96764
544. Tsujimoto N, Tsukahara M (1998) Jap Patent 1060020
545. Tsujimoto N, Tsukahara M (1998) Chem Abstr 128:180791

546. Reichert KH, Marquardt P, Eberstein C, Garmatter B, Sylvester G (1996) Int Patent 9631543
547. Reichert KH, Marquardt P, Eberstein C, Garmatter B, Sylvester G (1996) Chem Abstr 125:331272
548. Zhang YF, Ni XF, Li JF, Li WS, Shen ZQ (2003) Gaodeng Xuexiao Huaxue Xuebao 24:1499
549. Zhang YF, Ni XF, Li JF, Li WS, Shen ZQ (2003) Chem Abstr 140:59991
550. Jentsch J, Schulze-Tilling A, Steinhauser N, Sylvester G, Mersmann F, Schneider JM (1999) Ger Patent 19801858
551. Jentsch J, Schulze-Tilling A, Steinhauser N, Sylvester G, Mersmann F, Schneider JM (1999) Chem Abstr 131:102665
552. Buysch HJ, Mendoza-Frohn C, Notheis U, Sylvester G (1996) Eur Patent 727447
553. Buysch HJ, Mendoza-Frohn C, Notheis U, Sylvester G (1996) Chem Abstr 125:224281
554. Steinhauser N (1999) Ger Patent 19754789
555. Steinhauser N (1999) Chem Abstr 131:74108
556. Windisch H, Steinhauser N (1999) Eur Patent 903355
557. Windisch H, Steinhauser N (1999) Chem Abstr 130:252795
558. Dauben M, Schneider J, Sylvester G, Steinhauser N (1999) Ger Patent 19801857
559. Dauben M, Schneider J, Sylvester G, Steinhauser N (1999) Chem Abstr 131:102664
560. Cann KJ, Williams GH, Apecetche MA, Moorhouse JH, Murugananadam N, Smith GG (1996) Int Patent 9604323
561. Cann KJ, Williams GH, Apecetche MA, Moorhouse JH, Murugananadam N, Smith GG (1995) Chem Abstr 123:341350
562. Bernier RJN, Boysen RL, Brown RC, Scarola LS, Williams GH (1995) US Patent 5453471
563. Bernier RJN, Boysen RL, Brown RC, Scarola LS, Williams GH (1995) Chem Abstr 123:341350
564. Sylvester G, Vernaleken H (1999) US Patent 5914377
565. Sylvester G, Vernaleken H (1999) Chem Abstr 131:32866
566. Sylvester G (1996) Eur Patent 736549
567. Sylvester G (1996) Chem Abstr 125:302977
568. Eberstein C, Garmatter B, Reichert KH, Sylvester G (1996) Chem Ing Tech 68:820
569. Eberstein C, Garmatter B, Reichert KH, Sylvester G (1996) Chem Abstr 125:224220
570. Sun J, He S, Zhou Q (2001) Chin J Chem Eng 9:217
571. Sun J, He S, Zhou Q (2001) Chem Abstr 135:196001
572. Zöllner K, Reichert KH (2001) Chem Eng Sci 56:4099
573. Zöllner K, Reichert KH (2001) Chem Abstr 135:289084
574. Sun J, Zhao J, He S, Zhou Q (2001) Chin J Chem Eng 9:367
575. Sun J, Zhao J, He S, Zhou Q (2001) Chem Abstr 136:218085
576. Fang D, Sun J, Zhou Q, Feng B (2002) Chin J Chem Eng 10:328
577. Fang D, Sun J, Zhou Q, Feng B (2002) Chem Abstr 137:263344
578. Taube R, Maiwald S, Ruehmer T, Windisch H, Giesemann J, Sylvester G (1996) Int Patent 9631544
579. Taube R, Maiwald S, Ruehmer T, Windisch H, Giesemann J, Sylvester G (1996) Chem Abstr 125:331273
580. Berndt H, Landmesser H (2003) J Mol Catal A Chem 197:245
581. Windisch H, Obrecht W, Michels G, Steinhauser N, Schnieder T (2000) Ger Patent 19832446
582. Windisch H, Obrecht W, Michels G, Steinhauser N, Schnieder T (2000) Chem Abstr 132:123051

583. Windisch H, Schnieder T, Michels G, Sylvester G, Vanhoorne P, Brandt HD (2001) Ger Patent 10032876
584. Windisch H, Schnieder T, Michels G, Sylvester G, Vanhoorne P, Brandt HD (2001) Chem Abstr 135:258403
585. Juengling S, Schade C, Gausepohl H, Warzelhan V (2000) Ger Patent 19926283
586. Juengling S, Schade C, Gausepohl H, Warzelhan V (2000) Chem Abstr 134:42888
587. Michels G, Windisch H, Krueger P, Vanhoorne P, Brandt HD (2000) Int Patent 2000069940
588. Michels G, Windisch H, Krueger P, Vanhoorne P, Brandt HD (2000) Chem Abstr 134:5508
589. Michels G, Windisch H, Steinhauser N (2000) Int Patent 2000069939
590. Michels G, Windisch H, Steinhauser N (2000) Chem Abstr 134:5507
591. Gandon PS (2004) Fr Patent 2850101
592. Gandon PS (2004) Chem Abstr 141:124923
593. Hu J, Zhou C, Ouyang J (1982) Zhonguo Kexueyan Changchum Yingyong Huasue Yanjiuso Jikan 19:63
594. Hu J, Zhou C, Ouyang J (1984) Chem Abstr 101:153246
595. Pan E, Zhong C, Xie D, Ouyang J (1982) Huaxue Xuebao 40:301
596. Pan E, Zhong C, Xie D, Ouyang J (1982) Chem Abstr 97:128111
597. Pan E, Zhong C, Xie D, Ouyang J (1982) Huaxue Xuebao 40:395
598. Pan E, Zhong C, Xie D, Ouyang J (1982) Chem Abstr 97:145320
599. Sun T, Yang J, Pang S, Ouyang J (1985) Yingyong Huaxue 2:47
600. Sun T, Yang J, Pang S, Ouyang J (1985) Chem Abstr 103:142442
601. Yu GQ, Hu ZY (1980) Gaofenzi Tongxun 321
602. Yu GQ, Hu ZY (1981) Chem Abstr 95:81614
603. Rafikov SR, Monakov YB, Bieshev YK, Valitova IF, Murinov YI, Tolstikov GA, Nikitin YE (1976) Dokl Akad Nauk SSSR 229:1174
604. Rafikov SR, Monakov YB, Bieshev YK, Valitova IF, Murinov YI, Tolstikov GA, Nikitin YE (1976) Chem Abstr 85:160632
605. Iovu H, Hubca G (2001) Roum Chem Q Rev 8:181
606. Iovu H, Hubca G (2001) Chem Abstr 135:46467
607. Sigaeva NN, Usmanov TS, Budtov VP, Monakov YB (2001) Russ J Appl Chem 74:1141
608. Sigaeva NN, Usmanov TS, Budtov VP, Monakov YB (2001) Chem Abstr 136:295134
609. Urazbaev VN, Efimov VP, Sabirov ZM, Monakov YB (2003) J Appl Polym Sci 89:601
610. Sigaeva NN, Usmanov TS, Budtov VP, Spivak SI, Zaikov GE, Monakov YB (2003) J Appl Polym Sci 87:358
611. Monakov YB, Sabirov ZM, Urazbaev VN, Efimov VP (2002) Vysokomol Soedin Ser A Ser B 44:389
612. Monakov YB, Sabirov ZM, Urazbaev VN, Efimov VP (2002) Chem Abstr 137:63510
613. Sigaeva NN, Usmanov TS, Budtov VP, Spivak SI, Zaikov GE, Monakov YB (2001) Russ Polym News 6:11
614. Sigaeva NN, Usmanov TS, Budtov VP, Spivak SI, Zaikov GE, Monakov YB (2001) Chem Abstr 136:217087
615. Usmanov TS, Maksyutova ER, Spivak SI (2002) Dokl Phys Chem 387:331
616. Usmanov TS, Maksyutova ER, Spivak SI (2002) Chem Abstr 138:272827
617. Kwag G, Lee H, Kim S (2001) Macromolecules 34:5367
618. Kwag G, Kim A, Lee S, Jang Y, Kim P, Baik H, Yoon DI, Jeong H, Lee JG, Lee H (2002) Rubber Chem Technol 75:907
619. Yoon NM, Gyoung YS (1985) J Org Chem 50:2443
620. Shan C, Lin Y, Jun O, Fan Y, Yang G (1987) Makromol Chem 188:629

621. Gromada J, Mortreux A, Nowogrocki G, Leising F, Mathivet T, Carpentier JF (2004) Eur J Inorg Chem 3247
622. Jun O, Fosong W, Baotong H (1983) MMI Press Symp Ser 4:265
623. Farina M (1987) Top Stereochem 17:1
624. Cossee P (1964) J Catal 3:80
625. Arlman EJ (1964) J Catal 3:89
626. Arlman EJ, Cossee P (1964) J Catal 3:99
627. Patat P, Sinn H (1958) Angew Chem 70:496
628. Natta G, Mazzanti G (1960) Tetrahedron 8:86
629. Klepikova VI, Erusalimskii GB, Lobach MI, Churlayeva LA, Kormer VA (1976) Macromolecules 9:214
630. Dolgoplosk BA, Oreshkin IA, Tinyakova EI, Yakovlev VA (1967) Izv Akad Nauk SSSR Ser Khim 2130
631. Dolgoplosk BA, Oreshkin IA, Tinyakova EI, Yakovlev VA (1967) Chem Abstr 67:109053
632. Oreskhin IA, Chernenko GM, Tinyakova EI, Dolgoplosk BA (1966) Dokl Akad Nauk SSSR 169:1102
633. Oreskhin IA, Chernenko GM, Tinyakova EI, Dolgoplosk BA (1966) Chem Abstr 65:107397
634. Martin HA, Jellinek F (1964) Angew Chem 76:274
635. Martin HA, Jellinek F (1964) Angew Chem Int Ed Engl 3:311
636. Ti V, Pattiasina JW, Teuben JH (1977) J Organomet Chem 262:157
637. Wilke G (1977) Ger Patent 1793788
638. Wilke G (1977) Chem Abstr 87:135956
639. Wilke G, Bogdanovic B (1961) Angew Chem 73:756
640. Porri L, Natta G, Gallazzi MC (1967) J Polym Sci Polym Symp 16:2525
641. Sabirov ZM, Monakov YB, Tolstikov GA (1989) J Mol Catal 56:194
642. Li F, Jin Y, Pei F, Wang F (1991) Yingyong Huaxue 8:81
643. Li F, Jin Y, Pei F, Wang F (1992) Chem Abstr 116:106437
644. Li F, Jin Y, Pei F, Wang F (1994) Gaofenzi Xuebao 257
645. Li F, Jin Y, Pei F, Wang F (1995) Chem Abstr 122:240511
646. Taube R, Gehrke JP, Schmidt U (1986) Makromol Chem Macromol Symp 3:389
647. Druz N, Zak AV, Lobach M, Shpakov PP, Kormer VA (1977) Eur Polym J 13:875
648. Pellecchia C, Proto A, Zambelli A (1992) Macromolecules 25:4450
649. Sabirov ZM, Minchenkova NK, Monakov YB (1989) Inorg Chim Acta 160:99
650. Dolgoplosk BA, Makovetskii KL, Redkina LI, Soboleva TV, Tinyakova EI, Yakovlev VA (1972) Dokl Akad Nauk SSSR 205:387
651. Dolgoplosk BA, Makovetskii KL, Redkina LI, Soboleva TV, Tinyakova EI, Yakovlev VA (1972) Chem Abstr 77:140604
652. Penczek S (2002) J Polym Sci Part A Polym Chem 40:1665
653. Kwag G, Lee JG, Bae C, Lee SN (2003) Polymer 44:6555
654. Penczek S, Kubisa P, Szymanski R (1991) Macromol Rapid Commun 12:77
655. Coates GW, Hustad PD, Reinartz S (2002) Angew Chem 114:2340
656. Coates GW, Hustad PD, Reinartz S (2002) Angew Chem Int Ed 41:2236
657. Jun O, Fu-Song W, Baotong H (1983) In: Quirk RP (ed) Transition Metal Catalyzed Polymerizations, Part A. Harwood Academic, New York, USA, p 293
658. Barbier-Baudry D, Bonnet F, Dormond A, Finot E, Visseaux M (2002) Macromol Chem Phys 203:1194
659. Windisch H, Obrecht W, Stere C, Scholl T, Nuyken O, Friebe L, Vierle M (2002) Eur Patent 1245600

660. Windisch H, Obrecht W, Stere C, Scholl T, Nuyken O, Friebe L, Vierle M (2002) Chem Abstr 137:263766
661. Friebe L, Nuyken O, Windisch H, Obrecht W (2003) Macromol Mat Eng 288:484
662. Darling TR, Davis TP, Fryd M, Gridnev AA, Haddleton DM, Ittel SD, Matheson RR, Moad G, Rizzardo E (2000) J Polym Sci Part A Polym Chem 38:1706
663. Szwarc M (2000) J Polym Sci Part A Polym Chem 38:1710

Rare-Earth Metals and Aluminum Getting Close in Ziegler-Type Organometallics

Andreas Fischbach[1] · Reiner Anwander[2] (✉)

[1]Department of Chemistry, University of California at Berkeley, Berkeley, CA 94720, USA
[2]Department of Chemistry, Universitetet i Bergen, Allégaten 41, N-5007 Bergen, Norway
reiner.anwander@kj.uib.no

1	Introduction	158
2	**Lanthanide Halides**	**162**
2.1	"Homoleptic" Lanthanide Halides	162
2.1.1	Active Centers in $LnCl_3 \cdot 3TBP/R_3Al$	163
2.1.2	Divalent Lanthanide Halides	165
2.2	Heteroleptic Lanthanide Halides	165
2.2.1	Halogenoaluminates	165
2.2.2	Lanthanidocene Chlorides	169
3	**Lanthanide Carboxylates**	**172**
3.1	Lanthanide Acetates	173
3.2	Lanthanide Butyroates	175
3.3	Lanthanide Neodecanoates	176
3.4	Lanthanide Benzoates	178
3.4.1	Lanthanidocene Benzoates	178
3.4.2	Homoleptic Lanthanide Benzoates	182
4	**Lanthanide Alkoxides and Aryloxides**	**189**
4.1	Lanthanide Methanolates	190
4.2	Lanthanide Propanolates	190
4.3	Lanthanide Butanolates and Pentanolates	192
4.4	Lanthanide Phenolates	195
4.4.1	Lanthanide(III) Phenolates	196
4.4.2	Lanthanide(II) Phenolates	202
4.5	Other Systems	203
5	**Lanthanide Siloxides**	**205**
6	**Lanthanide Amides**	**207**
6.1	Lanthanide Alkylamides	209
6.2	Lanthanide Silylamides	210
6.2.1	Adduct Formation and [Amide]→[Alkyl] Exchange	210
6.2.2	[Amide]→[Chloride] Exchange	211
6.2.3	[Amide]→[Hydride] Exchange	212
6.2.4	Cation Formation	213
6.3	Lanthanide Anilides	213
6.4	Other Systems	216

7	**Lanthanide Alkyls**	218
7.1	Lanthanide Allyl Complexes	218
7.2	Lanthanide Tetraalkylaluminates	221
7.2.1	Lappert's Concept of Donor-Induced Aluminate Cleavage: Reversibility and Adduct-Formation	223
7.2.2	Alkane Elimination: The Tetramethylaluminate Route	227
7.2.3	[Alkyl]→[Chloride] Exchange	228
7.2.4	Cation Formation	230
7.3	Other Systems	233
8	**Lanthanide Hydrides**	234
8.1	Lanthanidocene Aluminohydrides	234
8.2	Lanthanide Borohydrides	234
9	**Heterogenized Lanthanide Complexes**	237
9.1	Organic (Co)Polymer Supported Complexes	237
9.2	Silica-Supported Complexes	238
10	**Structural Features of Ln/Al Heterobimetallic Moieties**	241
10.1	Solution Structure: Site Mobility in Tetraalkylaluminate Complexes	241
10.1.1	Monomer–Dimer Equilibrium	241
10.1.2	Alkyl Exchange in Terminal Tetramethylaluminates	242
10.2	Alkyl Exchange in Trimethylaluminum Adduct Complexes	245
10.3	Solid-State Structures	246
10.3.1	Tetraalkylaluminates	247
10.3.2	Trialkylaluminum Complexes	253
10.3.3	Dialkylaluminum Complexes	259
10.3.4	Monoalkylaluminum and "Non-Alkylaluminum" Complexes	261
10.3.5	"Mixed-Alkylated" Complexes	266
11	**Conclusion and Perspective**	270
	References	272

Abstract A prolific and synergetic interplay of rare-earth metal components and organoaluminum reagents is of fundamental importance in Ziegler-type catalysts used in diene polymerization. The present article surveys organoaluminum-promoted alkylation/cationization activation pathways which have been elaborated for various lanthanide compounds, i.e., halides, carboxylates, alkoxides, aryloxides, (silyl)amides, and hydrides. Special emphasis is put on the identification of stable discrete Ln/Al heterobimetallic complexes and the interpretation of their solution and solid-state structure (LnX$_n$Al coordination modes). Surpassing the pure model character of many structurally evidenced Ln/Al heterobimetallics their applicability and performance in the manufacture of polydienes will be highlighted and emerging structure-reactivity relationships will be addressed. Consideration is also given to various heterogenized variants involving organopolymeric and inorganic support materials.

Keywords Aluminum · Diene polymerization · Lanthanides · Organometallics · Rare-earth metals

Abbreviations

Ar	aryl
Bu	butyl
i-Bu	*iso*butyl
t-Bu	*tert*.butyl
CN	coordination number
cot	cyclooctatetraene
d	day(s)
DIBAH	di*iso*butylaluminum hydride
DME	dimethoxyethane
DMSO	dimethyl sulfoxide
Do	donor
E	heteroatom (non-carbon atom)
equiv	equivalent(s)
Et	ethyl
FT IR	Fourier transform infrared spectroscopy
h	hour(s)
Hex	hexyl
HIBAO	hexa*iso*butylalumoxane
IP	isoprene
L	liter(s)
Ln	rare-earth metal (i.e., Sc, Y, La–Lu)
MAO	methylaluminoxane
Me	methyl
Mes	mesityl
min	minute(s)
MMA	methyl methacrylate
MMAO	modified methylaluminoxane
NMR	nuclear magnetic resonance
Oct	octyl
Ph	phenyl
phen	1,10-phenanthroline
Pr	propyl
i-Pr	*iso*propyl
py	pyridine
rt	room temperature
SEC	size exclusion chromatography
TBP	tributylphosphate
TEA	triethylaluminum
THF	tetrahydrofuran
TMA	trimethylaluminum
TON	turn over number
VT	variable temperature

1
Introduction

The identification and characterization of organometallics featuring a transition metal and aluminum in the same molecule has been pursued since Ziegler's epoch-making discovery of novel polymerization catalysts (e.g., $TiCl_4/Et_2AlCl$) [1, 2]. A crucial issue being still under investigation is the nature of the interaction of the organoaluminum co-catalyst component(s) with the transition metal center. Metallocene complexes gained interest as discrete and more soluble variants of the classic Ziegler-type "inorganics" as early as 1957 when the applicability of $(C_5H_5)_2TiCl_2/Et_2AlCl$ for ethylene polymerization and concomitant mechanistic studies was proposed [3, 4]. The "stabilizing effect" of the rigid cyclopentadienyl ligands was also exploited by R. Pearce and M.F. Lappert in a comparative study on the alkylation behavior of Ti(III) and Ln(III) centers [5–11]. Their important findings during the period 1975–1980 uniquely emphasize distinct reaction pathways of the various metal centers with organoaluminum reagents, at the same time launching the *Lanthanide Ziegler–Natta Model*. Because of their relevance these fundamental investigations will be highlighted in the following (cf., Scheme 1 and 2). Accordingly, metallocene chloride complexes $[(C_5H_5)_2MCl]_2$ of Ti(III) and the smaller Ln(III) centers Sc, Y, and Gd – Yb were reacted with $LiAlR_4$ (R = Me, Et) to give the corresponding tetraalkylaluminate derivatives via salt metathesis. While complexes $(C_5H_5)_2Ti(AlR_4)$ "decomposed slowly in the solid state ... or rapidly in toluene solution ... at ambient temperature", the rare-earth metal derivatives exhibited enhanced thermal stability. For example, complexes $(C_5H_5)_2Sc(AlMe_4)$ and $(C_5H_5)_2Y(AlMe_4)$ could be sublimed without decomposition at 100 °C/0.1 Torr and 120 °C/0.05 Torr, respectively. However, the tetraethylaluminate derivatives $(C_5H_5)_2Ln(AlEt_4)$ (Ln = Y, Ho) "were much less stable, slowly decomposing at room temperature, in solution or in the solid state". X-ray structural analysis of $(C_5H_5)_2Ln(AlMe_4)$ (Ln = Y, Yb) revealed a "$Ln(\mu\text{-Me})_2AlMe_2$" ($[\eta^2]$) coordination mode for the aluminate ligand (Scheme 1), which could also be observed in solution by NMR spectroscopy. Interestingly, a variable-temperature NMR study revealed a rapid "bridge and terminal alkyl ligand site exchange process" depending on the size of the metal center. While complex $(C_5H_5)_2Sc(AlMe_4)$ was non-fluxional at ambient temperature (site exchange > 100 °C), the yttrium complexes $(C_5H_5)_2Y(AlR_4)$ (R = Me, Et) stayed fluxional as low as $-40\,°C$ ($^2J_{YH} = 4$–5 Hz, $^1J_{YC} = 12.2$ Hz, $\Delta G^\ddagger = 15.9$ kcal mol^{-1} at 392 K for [AlMe$_4$]).

Lappert and coworkers also found that such rare-earth metal aluminate complexes can be readily transformed into methyl derivatives by addition of equimolar amounts of donor Lewis base molecules such as pyridine or tetrahydrofuran (*Lappert's concept of donor-induced aluminate cleavage*, Scheme 1) [10]. Depending on the Lewis acidity of the Ln(III) center homo-

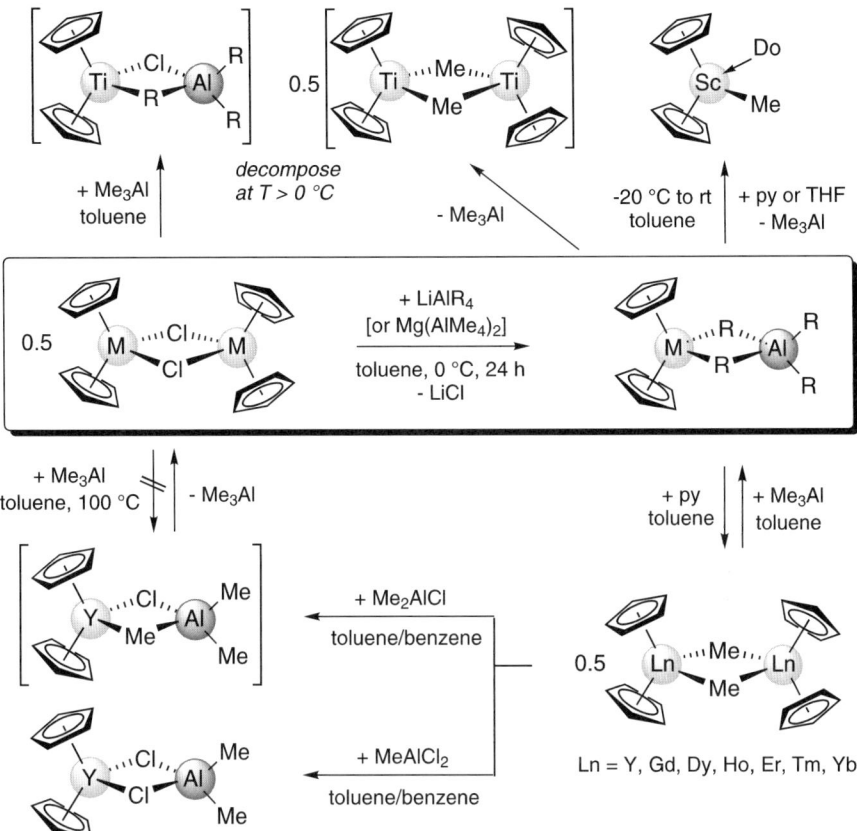

Scheme 1 Major findings of the comparative alkylation study by Pearce and Lappert. Non-isolated labile metal complexes are shown in *square parentheses*

bridged (Ln = Y, Dy – Yb) and donor-coordinated (Ln = Sc) methyl complexes were obtained. Attempts to prepare (μ-chloro)(μ-alkyl) heterobridged lanthanidocene complexes failed, while the corresponding titanocene complex could be identified, albeit being as unstable as the tetramethylaluminate congener. Derivatization reactions of complexes [(C$_5$H$_5$)$_2$Ln(μ-Me)]$_2$ proved the reversibility of the aluminate cleavage as well as a favorable Me/Cl ligand exchange via addition of MeAlCl$_2$. Furthermore, complexes [(C$_5$H$_4$R$'$)$_2$Ln(μ-Me)]$_2$ (Ln = Y, Er, Yb; R$'$ = H, Me, SiMe$_3$) and (C$_5$H$_4$R$'$)$_2$Y(AlMe$_4$) (Ln = Y, Er, Ho, Yb; R$'$ = H, Me, SiMe$_3$) were shown to be active homogeneous ethylene polymerization catalysts [8]. Unlike the former homobridged dilanthanide complexes (10–200 g mol^{-1} atm^{-1} h^{-1}, 70–100 °C, M_w/M_n = 1.8–2.5), the latter aluminate congeners exhibited a "threshold effect". While inactive at 75 °C/5 bar, a temperature ($\geq$ 95 °C) or pressure increase (33 bar) initiated polymerization (24 g mol^{-1} atm^{-1} h^{-1}, 100 °C, M_w/M_n = 1.5). These findings

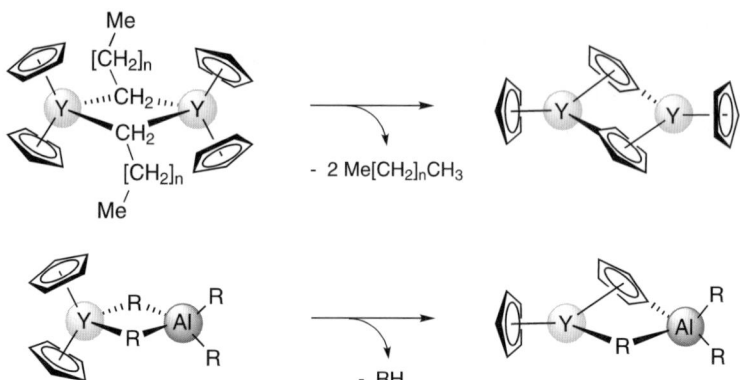

Scheme 2 Hydrogen abstraction as a viable deactivation process of ethylene polymerization catalysts: aluminate coordination as a beneficial effect?

could be corroborated five years later by P.L. Watson and T. Herskovitz for the permethylated catalyst precursor system $(C_5Me_5)_2Ln(AlMe_4)$ (Ln = Yb, Lu) [12]. Interestingly, the rate profile of the aluminate complexes did not reveal any complete catalyst deactivation as in the case of the aluminum-free dimer [8]. This enhanced lifetime was rationalized on the deactivation procedure shown in Scheme 2, emphasizing the non-availability of the second cyclopentadienyl ligand for H abstraction and hence a "remaining" active methyl ligand for the aluminate complex.

Clearly, these early investigations by Pearce and Lappert gave strong support for the presence of transient alkyl-bridged intermediates in classic Ziegler-type catalysts. This hypothesis was also in accordance with the idea that organoaluminum cocatalysts not only serve to alkylate M–Cl bonds but also to stabilize coordinatively unsaturated active centers via doubly alkyl-bridged intermediates with respect to both enhanced thermal stability and prolonged catalyst lifetime.

The relevance of Ln/Al heterobimetallic complexes for the emulation of zirconocene-based polymerization catalysis [13–15] was later on also stressed by the *Lanthanide Ziegler–Natta Model* [16]. Accordingly, lanthanidocene alkyl complexes were not only successfully employed for clarifying major initiation, propagation, and termination steps (Scheme 3) [17, 18].

In the course of these studies key features such as olefin coordination [19, 20], olefin insertion (propagation) [16, 21, 22], β-hydrogen elimination [16, 21, 22], and β-alkyl elimination [23] could be spectroscopically and structurally proven. In particular, yttrium aluminates have been proposed to model isoelectronic cationic homo- and heterobridged Zr/Al heterobimetallic complexes as dormant species and potential polymer chain transfer candidates [24]. Such zirconium aluminate complexes seem to be elusive [25], while the first structurally characterized titanium alumi-

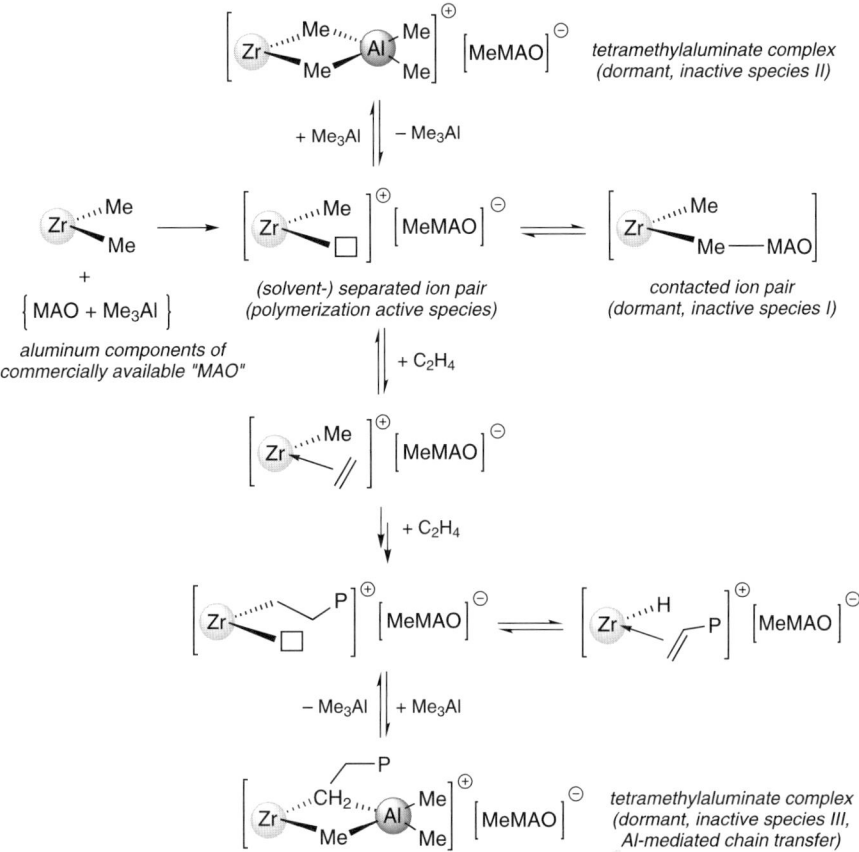

Scheme 3 Mechanistic scenario of zirconocene-promoted ethylene polymerization (according to [18]). Coordinated solvent molecules are omitted for clarity; P = polymer chain, □ = vacant coordination site

nate complex, [Ti(N*t*-Bu)(Me$_3$[9]aneN$_3$)(μ-Me)$_2$AlMe$_2$][B(C$_6$H$_5$)$_4$], has been reported recently [26]. Noteworthy, the interaction of Me$_3$Al with half-sandwich and postmetallocene complexes of the group 4 metal centers afforded methylidene, methylidine, and carbide moieties via multiple hydrogen abstraction, proposing an even more intricate catalyst deactivation scenario [27–39].

Beyond this "exclusive" lanthanide Ziegler–Natta model, Ziegler-type multicomponent systems ("*Mischkatalysatoren*") represent the only class of homogeneous rare-earth metal catalysts of considerable commercial relevance [40–43]. High-*cis*-1,4-polydienes are industrially produced from 1,3-dienes (butadiene and isoprene) in aliphatic or aromatic hydrocarbons by a number of *Mischkatalysatoren* based on the transition metals titanium, cobalt, and nickel, and the lanthanide element neodymium [40–47]. The formation of polymers with the highest cis-1,4-contents (> 98%) makes neodymium-based butadiene rubbers superior with respect to abrasion and cracking resistance, tack (the ability of the polymer to stick to itself), and raw polymer strength [48, 49]. Therefore, these polymers are particularly useful for the production of high-performance tires.

Since the early discoveries in the 1960s, numerous binary and ternary neodymium catalytic systems have been designed empirically utilizing Nd(III) carboxylates (also alkoxides or halides), aluminum alkyl halides, and aluminum alkyls or aluminum alkyl hydrides [43, 48, 50–52]. In situ formed cationic alkyl or hydrido Nd(III) sites were proposed as the active species promoting 1,3-diene polymerization via an allyl insertion mechanism. However, details of the polymerization mechanism and of the structure of the catalytically active center remained elusive. It was only the last five years, when organolanthanide research put more emphasis on organoaluminum-promoted alkylation reactions as well as the "cationization" phenomenon of rare-earth metal centers, supposedly the two key activation steps in Ziegler catalysts. The present account aims at a comprehensive survey of unequivocally identified Ln/Al organometallics elaborating synthesis, structural, and reactivity features of relevance for diene polymerization.

2
Lanthanide Halides

2.1
"Homoleptic" Lanthanide Halides

Initial reports on the stereospecific polymerization of dienes with binary lanthanide-based catalyst systems—composed of dehydrated lanthanide(III) chlorides $LnCl_3$ (Ln = Y, La, Ce, Pr, Nd, Sm, Gd, Er, and Yb) and trialkylaluminum reagents R_3Al (R = Et, i-Bu)—appeared as early as 1964 [53]. Because of the low solubilities of $LnCl_3$ in aliphatic and aromatic solvents catalytic activities were rather low, albeit high-cis-1,4-stereospecificities were observed. Addition of suitable amounts of alcohols to the original binary system led to enhanced activities without affecting the high stereospecificities for cis-1,4-polydienes (> 97%). For example, highly active ternary catalyst systems were obtained upon addition of excess of the alkylaluminum cocatalyst to a mixture of anhydrous $NdCl_3$ and ethanol in a 1 : 4 ratio [54]. In the following, a large variety of solubilizing O- and N-donor molecules (Do) was employed to generate more soluble and hence more efficient lanthanide precatalysts ($LnX_3 \cdot n$ Do), including, e.g., tetrahydrofuran [55–57], ethanol [54, 58, 59], *iso*propanol [57], pentan-n-ol (n = 1, 2, 3) [60], 2-ethylhexanol [61], dialkyl- and diarylsulfoxides [62], pyridine [57, 63–65], phenanthroline [63–65], ethylenediamine [63–65], tricresylphosphate [66], and tributylphosphate

(TBP) [67–69]. Although tremendous efforts were made to entirely elucidate the relationships between polymerization conditions on the one hand and catalytic activity and polymer properties on the other hand, proposals about the formation, number, and structure(s) of reactive intermediates and the catalytically active center(s) in these systems still lack final proof. As a matter of fact, the binary system $LnCl_3 \cdot 3TBP/i\text{-}Bu_3Al$ is one of the most comprehensively investigated diene polymerization catalysts. However, despite the numerous studies of Y.B. Monakov et al. [70–74] and H. Iovu et al. [75–79] no direct structural evidence of the proposed active species was achieved. In this context R.G. Bulgakov et al. recently claimed the generation of an $(i\text{-}Bu_2Al)_2O$ aluminoxane adduct of the terbium(III) derivative $TbCl_3 \cdot 3TBP$, evidenced only by ^{13}C NMR spectroscopy and an enhanced chemiluminescence [80].

2.1.1
Active Centers in $LnCl_3 \cdot 3TBP/R_3Al$

Numerous mechanistic studies involving the binary $NdCl_3 \cdot 3TBP/\, i\text{-}Bu_3Al$ catalyst mixture contributed to a better understanding of the stereospecific diene polymerization. A series of kinetic investigations revealed a first-order reaction with respect to the monomer, together with a strong influence of the initial monomer concentration and the polymerization temperature [76]. Compared to classic Ziegler-type catalysts ($E_a = 40\text{--}60\,\text{kJ}\,\text{mol}^{-1}$ [81]) a lower activation energy of $25.58\,\text{kJ}\,\text{mol}^{-1}$ was calculated and attributed to the higher catalytic activity of the neodymium-based systems. Bimodal molecular weight distributions and the influence of catalyst and cocatalyst concentrations suggested two preferential active centers being generated upon total displacement of the phosphate ligands and subsequent monoalkylation of the lanthanide metal center. Iovu and coworkers proposed the formation of butylated species $(i\text{-}Bu)NdCl_2$ as an intermediate [77]. Preformation, i.e., aging of the catalyst mixtures in the presence of a small amount of diene, enhanced the overall reactivity and was interpreted as formation of a much more stable η^3-butenyl unit at the expense of a "disfavored $Nd-C\,\sigma$-bond" [76, 78]. Furthermore, a bimetallic structure with neodymium and aluminum atoms bridged by chloride and alkyl ligands was assigned to one of the proposed active complexes [76, 82, 83]. The "*anti*" and "*syn*" arrangements of the polymer chain end are presented in Scheme 4.

Detailed investigations of the influence of the organoaluminum compound on the kinetic heterogeneity of active sites were performed by Y.B. Monakov and coworkers [84–91]. Butadiene was polymerized with $NdCl_3 \times 3TBP$ in the presence of different organoaluminum activators R_3Al ($R = Et$, i-Bu, Hex, Oct). As a first approximation broad bimodal molecular weight distributions were obtained for all of the cocatalysts, however, a more detailed analysis indicated the contribution of at least four types of structurally different active species. On the basis of mathematical methods (Tikhonov's regularization

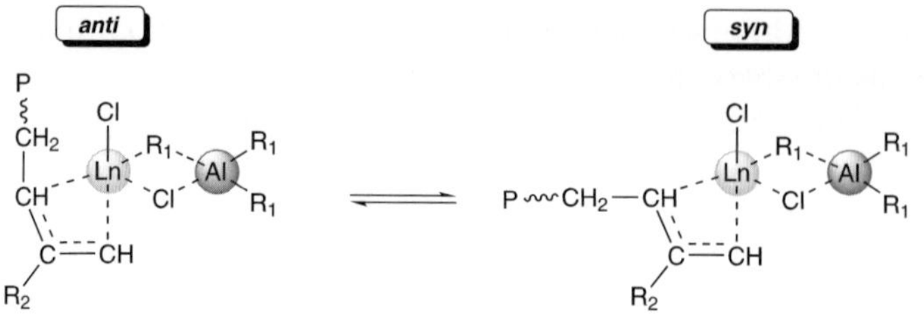

Scheme 4 Proposed "*anti*" and "*syn*" structures of the last chain unit of polydiene obtained from binary $LnCl_3 \cdot 3TBP/i\text{-}Bu_3Al$ catalysts (adopted from [76]). P= polymer chain, $R_1 = i\text{-}Bu$, $R_2 = H$, Me

method [92, 93]) the experimental curves of molecular weight distributions were transformed into kinetic activities as a function of the molecular weight. Polymodal molecular weight distributions were obtained, with the four distinct maxima representing four different types of active centers. The fact that even at the lowest butadiene conversions four maxima were observed led to the suggestion that the catalytic system initially is non-uniform, with all four types of active sites arising from the initial stage of the activation process [84].

In order to obtain distinct information about the ratio of lifetimes of active sites in their π- and σ-states quantum-chemical studies were performed, modeling the growing polymer chain by $CH_3CH=CHCH_2\text{-}CH_2CH=CHCH_2$ and substituting $i\text{-}Bu_3Al$ by Me_3Al. Thus, full energies E_n, calculated for π-(*anti*, *syn*) and σ-coordination modes (*cis*, *trans*), revealed the superior stability of π-coordinated polymer chains for all of the structures presented below (Fig. 1, Table 1). In contrast, formal peralkylation of the lanthanide

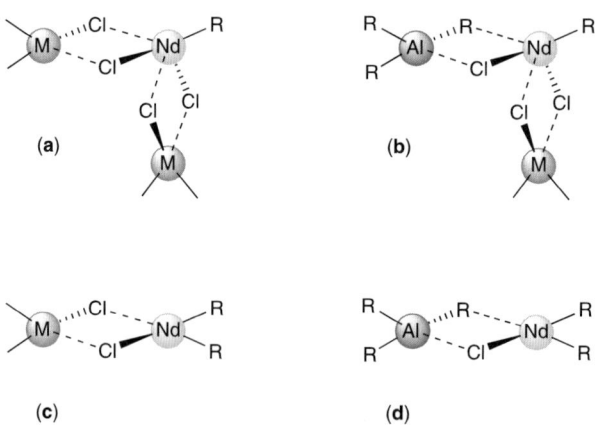

Fig. 1 Proposed active sites with *cis*-directing metal environments in binary $NdCl_3 \cdot 3TBP/R_3Al$ catalyst mixtures (adopted from [86]). M = Al or Nd; R = alkyl

Table 1 Full energies E_n calculated for different types of active centers with π- and σ-coordinated growing polymer chains (adopted from [94])

Structure[a]	Full energies E_n (kJ mol^{-1})[b] for structures				
	a	b	c	d	e
π-(anti-)	− 46.3	− 40.2	− 16.9	− 6.5	0
π-(syn-)	− 59.1	− 51.4	− 31.7	− 24.9	− 12.6
σ^α-(cis-)	− 5.4	− 4.7	− 4.7	− 4.9	− 19.0
σ^α-(trans-)	− 14.3	− 11.2	− 15.7	− 11.7	− 25.9
σ^γ-	0	0	0	0	− 14.4

[a] σ^α: Lanthanide atom is linked to the C_α atom of the terminal unit;
σ^γ: Lanthanide atom is linked to the C_γ atom of the terminal unit
[b] The energy of the least favorable structure is taken as the zero value

metal center, i.e., the experimentally not observed formation of a LnR$_3$ species (e), resulted in energetically more favorable σ-bound motifs [94].

2.1.2
Divalent Lanthanide Halides

W.J. Evans et al. reported the polymerization of isoprene with divalent lanthanide iodides LnI$_2$ and LnI$_2$(THF)$_x$ as initiators [95]. Accordingly, neodymium, samarium, dysprosium, and thulium derivatives gave high-*cis*-1,4-polyisoprenes with high molecular weights and relatively narrow molecular weight distributions. Although monomer conversions were significantly higher in the presence of excess of *i*-Bu$_3$Al active single component catalysts were observed both for unsolvated and solvated iodides. A direct correlation between the lanthanide metal reduction potential and the catalytic activity was not observed (Table 2).

2.2
Heteroleptic Lanthanide Halides

2.2.1
Halogenoaluminates

Lanthanide halide complexes free of coordinated Lewis bases, such as alcohols, phosphates, amines, dimethylsulfoxide, or THF, suffer from low solubilities in non-coordinating solvents. Therefore, catalytic systems based on LnCl$_3$ generally require preformation or aging in order to reach maximum activities. In contrast, lanthanide tetrahalogenoaluminate complexes are soluble in aromatic solvents. Such simple Ln/Al heterobimetallic halide

Table 2 Polymerization of isoprene with divalent lanthanide iodides LnI_2 (Ln = Nd, Sm, Dy, Tm) and $LnI_2(THF)_x$ (Ln = Sm, Tm) as initiators (from [95])

Catalyst precursor[a]	Cocatalyst	Al : Ln ratio	conv. %[b]	M_w 10^{-3} [c]	M_w/M_n^c
NdI_2	none	0	42	400	3.3
SmI_2	none	0	47	1900	2.2
$SmI_2(THF)_x$	none	0	44	1300	2.5
DyI_2	none	0	51	3000	1.5
TmI_2	none	0	29	200	2.4
$TmI_2(THF)_x$	none	0	39	1000	2.0
NdI_2	i-Bu_3Al	10	74	500	2.2
SmI_2	i-Bu_3Al	10	81	2100	1.2
$SmI_2(THF)_x$	i-Bu_3Al	10	76	1500	2.2
DyI_2	i-Bu_3Al	10	71	200	2.9
TmI_2	i-Bu_3Al	10	84	2300	4.0
$TmI_2(THF)_x$	i-Bu_3Al	10	89	2800	3.0

[a] General conditions: 5 mL hexane, 20 mg LnI_2, 3 mL isoprene (30 mmol), ambient temperature
[b] Gravimetrically determined
[c] Determined by size exclusion chromatography

complexes are accessible as η^6-coordinated arene molecules of the general formula $(\eta^6$-arene$)Ln(AlX_4)_3$ (X = Cl, Br, I). Since the first report on $(\eta^6$-$C_6Me_6)Sm(AlCl_4)_3$, which was synthesized by F.A. Cotton and W. Schwotzer in the presence of aluminum powder (reductive conditions) [96, 97], a variety of metal centers and arene molecules, such as benzene, toluene, xylene, and durene (1,2,4,5-tetramethylbenzene), have been employed [98, 99]. Isolable analogues were synthesized with $LnCl_3$, $LnBr_3$, and partially, with LnI_3, even in the absence of aluminum (non-reductive conditions) [100, 101]. They were employed as precatalysts for the polymerization of butadiene and isoprene in the absence of coordinating Lewis bases [102–106]. Furthermore, substitution of some of the halogen atoms in the [AlX_4] tetrahedrons by R (alkyl) led to a series of well-soluble alkylhalogenoaluminate complexes [105, 106]. Accordingly, addition of 1.5 equivalents of trialkylaluminum reagents to toluene mixtures of the parent tetrahalogenoaluminates gave η^6-arene lanthanide(III) alkyltrishalogenoaluminate compounds of the general formula $(\eta^6$-arene$)Ln(AlX_3R)_3$ (Scheme 5). Yttrium, praseodymium, neodymium, samarium, and gadolinium derivatives were isolated in yields between 56 and 90%, both for Me_3Al and Et_3Al, as well as for chloro-, bromo-, and iodoaluminates. Their overall composition was proven by elemental analysis and IR spectroscopy. For the neodymium derivative

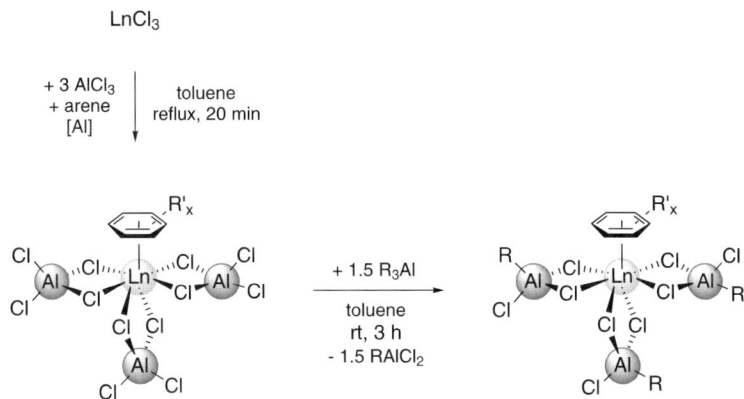

Scheme 5 Synthesis of (η^6-arene)Ln(AlCl$_4$)$_3$ (Ln = Y, Pr, Nd, Sm, Gd); subsequent reaction with trialkylaluminum reagents Me$_3$Al or Et$_3$Al gives alkylated (η^6-arene) Ln(AlCl$_3$R)$_3$ (R = Me, Et) compounds in good yields [105]

(η^6-C$_6$H$_5$Me)Nd[(μ-Cl)$_2$AlClMe]$_3$ the solid state structure was confirmed by X-ray crystallography (Fig. 22) [105].

The more bulky *iso*butyl organoaluminum reagent *i*-Bu$_3$Al did not produce stable η^6-arene lanthanide complexes with partially alkylated chloroaluminate environments. Instead, elemental analysis and IR spectroscopy indicated the loss of the coordinated arene molecules in aliphatic solvents associated with the formation of LnCl$_3$(*i*-BuAlCl$_2$)$_3$ complexes (Scheme 6a). Attempts to isolate higher alkylated η^6-arene lanthanide compounds were not successful. For example, the reaction of (η^6-C$_6$Me$_6$)Ln(AlCl$_4$)$_3$ with larger amounts of different trialkylaluminum reagents gave materials insoluble in toluene. The formation of polymeric {LnCl$_3$}$_x$ with variable amounts of surface chemisorbed alkylaluminum derivatives was proposed (Scheme 6b) [105].

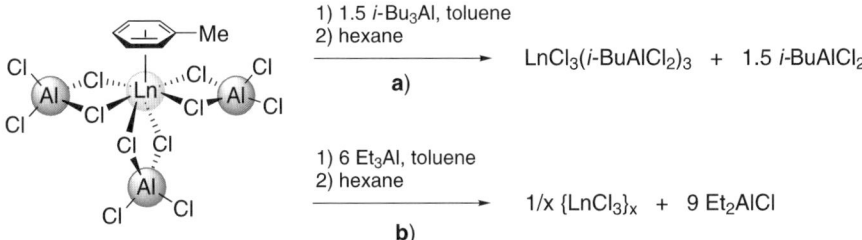

Scheme 6 Reaction of (η^6-arene)Ln(AlCl$_4$)$_3$ with 1.5 equivalents of *i*-Bu$_3$Al (a) or 6 equivalents of Et$_3$Al (b) [105]

Table 3 Polymerization of butadiene with $(\eta^6\text{-}C_6H_5Me)Ln(AlX_3R)$ (from [105])

Ln/X [a]	Cocatalyst [b]	Time min	conv. %	cis-1,4 %	trans-1,4 %	1,2 %
Pr/Cl	DIBAH	120	30	95.8	1.8	2.6
Pr/I	DIBAH	60	46	98.8	1.0	0.2
Nd/Cl	DIBAH	120	80	98.1	1.6	0.3
Nd/Br	DIBAH	120	34	98.6	1.1	0.3
Gd/Cl	DIBAH	120	30	96.6	1.7	1.7
Pr/Cl	Bu_2Mg	20	92	96.9	2.2	0.9
Pr/I	Bu_2Mg	30	94	88.2	10.8	1.0
Nd/Cl	Bu_2Mg	15	73	97.4	1.8	0.8
Gd/Cl	Bu_2Mg	15	70	86.8	12.8	0.4
Y/Cl	Bu_2Mg	20	87	0.4	99.5	0.1

[a] General conditions: 150 mL hexane, 15 g butadiene, $[Ln] = 3 \times 10^{-4}$ mol L^{-1}, $T = 50\,°C$. Ln/X indicates a toluene catalyst solution obtained in situ from the reaction of parent $(\eta^6\text{-}C_6H_5Me)Ln(AlX_4)_3$ and 1.5 equivalents of $i\text{-}Bu_3Al$
[b] $n(Ln):n(DIBAH) = 1:50$ or $n(Ln):n(Bu_2Mg) = 1:6$

Such tetrahalogenoaluminate complexes were employed as precatalysts in the homo- and copolymerization of butadiene and isoprene. When combined with alkylaluminum cocatalysts, most favorably $i\text{-}Bu_3Al$, high-*cis*-1,4-polymers were obtained, with maximum activities in petrol ether at 50 °C [102–104]. Increasing amounts of $i\text{-}Bu_3Al$ enhanced the catalytic activity albeit decreasing the molecular weights [103]. This suggests the generation of more active sites within the binary catalyst mixtures, and given the observation that larger amounts of catalytically inactive $LnCl_3$ are present, the occurrence of fast chain transfer between Ln and Al centers. Furthermore, high molecular weight high-*cis*-polybutadienes were obtained with monoalkylated halogenoaluminates, prepared in situ from lanthanide tetrahalogenoaluminates and 1.5 equivalents of $i\text{-}Bu_3Al$ (Table 3). In order to control the molecular weight a 50-fold excess of DIBAH was added as a cocatalyst, resulting in highly active catalyst mixtures. It should be mentioned, that similar reactivities have been observed in the presence of smaller amounts of the cocatalyst [105]. For comparison, lower molecular weights albeit higher catalyst activities were reported in combination with magnesium-based cocatalysts. Binary mixtures of $(\eta^6\text{-}C_6H_5Me)Ln(AlCl_4)_3$ and Bu_2Mg (1 : 6) afforded high-*cis*-polybutadienes with molecular weights sensitive to the type of metal center (Ln = Pr, Nd, Gd) and halogen atom X (X = Cl, Br, I). In contrast, yttrium-based catalyst systems gave high-*trans*-polybutadiene (99.5% *trans*) [105].

2.2.2
Lanthanidocene Chlorides

It is a generally accepted view that the formation of alkylated rare-earth metal centers in halide-based multicomponent mixtures is crucial for the initiation of catalytic diene polymerization. Inspired by the *Lanthanide Ziegler–Natta Model* and the stability of a variety of isolated and structurally fully characterized lanthanidocene(III) alkylaluminate complexes of general type $Cp^{R'}_2Ln[(\mu\text{-}R)_2AlR_2]$ ($Cp^{R'}$ = substituted cyclopentadienyl/indenyl ligand, e.g., C_5Me_5, C_5Me_4H, indenyl, 2-methylindenyl; R = Me, Et, *i*-Bu) [5, 6, 9, 10, 16, 24, 107–111] the interaction of discrete metallocene chloride complexes with organoaluminum reagents was examined. Such lanthanidocene-derived adduct complexes display valuable models and have not been examined in diene polymerization reactions. Catalytic activity might be achieved under special reaction conditions as shown for cationic species (Sect. 7.2.4). Alternatively, one can speculate about active initiators originating from oversaturated lanthanidocene molecules, which favor an $\eta^5\text{-}\eta^3\text{-}\eta^1$ coordination shift of the otherwise innocent cyclopentadienyl spectator ligand [112].

2.2.2.1
Adduct Formation with AlCl₃

Already 25 years ago P.L. Watson et al. described a high-yield reaction of the strong Lewis acid $AlCl_3$ with ate complexes $Li[Ln(C_5Me_5)_2Cl_2]$ (Ln = Yb, Lu; Scheme 7) and the chemistry of the Ln/Al heterobimetallic addition products $(C_5Me_5)_2Ln[(\mu\text{-}Cl)_2AlCl_2]$ [113, 114]. The molecular structure of the ytterbium derivative was confirmed by X-ray crystallography. Addition of THF did not promote ionic dissociation, but quantitatively generated the neutral monochloride complex $(C_5Me_5)_2YbCl(THF)$ [113].

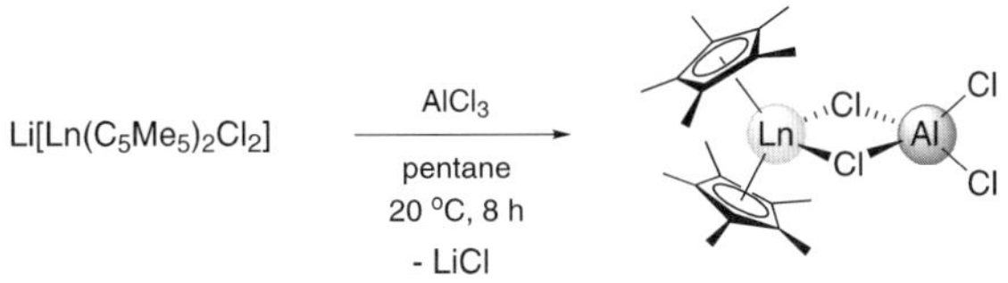

Scheme 7 Synthesis of lanthanidocene tetrachloroaluminates $(C_5Me_5)_2Ln(AlCl_4)$ (Ln = Yb, Lu) [113, 114]

2.2.2.2
Adduct Formation with R₃Al

Treatment of oligomeric lanthanidocene(III) chlorides $[(C_5Me_5)_2Ln(\mu\text{-}Cl)]_n$ (Ln = Y: *n* = 2; Ln = Sm: *n* = 3) with stoichiometric amounts of trialkylaluminum reagents R_3Al afforded heterobimetallic Ln(III)/Al complexes in good yields [115, 116]. Size differentiation on the basis of R was observed for both yttrium and samarium derivatives. As evidenced by X-ray crystallography, dimeric $[(C_5Me_5)_2Y(Cl)]_2$ produced monomeric (μ-alkyl)(μ-

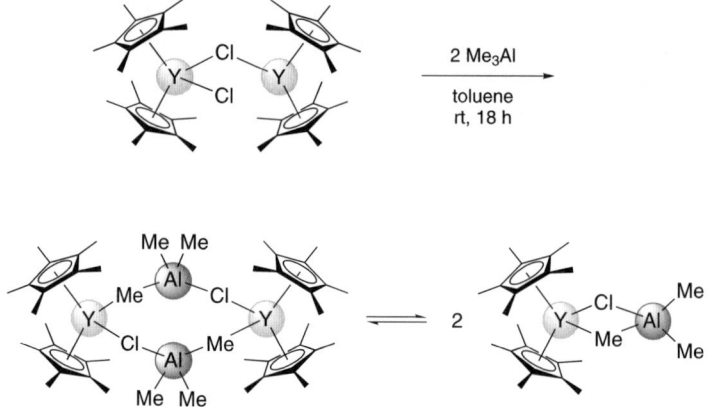

Scheme 8 Reactivity of $[Cp^*_2YCl]_2$ toward Me_3Al (monomer–dimer equilibrium) [115]

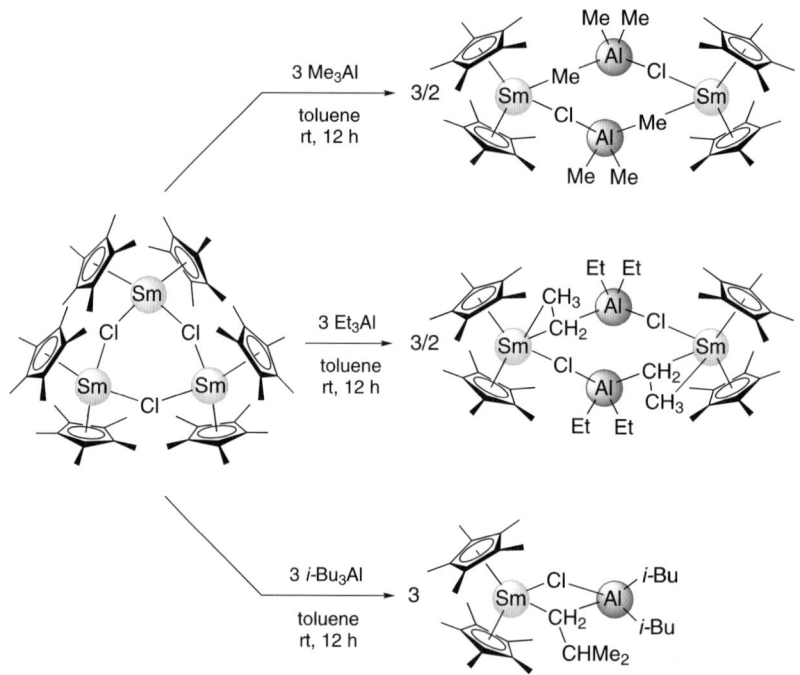

Scheme 9 Reactivity of $[Cp^*_2SmCl]_3$ toward R_3Al (R = Me, Et, i-Bu) [116]

chloride) bridged species [(C$_5$Me$_5$)$_2$Y(μ-Cl)(μ-R)AlR$_2$]$_2$ for triethyl- and tri*iso*butylaluminum. Ligand rearrangement was not observed even at elevated temperature. In analogy with the tetramethylaluminate complexes {(C$_5$Me$_5$)$_2$Ln[(μ-Me)$_2$AlMe$_2$]}$_2$ [109, 110], a monomer–dimer equilibrium was suggested for the trimethylaluminum yttrium derivative at ambient temperature (Scheme 8) [115].

Trimeric [(C$_5$Me$_5$)$_2$Sm(μ-Cl)]$_3$ was shown to react with trialkylaluminum reagents in a similar fashion. Monomeric (R = *i*-Bu) and dimeric (R = Me, Et) alkylaluminum adduct complexes were isolated and fully characterized by elemental analysis, NMR and IR spectroscopy, and X-ray crystallography (Scheme 9). It was suggested that the unusual asymmetric μ-η^1 : η^1(side-on) ethyl coordination mode of the bridging ethyl groups (Fig. 14) might facilitate special chemistry that is not available with methyl and *iso*butyl analogues (Fig. 13) [116].

2.2.2.3
Adduct Formation with R$_2$AlCl

Heterobimetallic Ln(III)/Al complexes, obtained from the reaction of lanthanidocene(III) chlorides with dialkylaluminum chlorides, were initially described by K.H. den Haan and J.H. Teuben [117]. Proton NMR spectroscopic data as well as the observation that the reaction of dimeric [(C$_5$H$_5$)$_2$Y(μ-Me)]$_2$ with Me$_2$AlCl exclusively generates (C$_5$H$_5$)$_2$Y[(μ-Cl)$_2$AlMe$_2$] [10], suggested the formation of a (μ-Cl)$_2$ bridged species also from the [(C$_5$H$_5$)$_2$Y(μ-Cl)]$_2$/Me$_2$AlCl reaction. Structural evidence was obtained for a variety of dif-

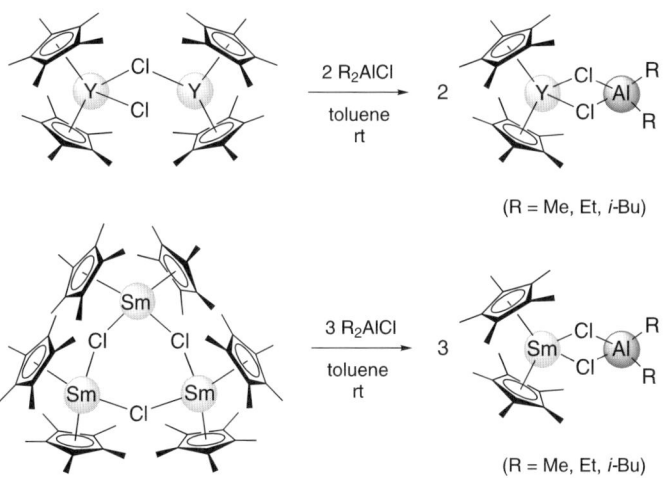

Scheme 10 Reactivity of [Cp*$_2$YCl]$_2$ and [Cp*$_2$SmCl]$_3$ toward R$_2$AlCl (R = Me, Et, *i*-Bu) [115]

ferent R_2AlCl reagents and $[(C_5Me_5)_2Ln(\mu\text{-}Cl)]_n$ precursors (Ln = Y: n = 2; Ln = Sm: n = 3) (Scheme 10) [115]. Monomer–dimer equilibria or ligand redistribution were excluded by NMR spectroscopy. Note that complexes $(C_5Me_5)_2Sm[(\mu\text{-}Cl)_2AlR_2]$ (R = Me, Et, i-Bu) are also obtainable from the reaction of carboxylate derivative $[(C_5Me_5)_2Sm(\mu\text{-}O_2CPh)]_2$ with R_2AlCl (see Sect. 3.4.1) [115].

3
Lanthanide Carboxylates

Numerous binary and ternary diene polymerization initiator systems with neodymium as the rare-earth metal component have been designed empirically and investigated since the early discoveries in the 1960s. Commercially used neodymium-based catalysts mostly comprise Nd(III) carboxylates, aluminum alkyl halides, and aluminum alkyls or aluminum alkyl hydrides [43, 48, 50–52]. Typically, the carboxylic acids, which are provided as mixtures of isomers from petrochemical plants carry solubilizing aliphatic substituents R. They are treated with the alkylaluminum reagents to generate the active catalysts in situ (Scheme 11).

Although heterobimetallic complexes with alkylated rare-earth metal centers were proposed to promote 1,3-diene polymerization via an allyl insertion mechanism, details of the polymerization mechanism and of the structure of the catalytically active center(s) are rare [58, 83, 118–125]. Moreover, until now, the interaction of the "cationizing" chloride-donating reagent with alkylated rare-earth metal centers is not well-understood. Lanthanide carboxylate complexes, which are used in the industrial-scale polymerization of butadiene and isoprene, are generally derived from octanoic, versatic, and

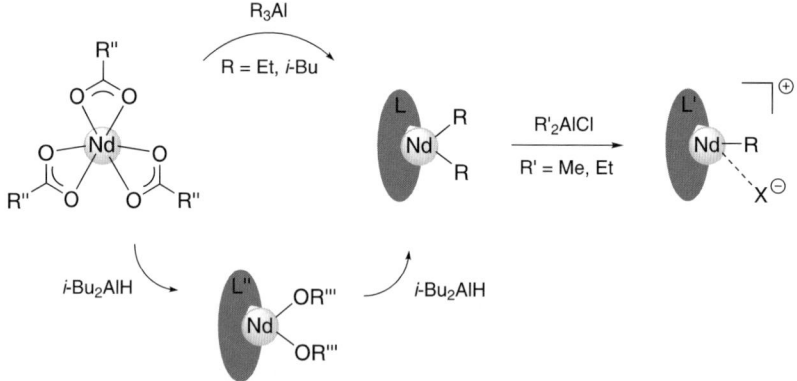

Scheme 11 Proposed activation sequence of oligomeric Ln(III) carboxylates via different organoaluminum cocatalysts

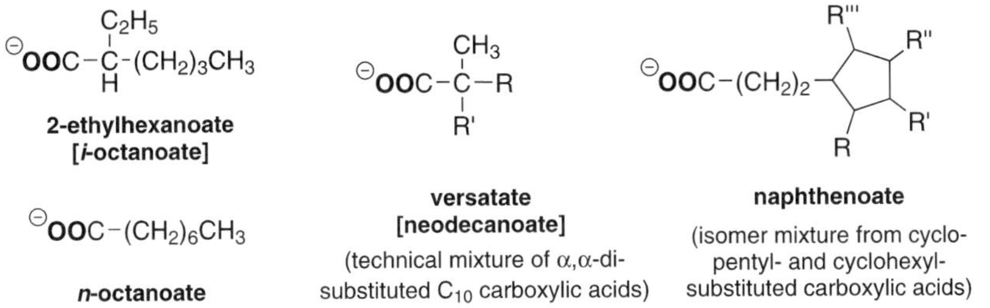

Chart 1

$^\ominus$OOC-C(C$_2$H$_5$)(H)-(CH$_2$)$_3$CH$_3$
2-ethylhexanoate
[*i*-octanoate]

$^\ominus$OOC-C(CH$_3$)(R')-R
versatate
[neodecanoate]
(technical mixture of α,α-di-
substituted C$_{10}$ carboxylic acids)

$^\ominus$OOC-(CH$_2$)$_2$- [cyclopentyl ring with R, R', R'', R''']
naphthenoate
(isomer mixture from cyclo-
pentyl- and cyclohexyl-
substituted carboxylic acids)

$^\ominus$OOC-(CH$_2$)$_6$CH$_3$
***n*-octanoate**

Chart 1 Carboxylate ligands employed in industrial scale lanthanide-based diene polymerization catalysts

naphthenonic acids (Chart 1). Because of the large "aliphatic tails", enhanced solubilities of the corresponding metal complexes in aliphatic solvents are imparted. Addition of alkylating organoaluminum reagents produces even more soluble components, however, such oily products exhibit poor crystallization behavior and NMR spectra, which are extremely difficult to interpret. In order to get a deeper insight into possible activation pathways and structure-reactivity relationships of these multicomponent mixtures model complexes based on more advantageous carboxylate components were studied.

3.1
Lanthanide Acetates

Diene polymerization with lanthanide acetate-based catalytic systems was first reported in 1979 [126]. Ouyang and coworkers reacted silver trifluoroacetate with neodymium(III) chloride to obtain $(CF_3CO_2)_2NdCl \cdot 2Et_2O$ as a precatalyst component that efficiently initiated the *cis*-1,4-stereospecific polymerization of butadiene and isoprene in the presence of the trialkylaluminum reagents *i*-Bu$_3$Al or Et$_3$Al. Slightly modified binary mixtures of general type $(CF_3CO_2)_2LnX \cdot EtOH/R_3Al$ (or R_2AlH) were studied with respect to catalytic activities and intrinsic viscosities of the polydienes obtained. In general, activities increased with higher electronegativities of the halogen atom X [127]. Comprehensive studies on the effect of acetate pK_a values and the *cis*-polymerization mechanism were reported by E. Kobayashi et al. using homoleptic neodymium and gadolinium acetates [128–131]. The homo- and copolymerization of butadiene and styrene was carried out with ternary catalytic systems of general type $\{Ln(O_2CR)_3\}_x/i$-Bu$_3$Al/Et$_2$AlCl (R = CH$_3$, CH$_2$Cl, CHCl$_2$, CCl$_3$, CF$_3$). As summarized in Table 4, gadolinium-based catalysts gave high-*cis*-polybutadienes in 22–85% yields with polymerization rates depending on the pK_a value of the acetate ligands. It was suggested that an increase of the acetate ligand acidity may lower the LUMO energy level of the gadolinium metal catalyst and result in higher reaction rates [129]. Neodymium-based catalysts gave highest activities in butadiene, isoprene,

Table 4 Polymerization of butadiene with ternary $Gd(O_2CR)_3/i\text{-}Bu_3Al/Et_2AlCl$ catalysts (from [129])[a]

R	Time h	Yield %	cis-1,4 %[b]	trans-1,4 %[b]	1,2 %[b]	M_w 10^{-4}
CH_3	24	48	99	0	1	84
CH_2Cl	18	22	99	0	1	55
$CHCl_2$	18	53	99	1	0	79
CCl_3	3	62	98	1	1	66
CF_3	2	85	97	2	1	72

[a] General conditions: polymerization in hexane, [butadiene] = 4 mol L^{-1}, [Gd] = 10^{-2} mol L^{-1}, $n(i\text{-}Bu_3Al):n(Gd) = 25:1$, $n(Et_2AlCl):n(Gd) = 2:1$; catalyst aging at 0 °C for 30 min
[b] Measured by the IR method in CS_2 solution

2,3-dimethylbutadiene, and styrene polymerization with trichloroacetate ligands (R = CCl_3) [130, 131].

First structural evidence for the formation of heterobimetallic Ln/Al complexes in carboxylate-based catalytic systems was obtained from the reaction of homoleptic rare-earth metal trifluoroacetates with equimolar amounts of $i\text{-}Bu_2AlH$ and Et_3Al, respectively [132]. Alkylated yttrium, neodymium, and

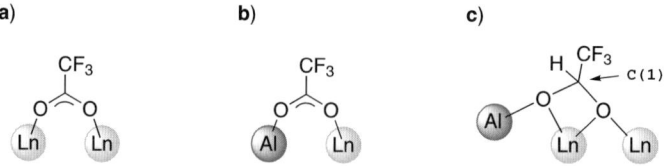

Scheme 12 Synthesis of heterobimetallic Ln/Al carboxylate complexes from $Ln(O_2CCF_3)_3$ (Ln = Y, Eu, Nd) and observed ligand coordination patterns (a–c) [132]

europium carboxylate complexes were generated and isolated in low yields (< 20%) from THF solutions via crystallization. Elemental analysis, X-ray crystallography, and 1D and 2D NMR spectroscopy suggested the formation of dimeric complexes of general formula $[(CF_3CO_2)_2Ln(CF_3CHO_2)AlR_2 \cdot 2THF]_2$ (Ln = Y, Nd, R = i-Bu; Ln = Eu, R = Et), featuring three types of ligand coordination modes (Scheme 12).

Interestingly, carbon and oxygen atoms of one of the original acetate ligands (motifs **a** and **b**) lost coplanarity with the former sp^2-hybridized carboxylic carbon atom (C(1), motif **c**). Although the X-ray crystallographic study did not reveal the presence of a C(1)-bonded hydrogen atom, a 1D and 2D NMR spectroscopic investigation of the isomorphous diamagnetic yttrium derivative finally confirmed the existence of a sp^3-hybridized carbon atom.

3.2
Lanthanide Butyroates

Model compounds featuring larger butyroate-based metal environments were reported by W.J. Evans. By reacting lanthanide metals with 2,2-dimethyl- or 2-phenylbutyric acid in the presence of catalytic amounts of $HgCl_2$ and $Hg(OAc)_2$ [133], homoleptic and carboxylic acid-coordinated carboxylate complexes $\{Ln(O_2CR)_3\}_x$ (R = $C(CH_3)_2CH_2CH_3$) and $Ln_2(O_2CR')_6(HO_2CR')_4$ (R' = $CH(Ph)CH_2CH_3$) were obtained and fully characterized [134]. In order to simulate the activation procedure, Et_2AlCl was added to the homoleptic neodymium and lanthanum butyroate complexes in a $n(Ln):n(Al)$ ratio of 1:5, reproducibly generating hexane insoluble materials of net compositions $[Nd_2AlCl_5C_{13}H_{22}O_x]$ and $[La_2AlCl_5C_{12}H_{21}O_y]$, respectively (Scheme 13). As

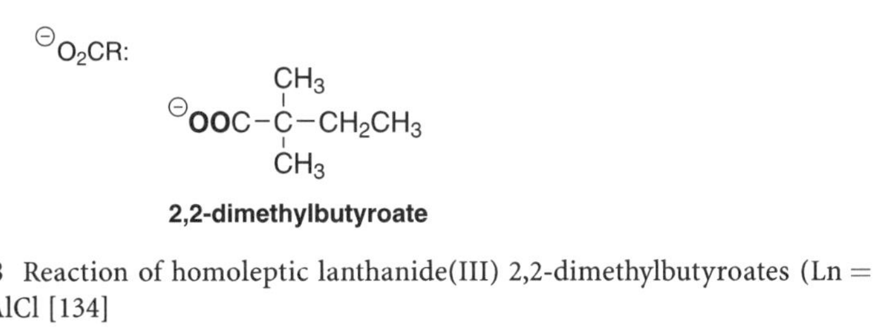

$\{Nd(O_2CR)_3\}_x$ + 5 Et_2AlCl $\xrightarrow[\text{rt, 1 h}]{\text{hexane}}$ " $Nd_2AlCl_5C_{13}H_{22}O_x$ "

$\{La(O_2CR)_3\}_x$ + 5 Et_2AlCl $\xrightarrow[\text{rt, 1 h}]{\text{hexane}}$ " $La_2AlCl_5C_{12}H_{21}O_y$ "

$^\ominus O_2CR$:

$^\ominus OOC-\underset{\underset{CH_3}{|}}{\overset{\overset{CH_3}{|}}{C}}-CH_2CH_3$

2,2-dimethylbutyroate

Scheme 13 Reaction of homoleptic lanthanide(III) 2,2-dimethylbutyroates (Ln = La, Nd) with Et_2AlCl [134]

a byproduct, the formation of $Et_2Al(O_2CR)$ was observed, corroborating the previously proposed cleavage of lanthanide carboxylate bonds upon addition of the organoaluminum chloride cocatalyst [50, 135, 136]. Noteworthy, perhalogenated $\{LnCl_3\}_x$ did not form as the predominant product in the presence of up to five equivalents of the chloride source.

Attempts to structurally characterize the generated Ln(III)/Al heterobimetallic complexes were not successful. Addition of coordinating solvents, such as THF or pyridine, afforded complex mixtures from which only lanthanide(III) chloride donor-adducts crystallized. The presence of reactive ethyl groups was confirmed by the reaction with D_2O (generating CH_3CH_2D) [134]. The catalytic relevance of such heterobimetallic complexes was confirmed by the quantitative conversion of a 7500-fold excess of isoprene into polyisoprene within 5–10 minutes at ambient temperature, after the addition of one equivalent of i-Bu_3Al cocatalyst.

3.3
Lanthanide Neodecanoates

For more than three decades, lanthanide complexes with long-chained aliphatic carboxylate ligands were employed in diene polymerization catalysis. Except for a small number of aliphatic model systems (Sects. 3.1, 3.2) there is still little known about the coordination environment of the neodymium centers and the changes caused by the addition of different types of organoaluminum cocatalysts. By using matrix-assisted laser desorption ionization time-of-flight (MALDI-TOF) mass spectrometry G. Kwag was able to identify a series of oligomeric and hydrated carboxylate complexes in parent neodymium(III) neodecanoates (i.e., 2,2-dimethyloctanoates). These neodecanoate complexes were synthesized from hydrated $NdCl_3$ (or nitrate and acetate) and sodium neodecanoate, according to Scheme 14 [124]. Because of the higher coordination numbers of the larger lanthanide metals ($CN = 8-12$) [137] and the aqueous synthesis protocol, neodymium neodecanoate (ND) does not exist as a monomer. Synchrotron X-ray absorption spectroscopy (XAS) confirmed the all-oxo environment and the strong ionic character of the trivalent neodymium center [121, 124, 138, 139]. In situ investigation of the activation procedure, i.e., subsequent addition of i-Bu_3Al and Et_2AlCl, confirmed a non-surprising redox stability of the neodymium(III) metal centers in the presence of larger amounts of cocatalysts. Under similar conditions structurally related nickel- or titanium-based carboxylate catalyst components are rapidly reduced to low-valent metal species [140].

A monomeric (and anhydrous) neodecanoate (= NDH) was obtained from the ligand exchange reaction between hydrated neodymium(III) acetate $Nd(OAc)_3(H_2O)_6$ and neodecanoic acid (Scheme 14). The coordination sphere of the large neodymium cation is saturated by an additional molecule of neodecanoic acid, as evidenced by MALDI-TOF mass spec-

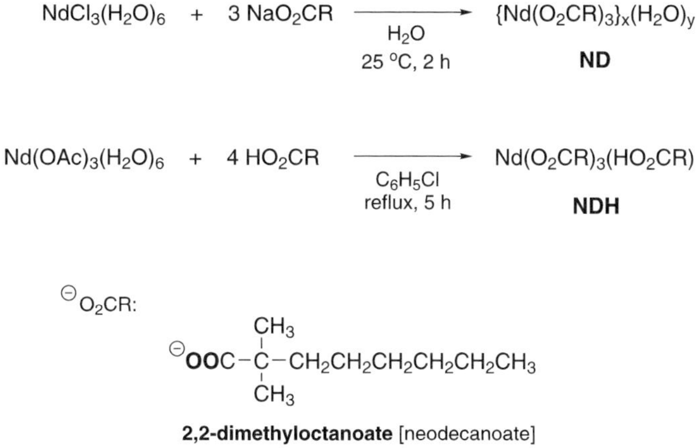

$$NdCl_3(H_2O)_6 + 3\,NaO_2CR \xrightarrow[25\,°C,\,2\,h]{H_2O} \{Nd(O_2CR)_3\}_x(H_2O)_y$$
ND

$$Nd(OAc)_3(H_2O)_6 + 4\,HO_2CR \xrightarrow[reflux,\,5\,h]{C_6H_5Cl} Nd(O_2CR)_3(HO_2CR)$$
NDH

$$^{\ominus}O_2CR:$$

$$^{\ominus}OOC-\underset{\underset{CH_3}{|}}{\overset{\overset{CH_3}{|}}{C}}-CH_2CH_2CH_2CH_2CH_2CH_3$$

2,2-dimethyloctanoate [neodecanoate]

Scheme 14 Synthesis of oligomeric (ND) and monomeric neodymium(III) neodecanoates (NDH) from hydrated neodymium chlorides and acetates [124]

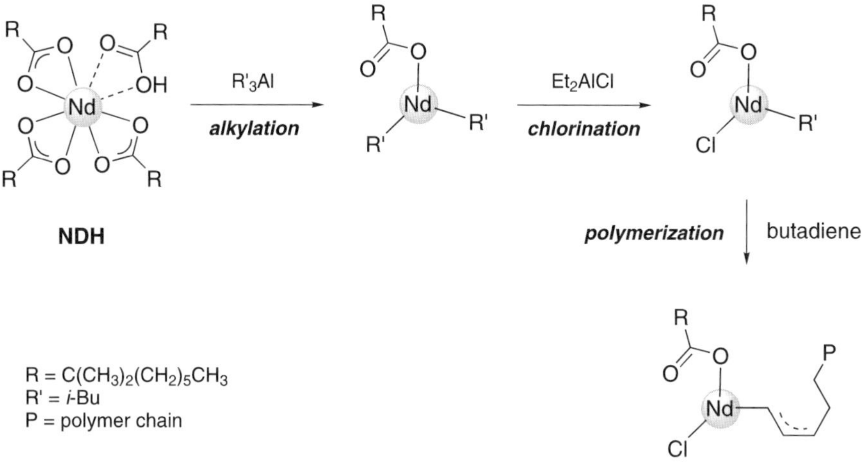

NDH

$R = C(CH_3)_2(CH_2)_5CH_3$
$R' = i\text{-}Bu$
$P = $ polymer chain

Scheme 15 Proposed activation and propagation mechanism involving monomeric neodymium(III) neodecanoate NDH (adopted from [124])

trometry, IR spectroscopy, and the X-ray absorption near edge structure (XANES) [124]. On the basis of these findings, a possible activation pathway (Scheme 15) and the structure of an active center was proposed and optimized by density functional methods, with Nd^{3+} coordinated by carboxylate, chloride, active allylic end, and the penultimate double bond of the growing unsaturated polymer chain [124, 141]. Although EXAFS spectroscopy indicated the presence of Nd – C and Nd – Cl bonds after alkylation and chlorination of the parent neodecanoate NDH, none of the proposed intermediates have been isolated and fully characterized.

3.4
Lanthanide Benzoates

Lanthanide carboxylates derived from a variety of symmetrically substituted benzoic acids (see Chart 2) were synthesized according to standard procedures and studied as realistic model systems in alkylation and diene polymerization scenarios. Particularly, alkylation and ligand exchange reactions were investigated using organoaluminum and organoaluminum chloride reagents. Several heterobimetallic Ln(III)/Al complexes were isolated and fully (structurally) characterized.

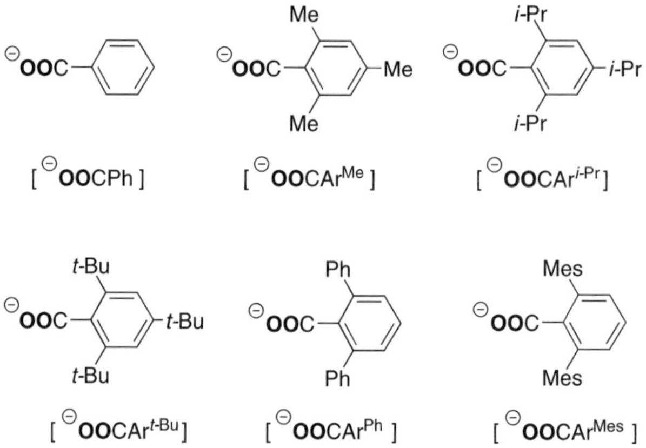

Chart 2 Symmetrically substituted benzoate ligands used as model ligands in order to investigate the alkylation behavior of lanthanide carboxylates

3.4.1
Lanthanidocene Benzoates

Once more, the well-defined and robust coordination environment of the $[(C_5Me_5)_2Ln]^+$ unit was exploited for model reactions, facilitating the isolation of a variety of heterobimetallic Ln(III)/Al complexes [115, 116, 142, 143]. As mentioned earlier, trialkylaluminum reagents R_3Al (R = Me, Et, i-Bu) easily react with lanthanidocene(III) chlorides $\{(C_5Me_5)_2LnCl\}_x$ to give mixed alkyl monochloride complexes, $(C_5Me_5)_2Ln(\mu\text{-Cl})(\mu\text{-R})AlR_2$, in good yields [115]. Similarly, $(\mu\text{-R})_2$-bridged lanthanidocene(III) tetraalkylaluminate complexes were obtained from lanthanidocene(III) benzoates, while the formation of various byproducts strongly depended on the nature of the trialkylaluminum compound. Tetramethylaluminate complexes $Cp^R{}_2Ln[(\mu\text{-Me})_2AlMe_2]$ (Ln = Y, La; $Cp^R = C_5Me_5$, 2-Me-C_9H_6) were obtained via [carboxylate]→[alkyl] interchange reactions from lanthanidocene carboxylates, $Cp^R{}_2Ln(O_2CAr^{Me})$ (Ln = Y, La; $Cp^R = C_5Me_5$) and

$Cp^R_2Y(O_2CAr^{i\text{-}Pr})$ (Cp^R = 2-Me-C_9H_6), and excess trimethylaluminum in good yields (Scheme 16) [142]. The addition of six equivalents of Et_3Al to a solution of dimeric $[(C_5Me_5)_2Sm(\mu\text{-}O_2CPh)]_2$ afforded a mixture of alkylated products (Scheme 17). Monomeric $(C_5Me_5)_2Sm[(\mu\text{-}Et)_2AlEt_2]$ [108] was isolated via crystallization from hexane in 47% yield. As a second fraction, a dinuclear ethylaluminum oxide complex was crystallized from the mother

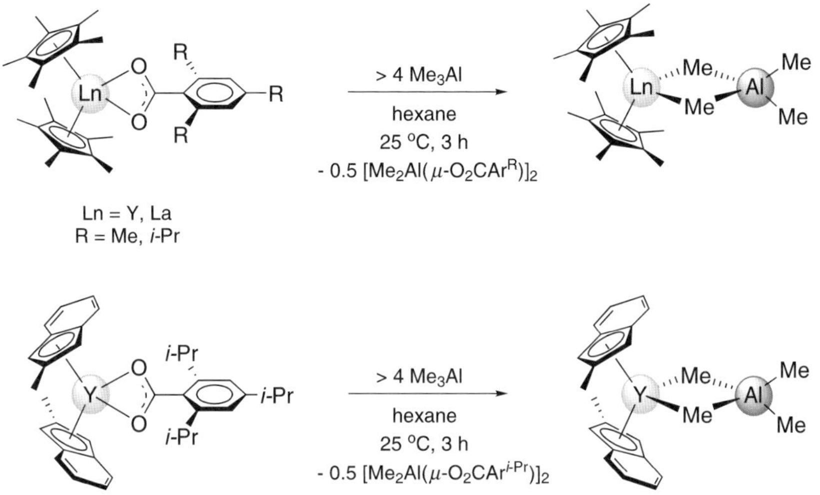

Scheme 16 Formation of lanthanidocene tetramethylaluminate complexes via [carboxylate]→[alkyl] interchange [142]

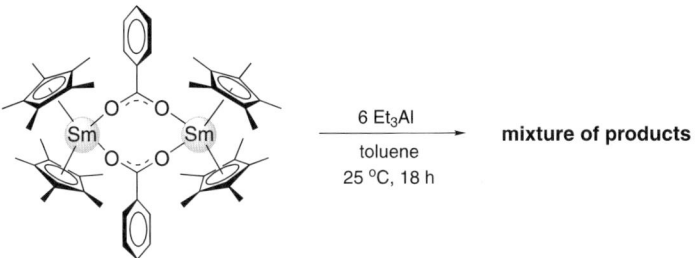

Scheme 17 Reactivity of $[(C_5Me_5)_2Sm(\mu\text{-}O_2CPh)]_2$ toward triethylaluminum. Two heterobimetallic Sm(III)/Al complexes were isolated via fractionated crystallization [116]

liquid in moderate, however, reproducible yields of about 30% and fully characterized by elemental analysis, and IR and NMR spectroscopy. Consistent with this C–O bond reaction, the GC-MS analysis of the organic byproducts revealed the formation of PhC(Et)$_2$(OH) and PhC(Et)=CHMe [116].

The solid-state structure of the ethylaluminum oxide metallocene complex (Fig. 30) shows two trivalent samarocene units connected by an [(Al$_2$Et$_5$O)$_2$]$^{2-}$ ethylalumoxane ("EAO") unit. The latter was described as an adduct of two molecules of triethylaluminum with two [AlEt$_2$O]$^-$ anions [116]. An unusual asymmetric μ-η^1 : η^1 (side-on) ethyl coordination mode was observed, which previously has been found only in a small number of lanthanide complexes, i.e., (C$_5$Me$_5$)$_2$Yb(μ-η^1 : η^1-Et)AlEt$_2$(THF) [144], (C$_5$Me$_5$)$_2$Sm(THF) (μ-η^1 : η^1-Et)AlEt$_3$ (Fig. 9) [115] and (C$_5$Me$_5$)$_2$Sm(THF) (μ-η^1 : η^1-Et) (μ-Cl)AlEt$_2$ [116].

The reaction of [(C$_5$Me$_5$)$_2$Sm(μ-O$_2$CPh)]$_2$ with trimethyl- and tri*iso*-butylaluminum also formed product mixtures and initially only tetraalkylaluminate complexes (C$_5$Me$_5$)$_2$Sm[(μ-R)$_2$AlR$_2$] (R = Me, *i*-Bu) were unequivocally identified. A more detailed investigation of the binary mixture [(C$_5$Me$_5$)$_2$Sm(μ-O$_2$CPh)]$_2$/R$_3$Al (R = Me, *i*-Bu) provided deeper insight into the alkylation scenario. While the 1 : 4-reaction produced mixtures with fully identified tetraalkylaluminate complexes, the 1 : 2-reaction gave adduct (C$_5$Me$_5$)$_2$Sm(μ-O$_2$CPh)(μ-*i*-Bu)Al(*i*-Bu)$_2$, so-to-speak as the first (most likely) intermediate of the precatalyst activation pathway. Despite the equilibrium shown in Scheme 18, addition of another two equivalents of *i*-Bu$_3$Al only led to a completion of the [carboxylate]→[aluminate] exchange. The

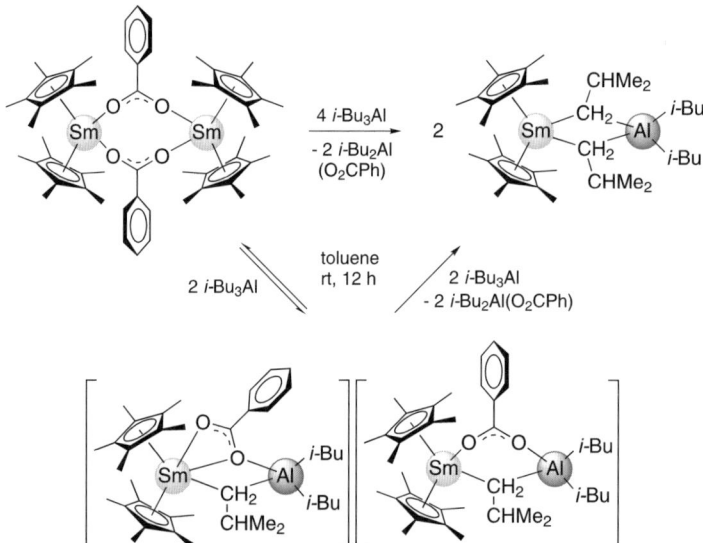

Scheme 18 Adduct and alkylaluminate formation in the reaction of [(C$_5$Me$_5$)$_2$Sm(μ-O$_2$CPh)]$_2$ with *i*-Bu$_3$Al [143]

fully characterized (μ-O$_2$CPh)(μ-i-Bu)-heterobridged compound crystallized as a 1 : 1-mixture of two isomers (Fig. 15). Surprisingly, the solid-state structure revealed no single preference of the carboxylate ligand, resulting in both dihapto ($CN = 9$) and monohapto coordination ($CN = 8$) to the samarocene metal center.

Modeling studies with aliphatic carboxylate complexes (see Sect. 3.3) indicated that their interaction with alkylaluminum chloride reagents, e.g., Et$_2$AlCl, most likely does not generate larger amounts of aluminum- and alkyl-free lanthanide(III) chlorides. Although carboxylate ligand exchange reactions were assumed to give terminal chloro- and/or alkyl species [145, 146], there is a dearth of structural data on a direct [carboxylate]→[chloride] transformation. Nevertheless, mixed chloride alkylaluminum complexes of general type (C$_5$Me$_5$)$_2$Sm[(μ-Cl)$_2$AlR$_2$] (R = Me, Et, i-Bu) were reported from the reaction of dimeric [(C$_5$Me$_5$)$_2$Sm(μ-O$_2$CPh)]$_2$ with four equivalents of the alkylaluminum chloride reagents R$_2$AlCl (Scheme 19) [115]. A mechanistic explanation of this ligand exchange reaction includes the initial breakup of the dimeric carboxylate-bridged samarocene by R$_2$AlCl and the generation of alkylaluminum adducts (C$_5$Me$_5$)$_2$Sm(μ-O$_2$CPh)(μ-Cl)AlR$_2$, which were not observed by means of NMR spectroscopy. Subsequent formation of intermediate "(C$_5$Me$_5$)$_2$SmCl" via alkylaluminum carboxylate dissociation and final R$_2$AlCl coordination were suggested to generate the (μ-Cl)$_2$-homobridged heterobimetallic reaction product [115].

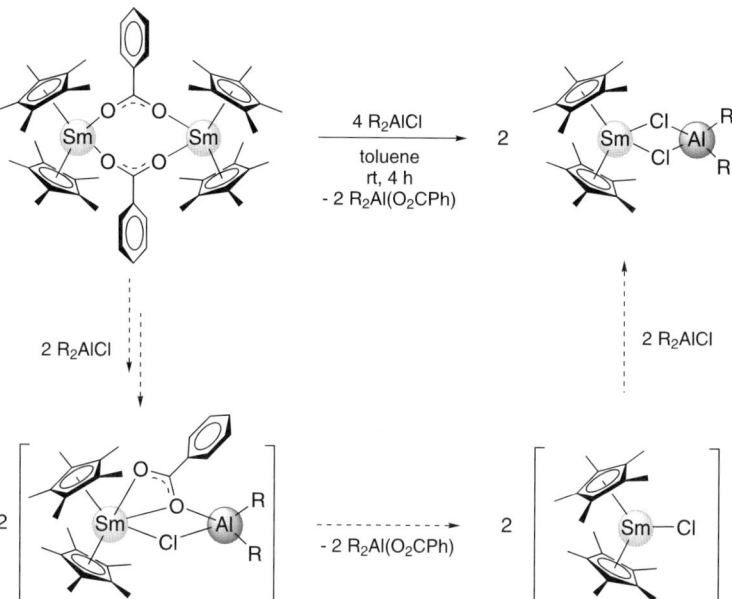

Scheme 19 Formation of (C$_5$Me$_5$)$_2$Sm[(μ-Cl)$_2$AlR$_2$] (R = Me, Et, i-Bu). Possible reaction intermediates are shown in parenthesis [115]

3.4.2
Homoleptic Lanthanide Benzoates

The organoaluminum-mediated alkylation and chlorination of tailor-made homoleptic rare-earth metal complexes with sterically encumbered, symmetrically substituted benzoate ligands (see Chart 2) was studied in detail. Oligomeric lanthanide carboxylate complexes, structurally characterized as monomeric or dimeric donor molecule adducts, were employed to investigate the implications of the degree of lanthanide metal alkylation and the organoaluminum chloride-mediated cationization for 1,3-diene polymerization [147, 148]. A series of homoleptic lanthanide carboxylate complexes was reacted with trialkylaluminum reagents R_3Al (R = Me, Et), and alkylated rare-earth metal complexes were isolated and identified. The structural chemistry of the alkylated products revealed a tremendous effect of the benzoate substitution.

3.4.2.1
[Carboxylate]→[Alkyl] Exchange in $Ln(O_2CAr^{i-Pr})_3$ and $Ln(O_2CAr^{t-Bu})_3$

Addition of a slight excess of trimethylaluminum (3–4 equiv) to $\{Ln(O_2CAr^{i-Pr})_3\}_x$ yielded hexane insoluble (Ln = Y, Lu) or hexane soluble (Ln = La, Nd, Gd) Ln(III)/Al heterobimetallic complexes in moderate isolated yields of 35–45% (Scheme 20). A detailed investigation of the alkylated (minor) byproducts failed due to their similar solubilities in aliphatic and aromatic solvents. According to the ^{1}H NMR spectra of the crude reaction mixtures at least three different alkylated byproducts formed, however, homoleptic tetramethylaluminates $Ln[(\mu-Me)_2AlMe_2]_3$ [149] were not observed. The hexane insoluble yttrium and lutetium derivatives revealed a net composition of $[LnAl_3Me_8(O_2CAr^{i-Pr})_4]$ as identified by elemental analysis, and IR and NMR spectroscopy. Under similar reaction conditions the larger metal centers lanthanum, neodymium, and gadolinium readily formed hexane soluble heterobimetallic complexes. Elemental analysis, NMR spectroscopy, and X-ray crystallography revealed a molecular connectivity of $Ln[(O_2CAr^{i-Pr})_2(AlMe_2)]_2[AlMe_4]$ (Fig. 25). One molecule of hexane entrapped in the crystal lattice could not be removed under reduced pressure ($\sim 10^{-2}$ mbar). For the lanthanum derivative, a decoalescence temperature of the tetramethylaluminate ligand of 213 K was determined by means of variable-temperature NMR studies. Consistent with the decreased steric unsaturation and the slower alkyl exchange rate at the smaller yttrium center an aluminate decoalescence temperature of 263 K was observed [147].

Similar products were obtained via the reaction of carboxylates $\{Ln(O_2CAr^{i-Pr})_3\}_x$ (Ln = La, Nd) with 3–4 equivalents of triethylaluminum. Highly soluble heterobimetallic lanthanide carboxylates were crystallized in low yields and fully characterized. In the case of the smaller metal centers

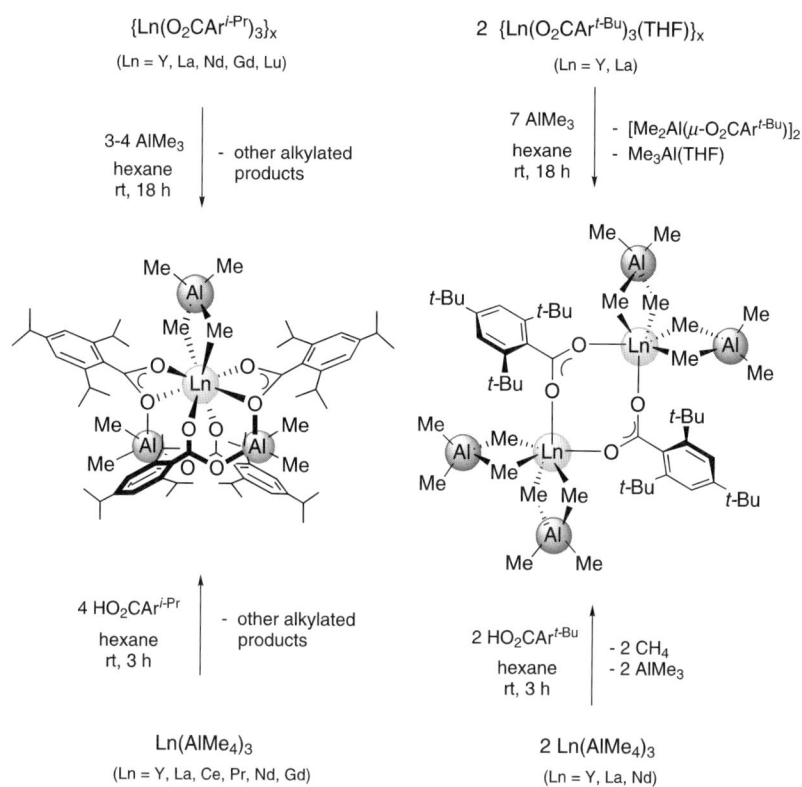

Scheme 20 Synthesis of heteroleptic tetramethylaluminate complexes via [carboxylate] → [alkyl] exchange (*top*) or [alkyl] → [carboxylate] exchange (*bottom*, "tetramethylaluminate route") [147, 148]

Y and Lu no alkylated products were isolated [148]. X-ray crystallography of both methyl and ethyl derivatives revealed the formation of eight-coordinate lanthanide metal centers with an unusual type of η^3-coordinated R$_2$Al-linked bis(carboxylate) ligand (Scheme 20).

Alkylation of complexes {Ln(O$_2$CAr$^{t\text{-}Bu}$)$_3$(THF)}$_x$ (Ln = Y, La) bearing the sterically more crowded *tert*.butyl-substituted carboxylate ligands produced exclusively hexane soluble products. Crystallization afforded "bis-alkylated" complexes with net composition [LnAl$_2$Me$_8$(O$_2$CAr$^{t\text{-}Bu}$)]. The soluble byproducts were investigated by proton NMR spectroscopy revealing the mere formation of two alkylated products [148]. The ambient temperature proton NMR spectra showed only one set of signals for the carboxylate ligands and broad resonances for the fluctuating tetramethylaluminate units. For the smallest metal center (Ln = Y), three well-resolved signals with an integral ratio of 24 : 12 : 12 were observed at 183 K, indicating three different types of aluminum-bonded methyl groups. The solid-state structures confirmed the formation of dimeric molecules $[Ln(AlMe_4)_2(\mu-O_2CAr^{t-Bu})]_2$ with a central eight-membered $\{LnOCO\}_2$-ring involving two bridging carboxylate ligands. Both of the lanthanide cations are 6-coordinated by two oxygen atoms of different carboxylate ligands and four methyl groups of the tetramethylaluminate units [148]. In order to synthesize larger amounts of the alkylated carboxylate complexes shown in Scheme 20, the "tetramethylaluminate route" was established (see Sect. 7.2.2). Accordingly, $Ln(AlMe_4)_3$ reacted with HO_2CAr^R (or $[Me_2Al(O_2CAr^R)]_2$) (R = i-Pr, t-Bu) to give the Ln(III)/Al heterobimetallic complexes in excellent yields [148].

3.4.2.2
[Carboxylate]→[Alkyl] Exchange in $Ln(O_2CAr^{Me})_3$ and $Ln(O_2CAr^{Ph})_3$

The alkylation reactions of lanthanide carboxylates bearing methyl-, phenyl-, or mesityl-substituted benzoate ligands with trialkylaluminum did not produce separable products, however, the tetramethylaluminate route facilitated the complete characterization of several of the generated and initially only spectroscopically identified species [148]. Combination of $Ln(AlMe_4)_3$ and $[Me_2Al(O_2CAr^{Me})]_2$ yielded two soluble compounds as indicated by proton NMR spectroscopy. Analysis of the ^1H and ^{13}C NMR spectra pointed at the formation of a lanthanide center that is surrounded by three dimethylaluminum-bridged biscarboxylate ligands (Scheme 21). Note that two of such in situ formed ancillary ligands were observed in complexes bearing *iso*propyl-substituted benzoates. In contrast, mono(carboxylate) bis(tetramethylaluminate) complexes $\{Ln[(\mu-Me)_2AlMe_2]_2(\mu-O_2CAr^{Ph})\}_2$ were generated in the reaction of phenyl-substituted benzoate complexes.

From these findings it was rationalized that not only electronic effects control the formation of differently alkylated products. The steric demand of the aliphatic and aromatic substituents in the 2,6-position of the benzoate

$Ln(AlMe_4)_3$ + 3 $[Me_2Al(O_2CAr^{Me})]_2$

$\xrightarrow[\text{rt, 18 h}]{\text{hexane}}$ $Ln[(O_2CAr^{Me})(\mu-AlMe_2)(Ar^{Me}CO_2)]_3$ + other products

(Ln = Y, La)

2 $Ln(AlMe_4)_3$ + 2 HO_2CAr^{Ph}

$\xrightarrow[\text{rt, 18 h}]{\text{hexane}}$ $\{Ln[(\mu-Me)_2AlMe_2]_2(\mu-O_2CAr^{Ph})\}_2$ + 2 CH_4 + 2 $AlMe_3$

(Ln = Y, La)

Scheme 21 Synthesis of heterobimetallic Ln(III)/Al carboxylate complexes via the "tetramethylaluminate route" [148]

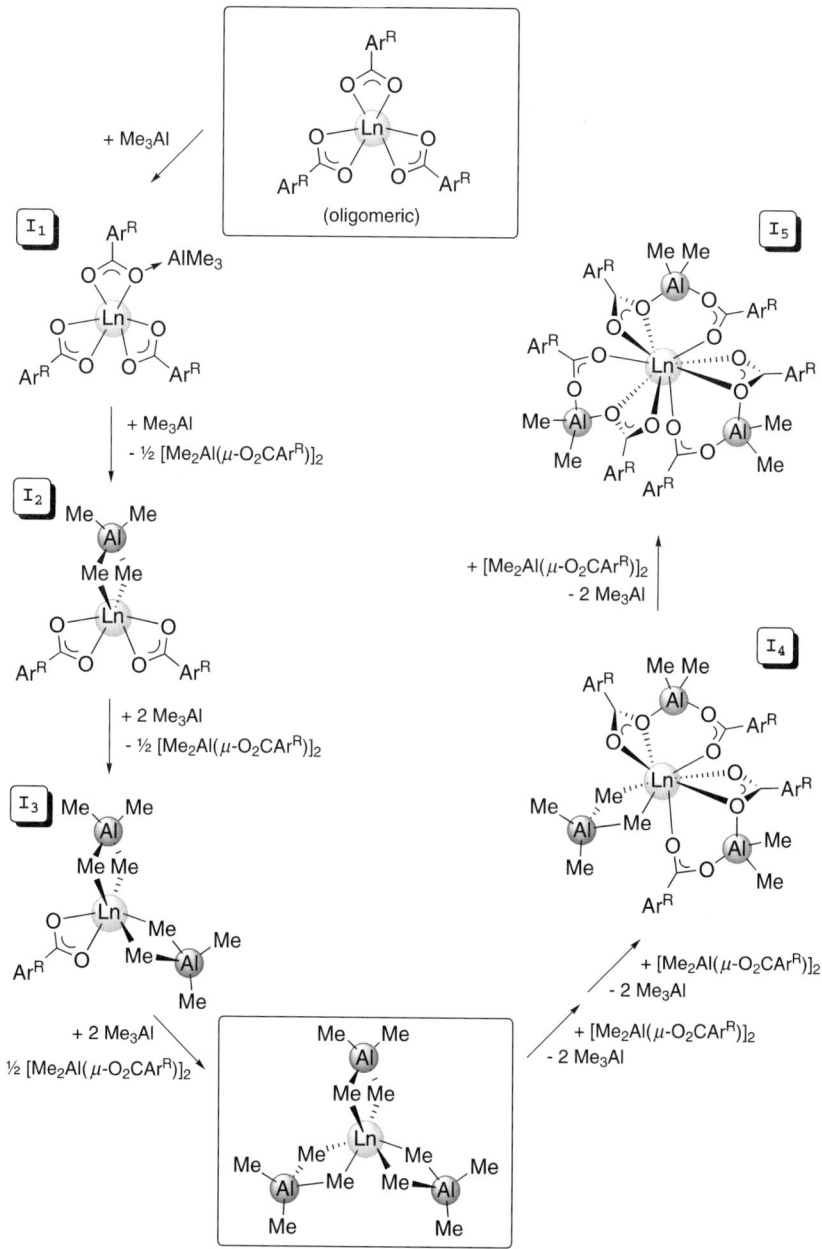

Scheme 22 [Carboxylate]→[alkyl] exchange: a mechanistic scenario of the formation of differently alkylated lanthanide carboxylate complexes (except for I_1, and I_2 all Ln complexes were completely characterized; $Ar^R = C_6H_2R'_3$-2,4,6 or $C_6H_3R'_2$-2,6) [148]

ligands seems to be a crucial factor determining the reaction of the homoleptic carboxylates with trialkylaluminum reagents. A mechanistic alkylation scenario is presented in Scheme 22. Accordingly, homoleptic tetramethylaluminate Ln(AlMe$_4$)$_3$ seem to be a key intermediate for the re-association reactions with separated dimethylaluminum carboxylate complexes. The active role of Ln(AlMe$_4$)$_3$ is supported by the "tetraalkylaluminate route" giving access to complexes I$_3$, I$_4$, and I$_5$. Furthermore, Ln(AlMe$_4$)$_3$ were found to display extraordinary catalytic activity in the *cis*-1,4-stereospecific polymerization of isoprene (Fischbach et al., 2006, personal communication) [150].

3.4.2.3
Reactivity toward R$_2$AlCl

In order to get a deeper insight into any structure-reactivity relationships of such rare-earth metal-based Ziegler catalysts, the interaction of the cationizing chloride-donating reagent R$_2$AlCl with the alkylated rare-earth metal centers was investigated. Accordingly, addition of equimolar amounts of Me$_2$AlCl to Ln[(O$_2$CAr$^{i\text{-}Pr}$)$_2$(AlMe$_2$)]$_2$(AlMe$_4$) immediately generated a white precipi-

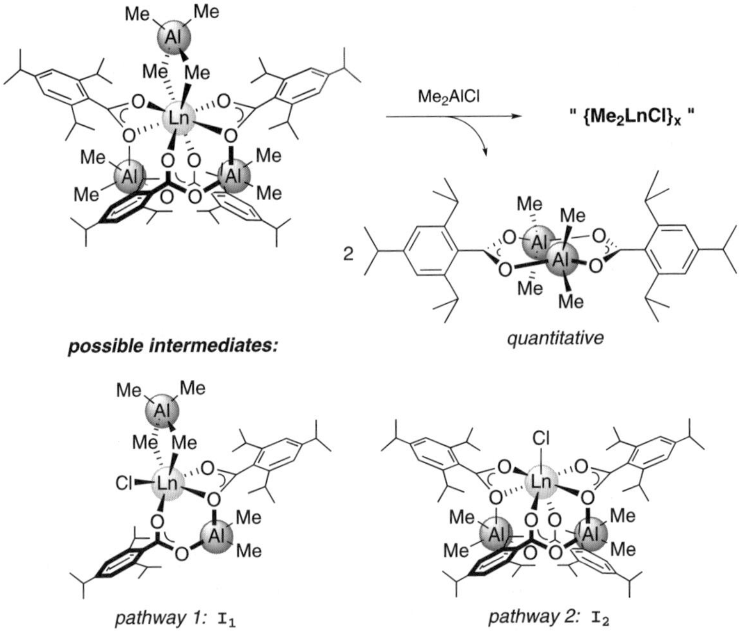

Scheme 23 Activation of Ln/Al heterobimetallic complexes Ln[(O$_2$CAr$^{i\text{-}Pr}$)$_2$ (AlMe$_2$)]$_2$ (AlMe$_4$) (Ln = Y, La, Nd) with Me$_2$AlCl and possible (*first*) intermediates (*bottom*) (adopted from [148])

tate. The NMR spectroscopic characterization revealed the formation of only one soluble product, which could be identified as the dimeric carboxylate-bridged dimethylaluminum complex $[Me_2Al(O_2CAr^{i-Pr})]_2$. The formation of two equivalents of $[Me_2Al(O_2CAr^{i-Pr})]_2$ per lanthanide metal center indicated that both chelating biscarboxylate ligands were cleaved by the addition of only one equivalent of Me_2AlCl. Although the microanalytical data of the insoluble lanthanide-containing products were not satisfactory, the amount of insoluble products pointed at the formation of polymeric/ionic "$\{Me_2LnCl\}_x$" as a possible polymerization initiating species. Two possible reaction pathways were presented. The first reaction intermediates of both pathways are shown in Scheme 23.

Given the thermal stability of "$LnMe_3$" [151] a possible ligand disproportionation reaction of "monocationic" "$\{Me_2LnCl\}_x$" into "$\{LnMe_3\}_x$" and "$\{LnCl_3\}_x$" was discussed. At $n(Cl):n(Ln)$ ratios ≥ 2, "dicationic" "$\{MeLnCl_2\}_x$" seems to be a chemically sound and realistic activation product [152–155]. However, given the ready availability of organoaluminum reagents in the reaction mixtures, equilibria reactions involving species "$\{Me_2LnCl\}_x(AlR_3)_y$" and "$\{MeLnCl_2\}_x(AlR_3)_z$" have to be taken into account as well.

3.4.2.4
Stereospecific Polymerization of Isoprene

Structure-reactivity studies were performed using binary Ziegler-type precatalyst mixtures (Fischbach et al., 2006, personal communication) [148]. Addition of various amounts of Et_2AlCl (as a cationizing reagent) to the aforementioned prealkylated Ln(III)/Al heterobimetallic benzoate complexes (Scheme 20) gave high-*cis*-1,4-polyisoprenes (Table 5). In order to elucidate the influence of the metal centers on the catalytic activity, the "monoalkylated" lanthanide carboxylate complexes $Ln[(O_2CAr^{i-Pr})_2(AlMe_2)]_2$ [$AlMe_4$] were used as precatalyst components at constant $n(Cl):n(Ln)$ ratios of 1:1. The yttrium and lutetium derivatives were completely inactive, whereas polyisoprene yields increased for the larger metal centers in the order La < Gd < Ce ~ Nd < Pr (Fischbach et al., 2006, personal communication) [148]. Tetraethylaluminate derivatives $Ln[(O_2CAr^{i-Pr})_2(AlEt_2)]_2[AlEt_4]$ displayed even higher activities and also gave almost quantitative conversions for the lanthanum derivative. Therefore, a distinct metal effect was not observed for $n(Cl):n(Ln)$ ratios of 1:1 and 2:1 within the reaction time of 24 h. This increased catalytic activity was rationalized on the enhanced solubility of the tetraethylaluminate versus the tetramethylaluminate Ln components. In general, stereospecificities and activities could be improved by increasing the amount of the halide-transferring cocatalyst [76, 125, 156–158]. At $n(Cl):n(Ln)$ ratios ≥ 3 polymer yields decreased again, due to the

Table 5 Polymerization of isoprene with "mono-alkylated" lanthanide carboxylate complex $Ln[(O_2CAr^{i-Pr})_2(AlR_2)]_2[AlR_4]$ (R = Me, Et)

Ln/R [a]	$n_{Cl}:n_{Ln}$ [b]	Yield [c] %	cis-1,4 [d] %	M_n [e] 10^{-3}	M_w [e] 10^{-3}	PDI [e]
Y/Me	1:1	0	–	–	–	–
La/Me	0:1	0	–	–	–	–
La/Me	1:1	12	95.1	137	558	4.08
La/Me	2:1	70	98.1	180	816	4.52
La/Me	3:1	21	>99	100	452	4.52
Nd/Me	0:1	0	–	–	–	–
Nd/Me	1:1	77	93.5	165	575	3.48
Nd/Me	2:1	>99	98.6	271	621	2.29
Nd/Me	3:1	98	>99	194	410	2.11
Gd/Me	1:1	24	92.0	112	589	5.27
Lu/Me	1:1	0	–	–	–	–
La/Et	1:1	97	94.3	78	290	3.73
La/Et	2:1	>99	96.6	91	454	4.98
La/Et	3:1	60	>99	66	217	3.27
Nd/Et	1:1	97	90.7	94	297	3.16
Nd/Et	2:1	>99	95.6	92	266	2.88
Nd/Et	3:1	>99	98.7	78	229	2.94

[a] General conditions: 8 mL hexane, 0.02 mmol precatalyst, 0–0.06 mmol Et_2AlCl (0 – 3 equiv), 20 mmol isoprene, 24 h, $T = 40\,°C$
[b] Catalyst preformation 30 min at ambient temperature
[c] Gravimetrically determined
[d] Measured by means of ^{13}C NMR spectroscopy
[e] Measured by means of size exclusion chromatography (SEC) against polystyrene standards

likely formation of higher halogenated and hence more insoluble lanthanide species [148].

The molecular weights of the isolated polymers showed relatively similar values for the tetramethylaluminate derivatives, however, the molecular weight distributions (MWD) strongly depend on the metal center (Table 5). All of the GPC elugrams showed polymodal mass distributions indicating two or three catalytically active species [148]. Monocarboxylate bis(tetramethylaluminate) complexes $[Ln(AlMe_4)_2(\mu\text{-}O_2CAr^{t-Bu})]_2$ displayed an even higher catalytic activity producing high-cis-polyisoprene in high to quantitative yields (Table 6). Molecular weights and molecular weight distributions were similar to those obtained via the mono(tetramethylaluminate) complexes, showing broader distributions for the less active lanthanum derivative(s) [148].

Table 6 Polymerization of isoprene with "bis-alkylated" lanthanide carboxylate complexes [Ln(AlMe$_4$)$_2$(μ-O$_2$CAr$^{t\text{-Bu}}$)]$_2$

Ln [a]	$n_{Cl} : n_{Ln}$ [b]	Yield [c] %	cis-1,4 [d] %	M_n [e] 10^{-3}	M_w [e] 10^{-3}	PDI [e]
La	1:1	83	97.5	125	593	4.76
La	2:1	> 99	> 99	175	564	3.23
La	3:1	86	> 99	57	307	5.37
Nd	1:1	> 99	98.7	277	787	2.84
Nd	2:1	> 99	> 99	250	705	2.82
Nd	3:1	> 99	> 99	165	462	2.79

[a] General conditions: 8 mL hexane, 0.02 mmol precatalyst, 0.02–0.06 mmol Et$_2$AlCl (1 – 3 equiv), 20 mmol isoprene, 24 h, $T = 40\,°C$
[b] Catalyst preformation 30 min at ambient temperature
[c] Gravimetrically determined
[d] Measured by means of ^{13}C NMR spectroscopy
[e] Measured by means of size exclusion chromatography (SEC) against polystyrene standards

4
Lanthanide Alkoxides and Aryloxides

Like the amide derivatives (see Sect. 6), rare-earth metal alkoxide (aryloxide) complexes Ln(OR)$_x$ are pseudo-organometallics exhibiting a Ln–

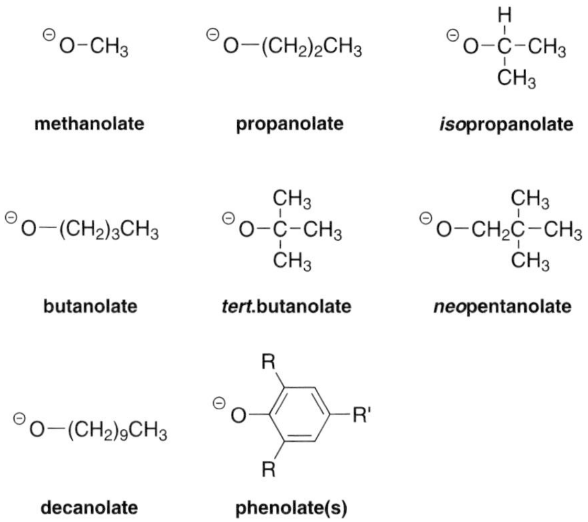

Chart 3 Alkoxide and aryloxide ligands used in lanthanide diene polymerization catalysts and model compounds

heteroatom bond which is readily hydrolyzed, in contrast to Ln–Cl and Ln–O(carboxylate) bond linkages. This well-known class of rare-earth metal compounds comprises highly soluble discrete complexes of well-defined nuclearity (mono-, di-, tri-, tetra-, penta-) depending on the bulkiness of the residue R [159, 160]. The diverse and multifaceted synthetic access combined with the intrinsic thermodynamically stable/kinetically labile bonding features make lanthanide alkoxides ideal components of Ziegler-type catalysts (Chart 3). The development and industrial application of diene polymerization catalysts based on lanthanide alkoxide precursors were mainly promoted by the Italian companies Anic, Montecatini, and Enichem.

4.1
Lanthanide Methanolates

The alkylation behavior of homoleptic lanthanide methanolate complexes is documented for a heterobimetallic Ln(III)/Al methoxide complex which was investigated by G. Lugli et al. [161]. Homoleptic neodymium(III) methanolate was reacted with trimethylaluminum in a n(Nd) : n(Al) ratio of 1 : 4 to give highly soluble Nd(OMe)$_3$(AlMe$_3$)$_4$ of well-defined stoichiometry. Elemental analysis and gas-volumetric analysis of the methyl group population via hydrolysis confirmed the overall composition, however, direct structural information was not provided. Nevertheless, the formation of a homoleptic tris(trimethylaluminum) adduct with "coordinated" Me$_3$Al, i.e., Nd[(μ-OMe)(μ-Me)AlMe$_2$]$_3$(AlMe$_3$), seems feasible. Polymerization of butadiene was performed in the presence of a series of Lewis acidic cocatalysts, including Me$_2$AlCl, EtAlCl$_2$, BCl$_3$, B(C$_6$F$_5$)$_3$, SnCl$_4$, and t-BuCl. Polybutadiene with predominantly cis-1,4-connectivity (70–99% cis) was obtained for all of the cocatalysts, except the "chloride-free" B(C$_6$F$_5$)$_3$ (24–70%) [161].

4.2
Lanthanide Propanolates

Diene polymerization via lanthanide propanolate-based catalytic systems was first reported by U. Pedretti et al. [162]. Several types of binary and ternary initiator systems have been developed since the late 1970s (see Table 7), including n- and isopropanolate ligands [162], homo- and heteroleptic lanthanide species [163–165], and different types of organoaluminum cocatalysts [162, 166–168]. Despite such extensive studies, structural evidence of the formation of alkylated rare-earth metal centers remained scarce.

Well-defined heterobimetallic complexes were isolated from Ln(Oi-Pr)$_3$/Et$_2$AlCl/Et$_3$Al and Ln(Oi-Pr)$_2$Cl/Et$_3$Al mixtures and characterized by means of elemental analysis, IR spectroscopy, and gas-volumetric analysis. The formation of complexes of general formula "(i-PrO)$_2$HLn$_2$Cl$_3$HAlEt$_2$" (Ln = Gd, Dy, Er, Tm) with a "two double bridged structure" was considered, how-

Table 7 Binary and ternary diene polymerization catalysts based on neodymium propanolates

Catalyst mixtures	Refs.
Nd(On-Pr)$_3$/i-Bu$_3$Al/AlBr$_3$	[162]
Nd(Oi-Pr)$_3$/i-Bu$_3$Al/EtAlCl$_2$	[162]
Nd(Oi-Pr)$_3$/Et$_3$Al/Et$_2$AlCl	[169]
Nd(Oi-Pr)$_3$/i-Bu$_3$Al/Et$_2$AlCl	[170]
Nd(Oi-Pr)$_{3-n}$Cl$_n$/Et$_3$Al	[163–165]
Nd(Oi-Pr)$_3$/i-Bu$_2$AlH/t-BuCl	[167]
Nd(Oi-Pr)$_3$/MAO/t-BuCl	[167]
Nd(Oi-Pr)$_3$/MAO	[166]
Nd(Oi-Pr)$_3$/i-Bu$_3$Al/[HNMe$_2$Ph][B(C$_6$F$_5$)$_4$]	[168]

ever, an unambiguous structural assignment was not possible [171]. According to Scheme 24 single crystals were isolated from a 2 : 20 : 3-mixture of Nd(Oi-Pr)$_3$, Et$_3$Al, and Et$_2$AlCl [170]. X-ray crystallography confirmed the formation of [Al$_3$Nd$_6$(μ_2-Cl)$_6$(μ_3-Cl)$_6$(μ_2-C$_2$H$_5$)$_9$(C$_2$H$_5$)$_5$(Oi-Pr)]$_2$ with alkylated rare-earth metal centers, showing three different types of seven-coordinated neodymium environments, with μ_2-Cl and μ_3-Cl bridges between the neodymium centers. Moreover, a very rare η^3-coordination mode of the AlEt$_4$ ligand could be identified. Butadiene polymerization studies performed with the isolated complex and the initial three-component mixture confirmed the "active species theory" [170]. Additional features of the heterobimetallic complex in butadiene polymerization were investigated, revealing the strong chain transfer ability of additional alkylaluminum reagents and higher reactivity for systems activated by methylaluminoxane [172–175].

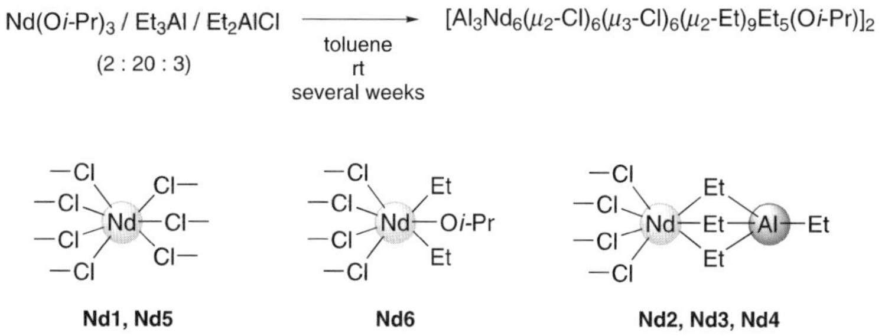

Scheme 24 Formation of a structurally characterized heterobimetallic Nd(III)/Al complex from Nd(Oi-Pr)$_3$ and alkylaluminum reagents. Three different types of metal environments are realized for six structurally independent neodymium cations Nd1–Nd6 [170]

It should be mentioned that S. Cesca and coworkers reported a heterobimetallic Ln/Al complex from a slow reaction of Ce(Oi-Pr)$_4$(HOi-Pr) with triethylaluminum as early as 1977 [176]. The crystalline cerium(III) complex (η^8-C$_8$H$_8$)Ce(μ-Oi-Pr)$_2$AlEt$_2$ was generated in the presence of cyclooctatetraene in low yields (30%). It was characterized by means of mass spectrometry, alcoholysis, deuterolysis, IR and NMR spectroscopy.

4.3
Lanthanide Butanolates and Pentanolates

A selection of multicomponent diene polymerization catalysts based on homo- and heteroleptic lanthanide alkoxide complexes with larger aliphatic groups, i.e., butanolates, pentanolates, and decanolates, is given in Table 8. Comprehensive studies were performed with the ternary system Nd(OR)$_3$/ i-Bu$_3$Al/Et$_2$AlCl showing high activities in isoprene polymerization. In the presence of a 30-fold excess of i-Bu$_3$Al and a n(Nd): n(Cl) ratio of 1 : 3 the catalytic activity decreased markedly depending on the alkoxy substituents R (n-C$_8$H$_{17}$ ≈ n-C$_6$H$_{13}$ ≥ n-C$_4$H$_9$ > i-C$_3$H$_7$ > i-C$_4$H$_9$) [170]. By using Gd(OR)$_3$/Et$_3$Al/Et$_2$AlCl catalyst mixtures, E. Kobayashi et al. observed higher activities for branched *neo*pentanolate ligands than linear n-pentanolates [177].

Structural evidence for the formation of alkylated lanthanide metal centers in the ternary catalyst mixtures stems predominantly from *tert*.butanolate complexes [178–181]. Fully characterized heterobimetallic *neo*pentanolate analogues were synthesized in recently [142].

Table 8 Binary and ternary diene polymerization catalysts based on lanthanide butanolates, pentanolates, and decanolates

Catalyst mixtures	Refs.
Nd(On-Bu)$_3$/i-Bu$_3$Al/AlBr$_3$	[162]
Nd(On-Bu)$_3$/i-Bu$_3$Al/Et$_2$AlCl	[170]
Nd(On-Bu)$_3$/i-Bu$_2$AlH/EtAlCl$_2$	[182]
Nd(Oi-Bu)$_3$/i-Bu$_3$Al/Et$_2$AlCl	[170]
Nd$_3$(Ot-Bu)$_9$(THF)$_2$/n-Hex$_2$Mg	[181]
Nd(Oi-C$_5$H$_{11}$)$_3$/i-Bu$_3$Al/AlBr$_3$	[162]
Gd(OCH$_2$$t$-Bu)$_3$/Et$_3Al/Et_2$AlCl	[177]
Nd(Oi-C$_{10}$H$_{21}$)$_3$/i-Bu$_3$Al/EtAlCl$_2$	[162]
Nd(Oi-C$_{10}$H$_{21}$)$_3$/i-Bu$_2$AlH/EtAlCl$_2$	[162]
Nd(Oi-C$_{10}$H$_{21}$)$_3$/i-Bu$_2$AlH/EtAlCl$_2$	[162]

A series of differently alkylated lanthanide metal complexes was obtained from the reaction of trimetallic $Y_3(Ot\text{-Bu})_7Cl_2(THF)_2$ with trimethylaluminum (Scheme 25) [178]. Heterobimetallic compounds **A**, **B**, and **C** were isolated via crystallization from hexane, toluene, or THF/DME mixtures, and characterized by means of elemental analysis, NMR spectroscopy, and X-ray crystallography. Thus, homo- and heteroleptic trimethylaluminum adducts $Y[(\mu\text{-}Ot\text{-Bu})(\mu\text{-Me})AlMe_2]_3$ (**A**, minor product), and $Y[(\mu\text{-}Ot\text{-Bu})(\mu\text{-Me})AlMe_2]_2(Ot\text{-Bu})(THF)$ (**B**) were identified as "chloride-free" products in the hexane soluble fraction. Additionally, crystallization of the hexane insoluble fraction from a THF/DME mixture revealed the generation of $Y[(\mu\text{-}Ot\text{-Bu})_2AlMe_2](Ot\text{-Bu})Cl(THF)_2$ (**C**), and $YCl_3(DME)_2$ (**D**). Similar reactions performed with both Et_3Al and Bu_3Al did not produce isolable and unequivocally identifiable products. Furthermore, alternative strategies to generate compounds **B** and **C** were not successful [178]. The quantitative formation of homoleptic trimethylaluminum adducts of general type $Ln[(\mu\text{-}OR)(\mu\text{-Me})AlMe_2]_3$, however, is achievable according to Scheme 26.

Correspondingly, treatment of homoleptic lanthanide *tert*.butanolate complexes $[Ln(Ot\text{-Bu})_3]_x$ (Ln = Y, Pr, Nd) [161, 179] and tetrameric *neo*pentanolates $[Ln(OCH_2t\text{-Bu})_3]_4$ (Ln = Y, La, Nd) [142] with stoichiometric amounts of Me_3Al (n(Ln):n(Al) = 1 : 3) generated highly soluble heterobimetallic complexes. Moreover, structurally related and fully characterized TMA adduct complexes $(C_5H_4SiMe_3) Y[(\mu\text{-}Ot\text{-Bu})(\mu\text{-Me})AlMe_2]_2$ [180] and

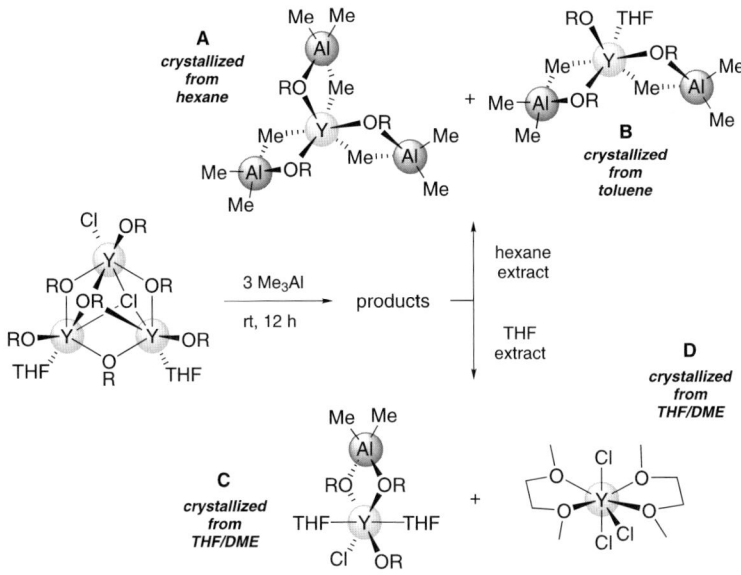

Scheme 25 Reactivity of trimetallic $Y_3(OR)_7Cl_2(THF)_2$ (Rt-Bu) with Me_3Al [178]

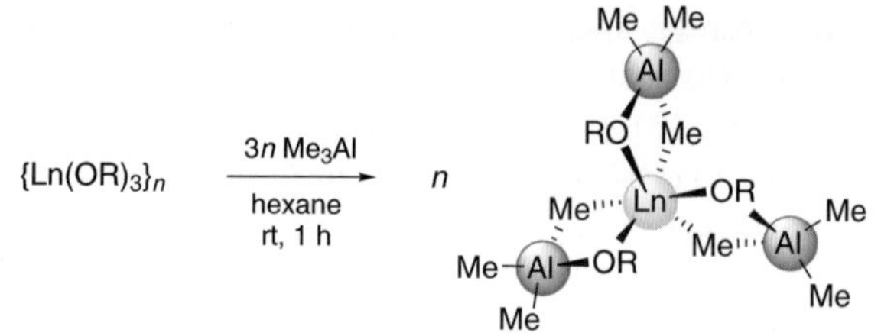

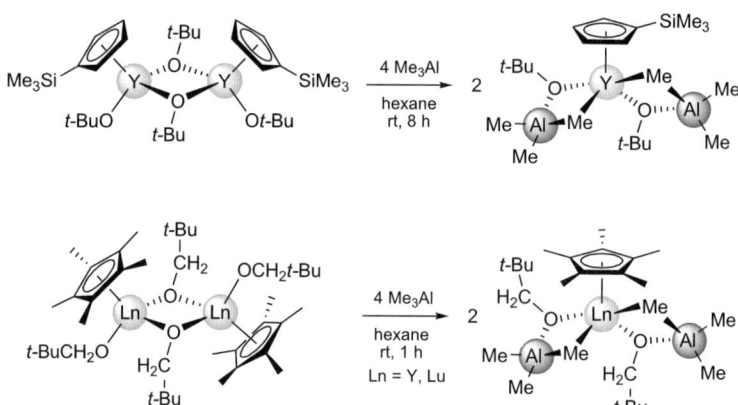

Scheme 26 Quantitative formation of homoleptic TMA-adducts from {Ln(OR)$_3$}$_n$ (R = t-Bu: Ln = Y, Pr, Nd; R = CH$_2$t-Bu: Ln = Y, La, Nd) and Me$_3$Al [142, 161, 179]

Scheme 27 Formation of half-lanthanidocene TMA-adduct complexes [142, 180]

(C$_5$Me$_5$)Ln[(μ-OCH$_2$t-Bu)(μ-Me)AlMe$_2$]$_2$ (Ln = Y, Lu) [142] were obtained from the reactions of cyclopentadienyl-substituted derivatives. According to Scheme 27, the addition of stoichiometric amounts of trimethylaluminum (n(Ln):n(Al) = 1:2) to the dimeric half-lanthanidocene bis(alkoxide) complexes quantitatively gave bis(TMA) adducts.

Activation of such homoleptic trimethylaluminum adduct complexes with 1–3 equivalents of the chloride source Et$_2$AlCl produced binary catalyst mixtures revealing the expected activity and cis-1,4-stereospecificity in isoprene polymerization (Table 9) [142]. High-yield polymeric materials with relatively narrow molecular weight distributions were obtained. No activity was observed for a n(Cl):n(Ln) ratio of 3:1.

Although alkoxide→alkylaluminum adduct formation seems to be the preferred reaction pathway of lanthanide alkoxides with trialkylaluminum reagents, the corresponding reaction of decamethylytterbocene derivative (C$_5$Me$_5$)$_2$Yb(Ot-Bu)(NH$_3$) takes a different course. Addition of three equivalents of Me$_3$Al yielded a purple 2:1-adduct as confirmed by elemental analysis and ^{1}H NMR spectroscopy. On the basis of the X-ray crys-

Table 9 Polymerization of isoprene with homoleptic lanthanide alkoxides $Ln(OCH_2t\text{-}Bu)_3(AlMe_3)_3$ (Ln = La, Nd)

Ln [a]	$n_{Cl} : n_{Ln}$ [b]	Yield [c] %	cis-1,4 [d] %	M_n [e] 10^{-3}	M_w [e] 10^{-3}	PDI [e]
La	1:1	78	98.9	270	641	2.37
La	2:1	> 99	> 99	517	954	1.85
La	3:1	0	–	–	–	–
Nd	1:1	90	98.6	339	693	2.05
Nd	2:1	> 99	> 99	384	670	1.74
Nd	3:1	0	–	–	–	–

[a] General conditions: 8 mL hexane, 0.02 mmol precatalyst, 0.02–0.06 mmol Et_2AlCl (1–3 equiv), 20 mmol isoprene, 24 h, T = 40 °C
[b] Catalyst preformation 30 min at ambient temperature
[c] Gravimetrically determined
[d] Measured by means of ^{13}C NMR spectroscopy
[e] Measured by means of size exclusion chromatography (SEC) against polystyrene standards

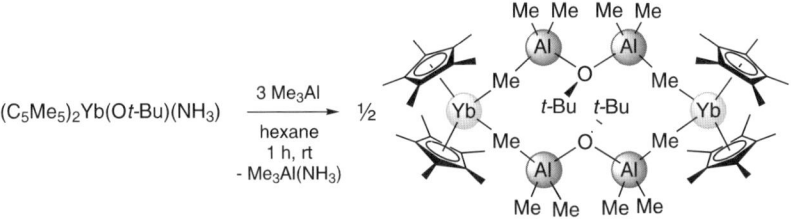

Scheme 28 Reaction of $(C_5Me_5)_2Yb(Ot\text{-}Bu)(NH_3)$ with Me_3Al [183]

tallographic characterization of the similarly behaving S-p-tolyl derivative $[(C_5Me_5)_2Yb(S\text{-}p\text{-}tolyl)(AlMe_3)_2]_2$ (Fig. 19) the formation of a dimeric molecule was rationalized, with two "cationic" metallocene fragments linked by two anionic $[Me_3Al - Ot\text{-}Bu - AlMe_3]^-$ units [183].

4.4
Lanthanide Phenolates

Aryloxide (phenolate) ligands provide rigid and versatile ancillary ligand sets in order to study the multifunctional reactivity of alkylaluminum and alkylmagnesium reagents toward Ln – OR moieties. Several types of symmetrically substituted phenolate ligands — summarized in Chart 4 — were employed for the synthesis of a variety of heterobimetallic lanthanide(III) and lanthanide(II) metal complexes. Alkylation reactions revealed the preferred

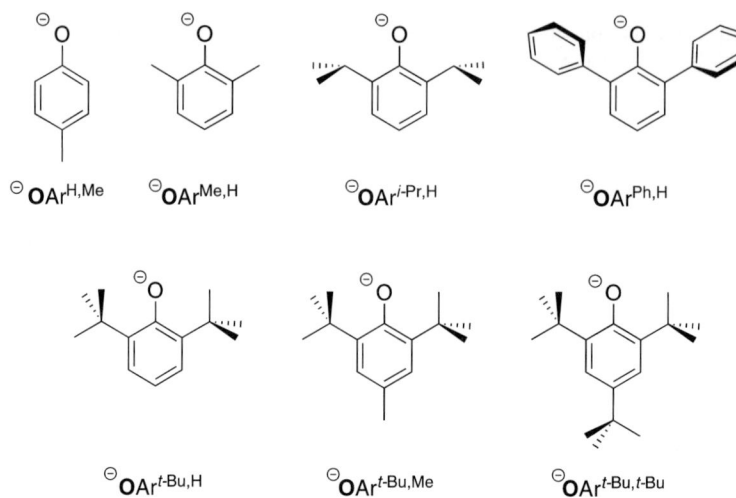

Chart 4 Symmetrically substituted phenolates used as model ligands to investigate the alkylation behavior of lanthanide alkoxides

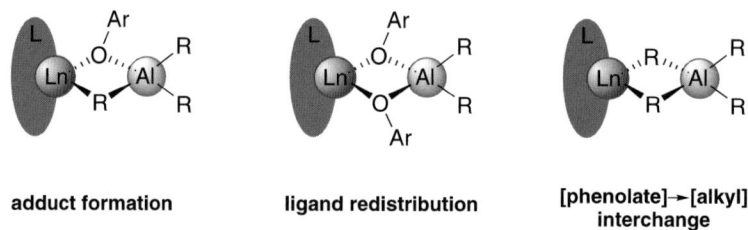

Fig. 2 Reactivity of lanthanide phenolates toward trialkylaluminum reagents R_3Al

formation of three types of lanthanide coordination environments as indicated by adduct formation, ligand redistribution, and [phenolate] → [alkyl] interchange (Fig. 2).

4.4.1
Lanthanide(III) Phenolates

4.4.1.1
Ligand Redistribution

Alkylated rare-earth metal complexes with rare-earth metal centers surrounded exclusively by oxygen donor ligands were reported from facile ligand redistribution processes in 2,6-dimethylphenolate/trialkylaluminum mixtures. As shown in Scheme 29 for the yttrium derivatives $Y(OAr^{Me,H})_2[(\mu\text{-}OAr^{Me,H})_2AlR_2](THF)_2$ (R = Me, Et), heterobimetallic 1 : 1-species were ac-

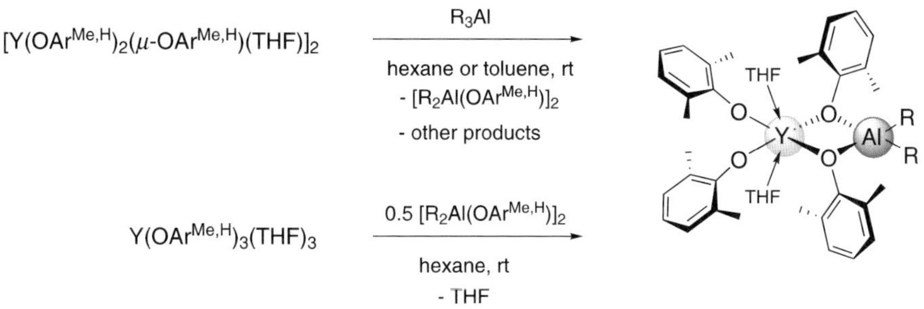

[Y(OAr^{Me,H})$_2$(μ-OAr^{Me,H})(THF)]$_2$

$\xrightarrow[\substack{\text{hexane or toluene, rt} \\ - [R_2Al(OAr^{Me,H})]_2 \\ - \text{other products}}]{R_3Al}$

Y(OAr^{Me,H})$_3$(THF)$_3$

$\xrightarrow[\substack{\text{hexane, rt} \\ - \text{THF}}]{0.5\,[R_2Al(OAr^{Me,H})]_2}$

Scheme 29 Formation of Y(OAr^{Me,H})$_2$[(μ-OAr^{Me,H})$_2$AlR$_2$](THF)$_2$ (R = Me, Et) [184]

cessible for small (Ln = Y, Yb) and large metal centers (Ln = Nd) [184]. NMR spectroscopy and X-ray structure analysis confirmed the formation of an anionic η^2-coordinated [(μ-OAr^{Me,H})$_2$AlR$_2$] ligand, previously observed for one of the products of the reaction of Y$_3$(Ot-Bu)$_7$Cl$_2$(THF)$_2$ with Me$_3$Al [178]. Furthermore, single crystals of the homoleptic derivative Nd[(μ-OAr^{H,Me})$_2$AlMe$_2$]$_3$ were obtained from the reaction of K$_3$Nd$_2$(OAr^{H,Me})$_9$(THF)$_7$ with excess trimethylaluminum (Fig. 20) [185]. Despite the formation of several alkylated byproducts in either reaction, only dimeric [R$_2$Al(OAr^{Me,H})]$_2$ and [Me$_2$Al(OAr^{H,Me})]$_2$, respectively, were unequivocally identified in the crude mixtures. Thus, initial adduct formation and subsequent cleavage of some of the Ln – OAr bonds, associated with the formation of truly alkylated metal centers (i.e., Ln – C species) seems to be plausible. Spectroscopic or structural evidence of both species was not obtained in these particular systems [184, 185].

4.4.1.2
Adduct Formation

The first structurally evidenced trialkylaluminum adducts of lanthanide phenolate complexes were reported by J.C. Gordon et al. by utilizing sterically encumbered 2,6-di*iso*propylphenolate derivatives [186–188]. Addition of four equivalents of trimethyl- or triethylaluminum to the π-arene-bridged homoleptic dimers [Ln(OAr$^{i\text{-}Pr,H}$)$_2$(μ-OAr$^{i\text{-}Pr,H}$)]$_2$ (Ln = Y, La, Nd, Sm) yielded Ln(III)/Al heterobimetallic compounds Ln(OAr$^{i\text{-}Pr,H}$)[(μ-OAr$^{i\text{-}Pr,H}$)(μ-R)AlR$_2$]$_2$ (R = Me, Et), with a n(Ln):n(Al) ratio of 1 : 2 (Scheme 30, Fig. 16). Preferentially, adduct formation stopped after the addition of two trialkylaluminum molecules. In the presence of larger amounts of AlMe$_3$, significant amounts of peralkylated Ln[(μ-Me)$_2$AlMe$_2$]$_3$ as well as dimeric [Me$_2$Al(OAr$^{i\text{-}Pr,H}$)]$_2$ and one additional unidentified alkylated species were observed for the yttrium derivative [188]. Furthermore, thermal instability was reported for the lanthanum and samarium derivatives, presumably leading to a spectroscopically not observed lanthanide dialkyl species

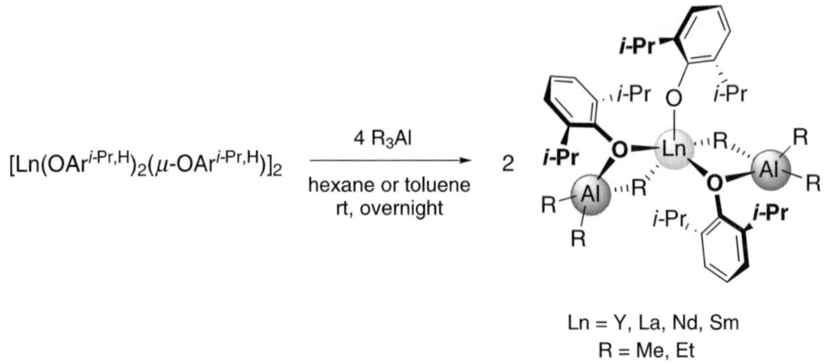

[Ln(OAr$^{i\text{-Pr,H}}$)$_2$(μ-OAr$^{i\text{-Pr,H}}$)]$_2$ $\xrightarrow[\text{rt, overnight}]{\text{4 R}_3\text{Al}}$ 2 [complex shown]

hexane or toluene
rt, overnight

Ln = Y, La, Nd, Sm
R = Me, Et

Scheme 30 Formation of bis(trialkylaluminum) adduct complexes Ln(OAr$^{i\text{-Pr,H}}$)[(μ-OAr$^{i\text{-Pr,H}}$)(μ-R)AlR$_2$]$_2$ (Ln = Y, La, Nd, Sm; R = Me, Et) [186–188]

"R$_2$Ln(OAr$^{i\text{-Pr,H}}$)". This was supported by the isolation of byproduct [Me$_2$Al(OAr$^{i\text{-Pr,H}}$)]$_2$ [187].

4.4.1.3
[Phenolate] → [Alkyl] Interchange

Similar adduct formation reactivity was observed for the bulkier 2,6-di-*tert*.butylphenolate ligands OAr$^{t\text{-Bu,R}'}$ (R' = H, Me, *t*-Bu) and the larger lanthanide metals lanthanum and neodymium. However, even the use of a large excess of trimethylaluminum yielded mono(TMA) adducts Ln(OAr$^{t\text{-Bu,R}'}$)$_3$(AlMe$_3$) exclusively (Scheme 31) [188]. The corresponding yttrium and lutetium phenolate complexes displayed a completely different reaction behavior. Under identical reaction conditions mono(tetramethylaluminate) complexes of type Ln(OAr$^{t\text{-Bu,R}'}$)$_2$[(μ-Me)$_2$AlMe$_2$] (Ln = Y, Lu; R' = H, Me, *t*-Bu) (Fig. 8) formed along with [Me$_2$Al(OAr$^{t\text{-Bu,R}'}$)]$_2$ and MeAl(OAr$^{t\text{-Bu,R}'}$)$_2$ (Fischbach et al., 2006, personal communication) [188]. The [aryloxide]→[alkyl] interchange reaction was found to be the predominant pathway for the smaller lanthanide metal centers. Addition of only one equivalent of trimethylaluminum did not produce isolable mono(TMA) adducts, and no "higher-alkylated" species were observed in the presence of larger amounts of the organoaluminum reagent. Furthermore, half-lanthanidocene complexes of type (C$_5$Me$_4$R'')Ln(OAr$^{t\text{-Bu,R}'}$)[(μ-Me)$_2$AlMe$_2$] were generated in the reaction of (C$_5$Me$_4$R'')Ln(OAr$^{t\text{-Bu,R}'}$)$_2$ (Ln = Y, Lu; R' = H, Me; R'' = H, Me) and excess Me$_3$Al (Scheme 32) [142, 188].

From the different alkylation products isolated and identified in binary Ln(OAr)$_3$/R$_3$Al mixtures, a mechanistic scenario with several reaction pathways was proposed (Scheme 33). The size of the metal center and the steric bulk of the aryloxide ligands were spotted as crucial factors [188]. Accordingly, initial attack of TMA onto homoleptic complexes Ln(OAr)$_3$ appeared

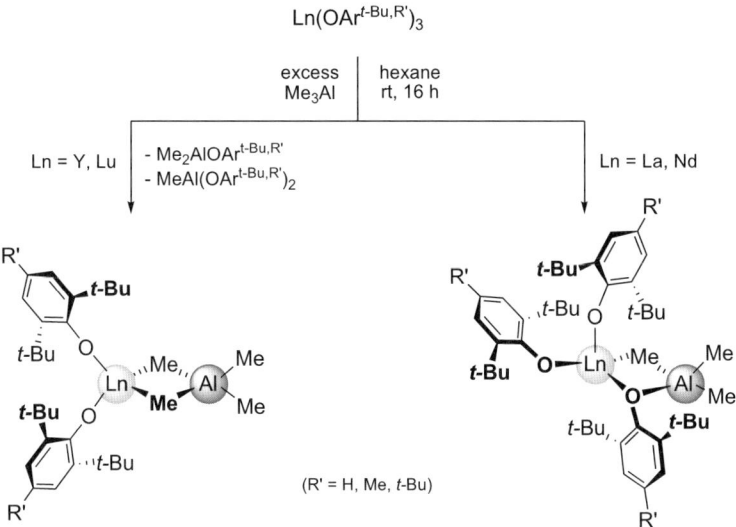

Scheme 31 Formation of mono(TMA) adducts and mono(tetramethylaluminate) complexes [188]

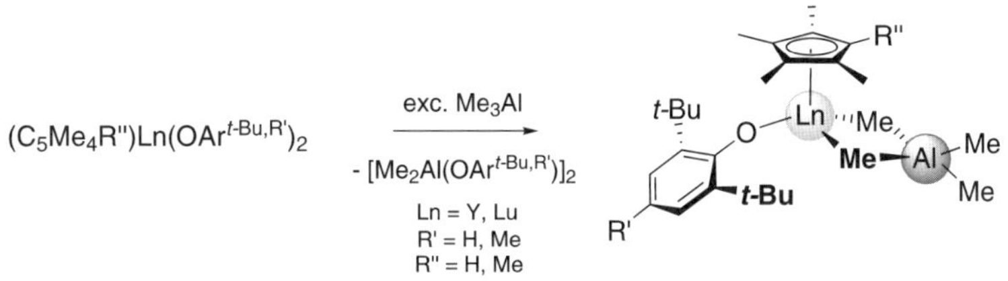

Scheme 32 Formation of heteroleptic half-lanthanidocene tetramethylaluminate complexes via [phenolate]→[alkyl] interchange [142, 188]

to be the rate-determining step, yielding stable mono(TMA) adducts I_1 for *ortho-t*-Bu$_2$-substituted ligands and large metal centers. The formation of mono(tetramethylaluminate) complexes I_2 via [phenolate]→[alkyl] interchange is the predominate pathway for the smaller lanthanide metals. Not only for steric reasons a dissociative mechanism was favored, with steric oversaturation forcing Me$_2$AlOAr dissociation. This is in accord with the decreased oxophilicity of the smaller rare-earth metal centers (disruption energy of metal monoxides, $D_0(\text{MO})$: La > Y > Lu ≫ Al) [189]. Furthermore, the formation of anionic [(μ-OAr)$_2$AlMe$_2$]$^-$ moieties can be explained by the re-addition of dimeric [Me$_2$Al(μ-OAr)]$_2$ to intermediates Ln(OAr)$_2$Me or Ln(OAr)$_2$(AlMe$_4$) (= I_3), respectively. In the presence of the less bulky *ortho-i*-Pr$_2$-substituted phenolates bis(TMA) adducts I_4 are preferentially formed. The kinetic control of the alkylation reaction was finally evidenced by the

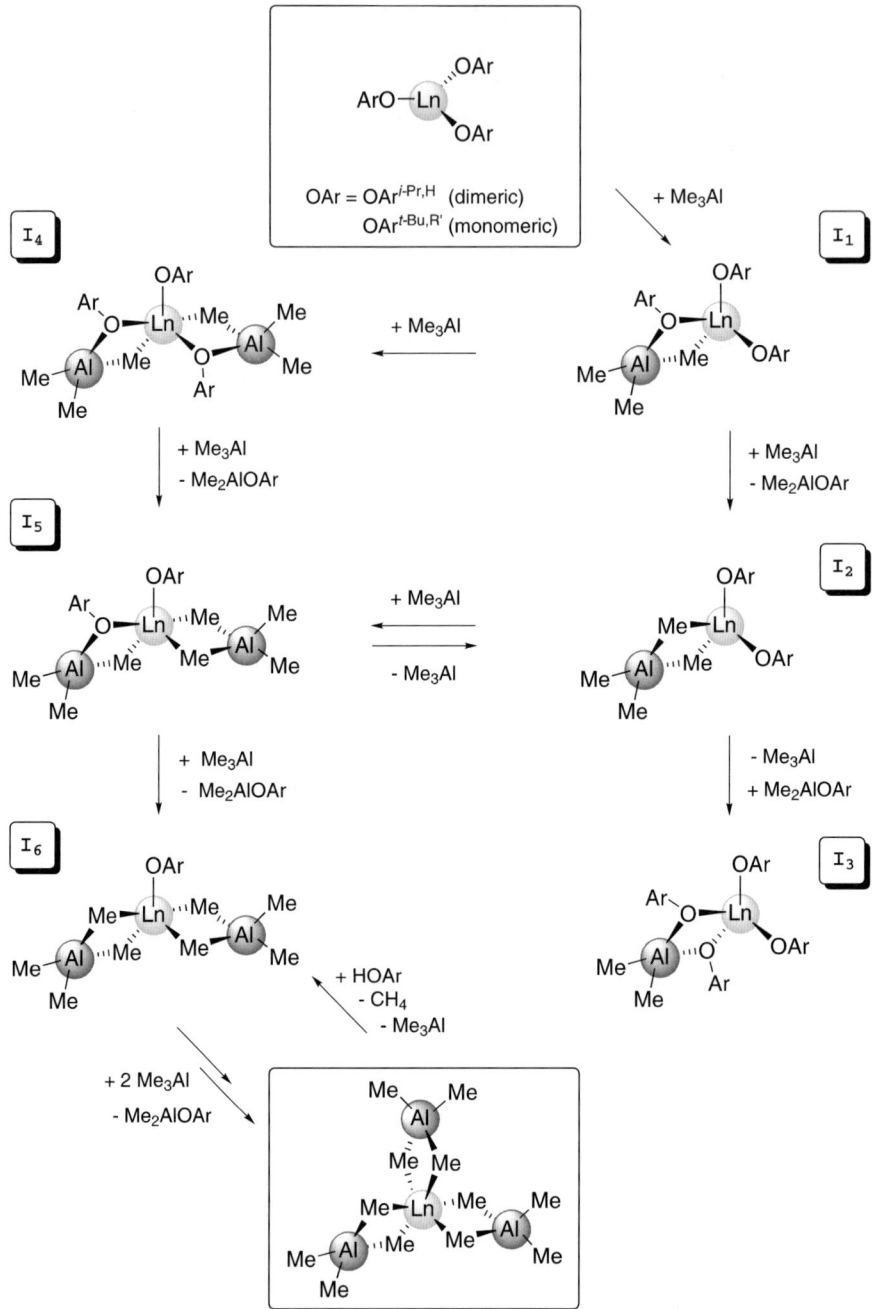

Scheme 33 Mechanistic scenario for the formation of differently alkylated lanthanide aryloxide complexes (adopted from [188])

generation of peralkylated species $Ln[(\mu\text{-Me})_2AlMe_2]_3$, probably via a series of other intermediates, including I_5 and I_6. Note, that intermediate I_6, i.e., the mono(aryloxide) bis(tetramethylaluminate) $Y(OAr^{t\text{-Bu,H}})(AlMe_4)_2$, was identified as one of the major products from a 1 : 1 reaction of homoleptic $Y[(\mu\text{-Me})_2AlMe_2]_3$ and $HOAr^{t\text{-Bu,H}}$ [142].

4.4.1.4
Precatalyst Activation and Stereospecific Isoprene Polymerization

Isoprene polymerization studies performed with discrete mono- and bis(trimethylaluminum) adducts, $Ln(OAr^{t\text{-Bu,Me}})_3(AlMe_3)$ and $Ln(OAr^{i\text{-Pr,H}})_3$ $(AlMe_3)_2$, revealed the importance of using excess alkylaluminum reagents in active catalyst mixtures. As summarized in Table 10, mono(TMA) adducts did not produce polyisoprene upon "activation" with 1–3 equivalents of Et_2AlCl. Although NMR spectroscopic studies pointed out the formation of several unidentified reaction products, the generation of catalytically in-

Table 10 Polymerization of isoprene with lanthanide aryloxides $Ln(OAr^{t\text{-Bu,Me}})_3(AlMe_3)$ and $Ln(OAr^{i\text{-Pr,H}})_3(AlMe_3)_2$

Ln [a]	$n_{Cl}:n_{Ln}$ [b]	Yield [c] %	cis-1,4 [d] %	M_n [e] 10^{-3}	M_w [e] 10^{-3}	PDI [e]
$Ln(OAr^{t\text{-Bu,Me}})_3(AlMe_3)$:						
La	1:1	0	–	–	–	–
La	2:1	0	–	–	–	–
La	3:1	0	–	–	–	–
Nd	1:1	0	–	–	–	–
Nd	2:1	0	–	–	–	–
Nd	3:1	0	–	–	–	–
$Ln(OAr^{i\text{-Pr,H}})_3(AlMe_3)_2$:						
La	1:1	53	98.0	76	325	4.26
La	2:1	54	> 99	390	870	2.23
La	3:1	33	> 99	76	318	4.17
Nd	1:1	92	98.9	355	744	2.10
Nd	2:1	> 99	> 99	223	453	2.03
Nd	3:1	38	> 99	107	371	3.46

[a] General conditions: 8 mL hexane, 0.02 mmol precatalyst, 0.02–0.06 mmol Et_2AlCl (1–3 equiv), 20 mmol isoprene, 24 h, $T = 40\,°C$
[b] Catalyst preformation 30 min at ambient temperature
[c] Gravimetrically determined
[d] Measured by means of ^{13}C NMR spectroscopy
[e] Measured by means of size exclusion chromatography (SEC) against polystyrene standards

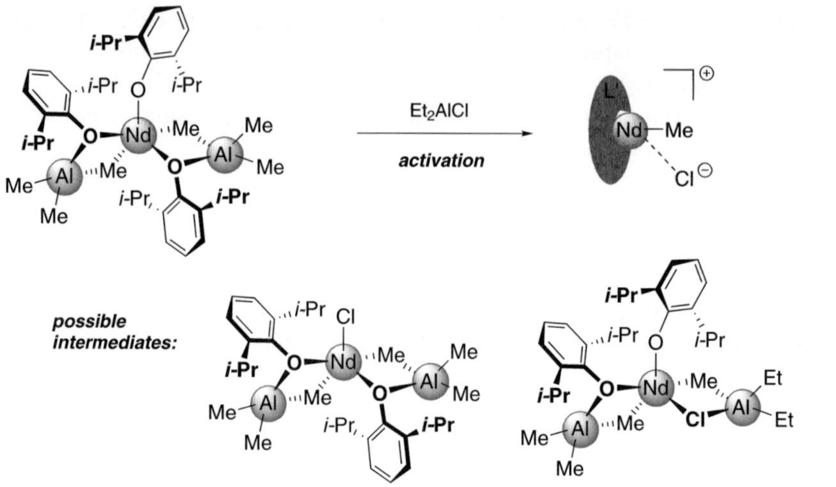

Scheme 34 Et$_2$AlCl-initiated [aryloxide]→[alkyl] transformation – a plausible "activation" step [142]

active [Ln(OAr$^{t\text{-Bu,Me}}$)$_2$Cl]$_x$ seems to be most likely. Thus, dimeric [R$_2$Al(OAr$^{t\text{-Bu,Me}}$)]$_2$ (R = Me, Et) were identified as major components in the 1:1-reactions with Me$_2$AlCl and Et$_2$AlCl, respectively. On the other hand, bis(TMA) adducts Ln(OAr$^{i\text{-Pr,H}}$)$_3$(AlMe$_3$)$_2$ gave active and highly cis-1,4-stereospecific catalyst mixtures. Once again, mechanistic NMR studies evidenced the formation of several activation products, with [R$_2$Al(OAr$^{i\text{-Pr,H}}$)]$_2$ being one of the major components. An [aryloxide]→[chloride] transfer (see Scheme 34) seems reasonable, however, was not structurally evidenced.

4.4.2
Lanthanide(II) Phenolates

Adduct formation and [phenolate]→[alkyl] interchange are the predominant reaction pathways observed in binary Ln(OAr)$_2$/R$_3$Al mixtures. As shown for a series of 2,6-di-*tert*.butyl- and diphenyl-substituted phenolate complexes of the divalent lanthanide metal centers samarium and ytterbium, homoleptic bis(trialkylaluminum) adducts Ln[(μ-OAr)(μ-R)AlR$_2$]$_2$ were generated in good yields (Scheme 35) [188]. The enhanced steric unsaturation of the larger divalent metal centers ensures the coordination of two R$_3$Al molecules, previously observed in Yb(II) and Sm(II) bis(trimethylsilyl)amide complexes, Ln[N(SiMe$_3$)$_2$]$_2$(AlMe$_3$)$_2$ (Fig. 18) [190].

Depending on the type of phenolate substituent in the 4-position (H vs. Me vs. *t*-Bu) varying amounts of peralkylated Yb(II) aluminate [Yb(AlMe$_4$)$_2$]$_x$ were obtained as byproduct. The sterically less crowded *i*-Pr$_2$-substituted

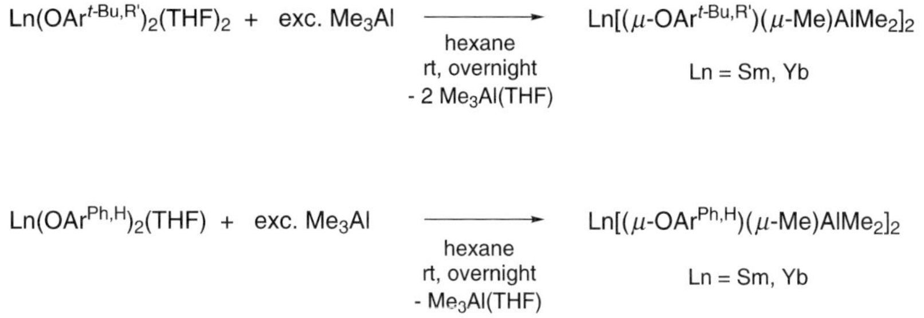

Ln(OAr$^{t\text{-Bu},R'}$)$_2$(THF)$_2$ + exc. Me$_3$Al $\xrightarrow[\text{- 2 Me}_3\text{Al(THF)}]{\substack{\text{hexane} \\ \text{rt, overnight}}}$ Ln[(μ-OAr$^{t\text{-Bu},R'}$)(μ-Me)AlMe$_2$]$_2$

Ln = Sm, Yb

Ln(OAr^{Ph,H})$_2$(THF) + exc. Me$_3$Al $\xrightarrow[\text{- Me}_3\text{Al(THF)}]{\substack{\text{hexane} \\ \text{rt, overnight}}}$ Ln[(μ-OAr^{Ph,H})(μ-Me)AlMe$_2$]$_2$

Ln = Sm, Yb

Scheme 35 Adduct formation in the reaction of lanthanide(II) phenolates (Ln = Sm, Yb; R' = H, Me, t-Bu) with excess of trimethylaluminum [188]

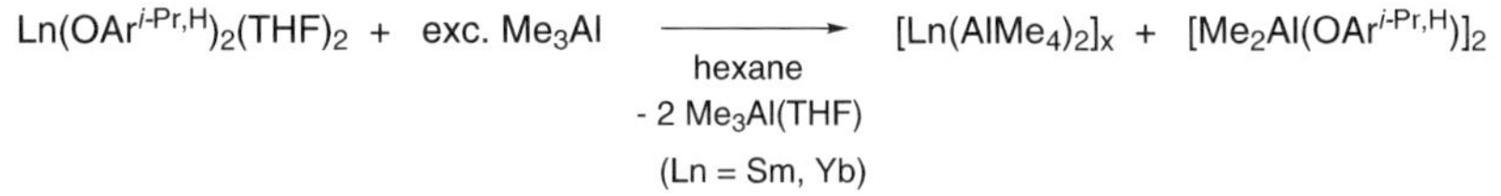

Ln(OAr$^{i\text{-Pr},H}$)$_2$(THF)$_2$ + exc. Me$_3$Al $\xrightarrow[\text{- 2 Me}_3\text{Al(THF)}]{\text{hexane}}$ [Ln(AlMe$_4$)$_2$]$_x$ + [Me$_2$Al(OAr$^{i\text{-Pr},H}$)]$_2$

(Ln = Sm, Yb)

Scheme 36 Peralkylation of lanthanide(II) phenolates [188]

phenolates afforded homoleptic tetramethylaluminates in quantitative yields (Scheme 36).

4.5
Other Systems

A mixed-alkylated heterobimetallic Ln/Al complex was reported by the Evans group [191]. In a two-step reaction the donor-functionalized europium(II) alkoxide [Eu(OCH$_2$CH$_2$OMe)$_2$]$_n$ was combined with [Me$_2$Al(OCH$_2$CH$_2$OMe)]$_2$ and trimethylaluminum to yield a polymetallic Eu(II)/Al(III) compound (Scheme 37, Fig. 31).

Binary neodymium alk(aryl)oxide/dialkylmagnesium diene polymerization catalysts were reported by J.-F. Carpentier and coworkers [181, 192]. The homopolymerization of butadiene, and copolymerization with styrene and

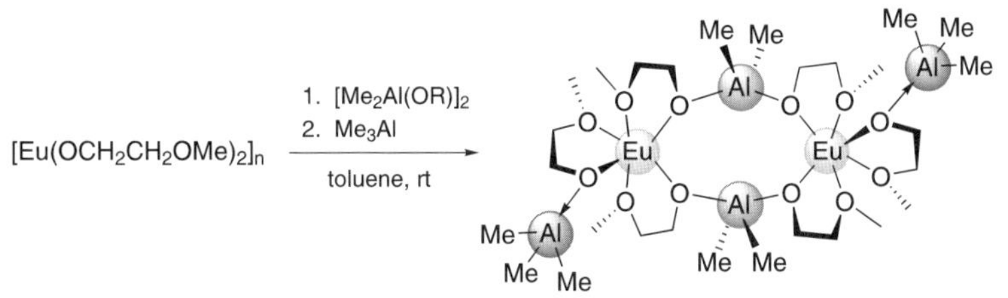

Scheme 37 Synthesis of a heterobimetallic Eu(II)/Al(III) compound from [Eu(OCH$_2$CH$_2$OMe)$_2$]$_n$ [191]

glycidyl methacrylate were investigated, revealing high-*trans*-1,4-stereospecificities, as well as higher activities and lower polydispersities for aryloxide-based catalyst precursors (Table 11). Since the in situ formation of lanthanide alkyl species in [Nd(OR')$_3$]$_x$/R$_2$Mg mixtures was evidenced [193–195], initial formation of alkylated compound [Nd(OR')$_2$R] and bimetallic Nd/Mg species [Nd(OR')$_2$(μ-R)$_2$MgR] were suggested and an activation and polymerization mechanism was presented [181].

The isolation and structural characterization of alkylated species was achieved when trinuclear Nd$_3$(μ_3-O*t*-Bu)$_2$(μ_2-O*t*-Bu)$_3$(μ-O*t*-Bu)$_4$(THF)$_2$ and monomeric Nd(OAr$^{t\text{-Bu,Me}}$)$_3$(THF) were reacted with Mg(CH$_2$SiMe$_3$)$_2$ (Et$_2$O) [192]. As shown in Scheme 38, a heterobimetallic Nd/Mg complex was crystallized from toluene/MMA mixtures (0.03 equiv MMA vs. Nd) in 30–40% yield after a few weeks. X-ray crystallography confirmed the

Table 11 Polymerization of butadiene with magnesium-based catalyst mixtures (adopted from [181])

$n_{Nd} : n_{Mg}$ [a]	$n_{BD} : n_{Mg}$ [a]	T °C	t h	yield [b] %	M_n [c]	PDI [c]
Nd$_3$(O*t*-Bu)$_9$(THF)$_2$/(*n*-Hex)$_2$Mg :						
1.0	100	20	17	5	1900	1.77
1.0	100	60	17	22	2700	2.40
0.2	20	60	17	95	1700	4.40
0.2	200	60	17	47	10 300	5.10
Nd(OAr$^{t\text{-Bu,Me}}$)$_3$(THF)/(*n*-Hex)$_2$Mg :						
1.0	100	20	17	99	5800	1.26
0.2	20	20	17	98	1100	1.15
1.0	200	20	2	60	6300	1.18
1.0	1000	20	17	99	49 900	1.86

[a] General conditions: [BD]$_0$ = 3 M in hexane, catalyst preformation 1 h at 0 °C
[b] Gravimetrically determined
[c] Measured by means of size exclusion chromatography (SEC) against polystyrene standards

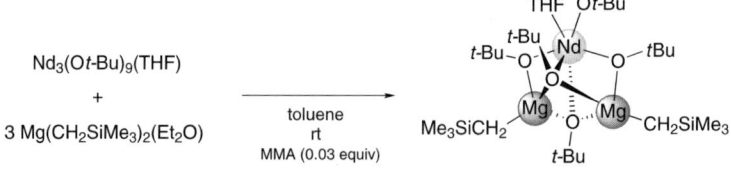

Scheme 38 Formation of a heterobimetallic Nd/Mg species [192]

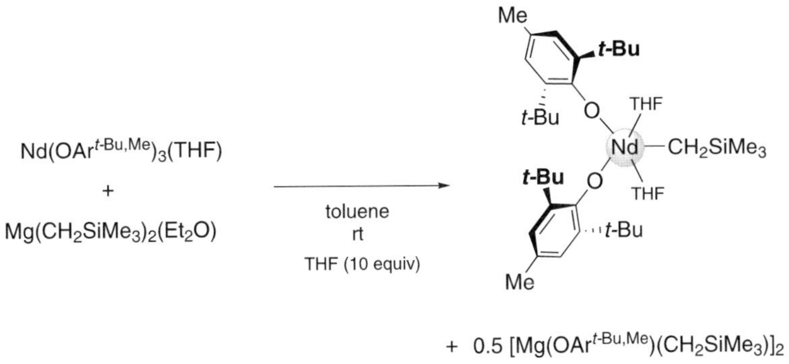

Scheme 39 Alkylation of monomeric Nd(OAr$^{t\text{-Bu,Me}}$)$_3$(THF) with Mg(CH$_2$SiMe$_3$)$_2$(Et$_2$O) [192]

formation of a six-coordinated neodymium center and an all-oxygen environment. Furthermore, [aryloxide]→[alkyl] exchange in equimolar mixtures of Nd(OAr$^{t\text{-Bu,Me}}$)$_3$(THF) and Mg(CH$_2$SiMe$_3$)$_2$(Et$_2$O) led to the isolation of a magnesium-free bis(aryloxide) mono(alkyl) complex (Scheme 39).

5
Lanthanide Siloxides

Heterogeneous diene polymerization catalysts based on modified and unmodified silica-supported lanthanide complexes are known as efficient gas-phase polymerization catalysts for a variety of support materials and activation procedures (see Sect. 9). Metal siloxide complexes M(OSiR$_3$)$_x$ are routinely employed as molecular model systems of such silica-immobilized/grafted metal centers [196–199]. Structurally authenticated alkylated rare-earth metal siloxide derivatives are scarce, which is surprising given that structural data on a considerable number of alkylated lanthanide alkoxide and aryloxide complexes with a variety of substitution patterns is meanwhile available.

Indeed, attempts to isolate heterobimetallic Ln/Al complexes from binary Ln(OSiR$_3$)$_3$ (SiR$_3$ = SiMe$_2$$t$-Bu, SiPh$_2$$t$-Bu, SiH($t$-Bu)$_2$, SiPh$_3$)/TMA mixtures were not successful. NMR spectroscopic studies showed the formation of peralkylated tetramethylaluminates Ln(AlMe$_4$)$_3$ and Ln – OSi cleavage products [Me$_2$Al(OSiR$_3$)]$_2$ in varying amounts, besides several unidentified alkylated species (Scheme 40) [142, 200].

Structural evidence of the formation of alkylated siloxide complexes was obtained from the reaction of homoleptic tetramethylaluminates Ln(AlMe$_4$)$_3$ with one equivalent of tris($tert$.butoxy)silanol HOSi(Ot-Bu)$_3$. This fast protonolysis reaction results in heterobimetallic Ln/Al complexes Ln[OSi(Ot-Bu)$_3$](AlMe$_4$)$_2$(AlMe$_3$) (Ln = Y, Ce, Pr, Nd, La) via quantitative

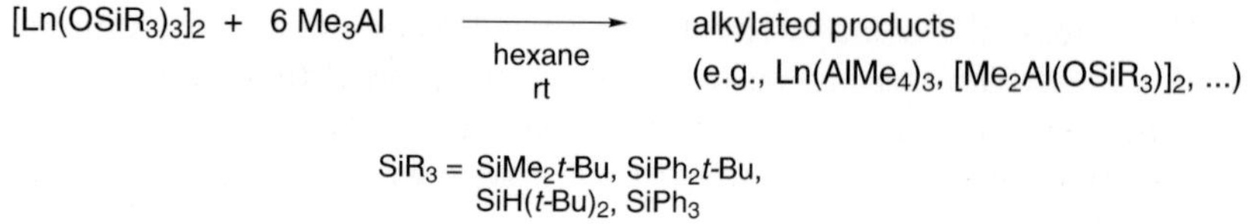

[Ln(OSiR₃)₃]₂ + 6 Me₃Al →(hexane, rt) alkylated products (e.g., Ln(AlMe₄)₃, [Me₂Al(OSiR₃)]₂, ...)

SiR₃ = SiMe₂t-Bu, SiPh₂t-Bu, SiH(t-Bu)₂, SiPh₃

Scheme 40 Formation of heterobimetallic Ln/Al complexes from lanthanide siloxide/TMA mixtures [142, 200]

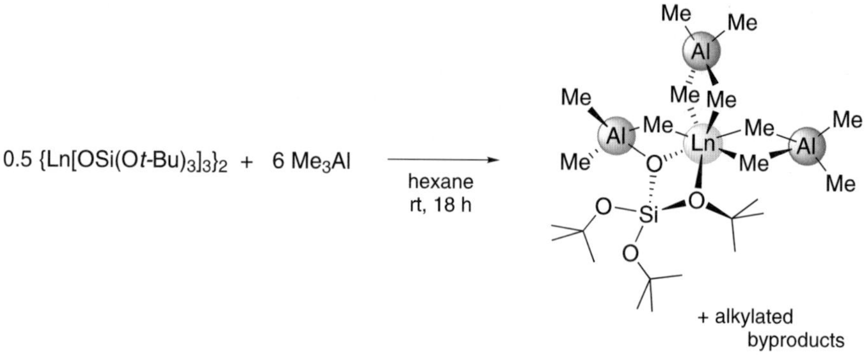

0.5 {Ln[OSi(Ot-Bu)₃]₃}₂ + 6 Me₃Al →(hexane, rt, 18 h) [structure] + alkylated byproducts

Scheme 41 Synthesis of structurally characterized heterobimetallic Ln/Al siloxide complexes Ln[OSi(Ot-Bu)₃](AlMe₄)₂(AlMe₃) (Ln = Y, La, Ce, Pr, Nd) [150, 200]

Table 12 Polymerization of isoprene with heterobimetallic Ln/Al siloxides Ln[OSi(Ot-Bu)₃] (AlMe₄)₂(AlMe₃) (Ln = Ce, Pr, Nd)

Ln [a]	$n_{Cl} : n_{Ln}$ [b]	Yield [c] %	M_n [d] 10^{-3}	M_w [d] 10^{-3}	PDI [d]
Ce	1:1	18	414	744	1.80
Ce	2:1	> 99	535	807	1.51
Ce	3:1	33	72	366	5.08
Pr	1:1	21	303	718	2.37
Pr	2:1	> 99	446	762	1.71
Pr	3:1	88	354	707	2.00
Nd	1:1	92	355	744	2.10
Nd	2:1	> 99	223	453	2.03
Nd	3:1	38	107	371	3.46

[a] General conditions: 8 mL hexane, 0.02 mmol precatalyst, 0.02–0.06 mmol Et₂AlCl (1–3 equiv), 20 mmol isoprene, 24 h, $T = 40\,°C$
[b] Catalyst preformation 30 min at ambient temperature
[c] Gravimetrically determined
[d] Measured by means of size exclusion chromatography (SEC) against polystyrene standards

[alkyl]→[siloxide] exchange (Fischbach et al., 2006, personal communication) [142, 150, 200]. As shown by NMR spectroscopy, these complexes are also generated as major products from the direct alkylation of homoleptic lanthanide siloxides {Ln[OSi(O*t*-Bu)$_3$]$_3$}$_2$ with excess trimethylaluminum (Scheme 41).

X-ray structure analysis revealed a 7-coordinate rare-earth metal center with two asymmetrically η^2-coordinating tetramethylaluminate ligands, an asymmetrically η^2-coordinating siloxide ligand and one methyl group of a trimethylaluminum donor molecule (Fig. 28). Such heteroleptic complexes can be regarded as molecular models of covalently bonded alkylated silica surface species. Moreover, isoprene was polymerized in the presence of 1–3 equivalents of diethylaluminum chloride, with highest activities observed for n(Cl):n(Ln) ratios of 2 : 1 (Table 12) (Fischbach et al., 2006, personal communication) [150].

6
Lanthanide Amides

Apart from inorganic halide and organometallic allyl components, essentially multicomponent mixtures based on oxygenated ligand systems, such as carboxylates and alkoxides, still dominate rare-earth metal-based Ziegler catalysis. As shown in Chart 5, only a few alkyl-, aryl-, and silylamide ligands were used to generate active catalyst mixtures. Recently, detailed investigations with commercially available homoleptic neodymium bis(trimethylsilyl)amide Nd[N(SiMe$_3$)$_2$]$_3$ were reported. Highly active ternary diene polymerization catalysts were generated in the presence of different types of organoaluminum (e.g., *i*-Bu$_3$Al/Et$_2$AlCl [158], MMAO/Et$_2$AlCl [201]), and organoboron cocatalysts (e.g., MMAO/B(C$_6$F$_5$)$_3$ [201], *i*-Bu$_3$Al/[NHMe$_2$Ph][B(C$_6$F$_5$)$_4$] [202]). In addition, the Dow Chemical Com-

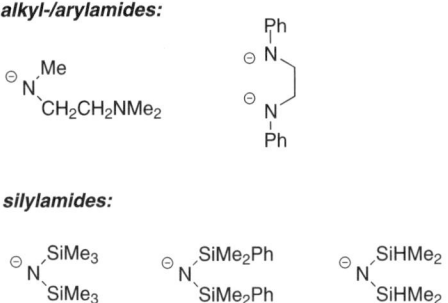

Chart 5 Amide ligands employed in lanthanide-based diene polymerization catalysts

pany described various additional supported and unsupported neodymium amide complexes [201].

By nature lanthanide amide complexes combine favorable steric factors—ease of ligand fine-tuning with respect to steric bulk/cone angle criteria—and concomitant high solubility with high ligand exchangeability [203]. In the peculiar case of organoaluminum cocatalyst interactions ligand exchange is uniquely driven by the much higher thermodynamic stability of the Al–N bond (the Al–N bond enthalpy in H_2N-AlH_2 was calculated as ca. 375 kJ mol^{-1} [204] while the absolute bond disruption energy of the amide ligand bond in $(C_5Me_5)_2Sm-NMe_2$ was determined as (201 ± 8) kJ mol^{-1} by anionic iodinolytic and alcoholytic isoperibol titration [205]). Although a large-scale industrial application of lanthanide amide complexes seems to be not feasible for cost factors, comprehensive studies on the alkylation behavior of alkyl-, aryl-, and silylamide complexes are meanwhile available. Five different types of activated lanthanide environments were accomplished upon reaction of homo- and heteroleptic amide complexes with typical cocatalysts used in diene polymerization, i.e., trialkylaluminum, R_3Al, dialkylaluminum hydride, R_2AlH, dialkylaluminum chloride, R_2AlCl, or borate reagents, [X][BR$_4$] (Scheme 42).

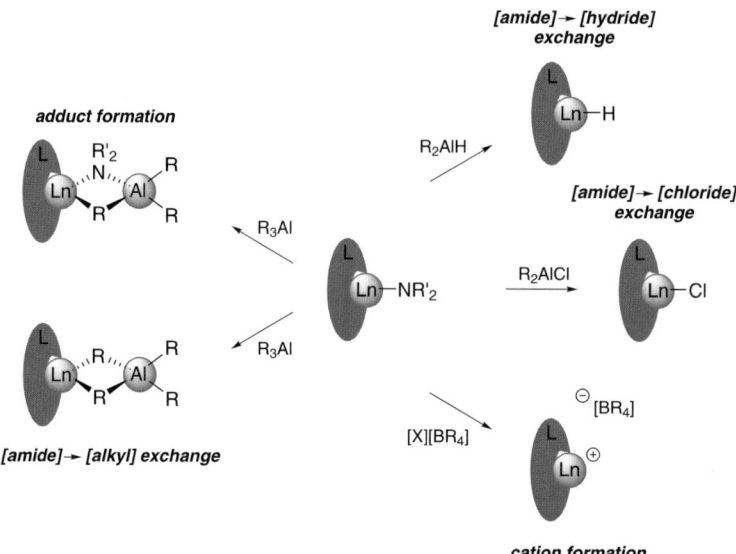

Scheme 42 Reactivity of lanthanide amide complexes with organoaluminum or borate reagents

6.1
Lanthanide Alkylamides

The reactivity of lanthanide alkylamides toward organoaluminum reagents was investigated by W.J. Evans et al. [149, 206–208]. Dimethylamide, Ln(NMe$_2$)$_3$(LiCl)$_3$, and di*iso*propylamide derivatives, Ln(N*i*-Pr$_2$)$_3$(THF), were reacted with different amounts of trimethylaluminum, Me$_3$Al, and trimethylgallium, Me$_3$Ga, to generate heterobimetallic Ln(III)/Al and Ln(III)/Ga species, respectively. While homoleptic Me$_3$M adducts Ln[(μ-NMe$_2$)(μ-Me)MMe$_2$]$_3$ were isolated with three equivalents of the Lewis acids [206], peralkylated tris(tetramethylmetallates) Ln[(μ-Me)$_2$MMe$_2$]$_3$ (M = Al, Ga) formed in the presence of excess Me$_3$M (Scheme 43) [149, 208]. Keeping this in mind, the ternary catalyst mixtures reported in the literature, depending on slight excess of trialkylaluminum reagents, most likely involve the forma-

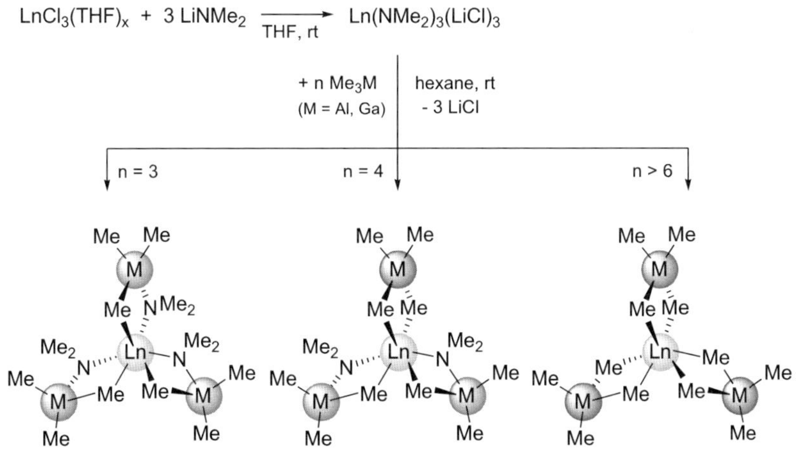

Scheme 43 Synthesis of heterobimetallic Ln(III)/M species from lanthanide amides Ln(NMe$_2$)$_3$(LiCl)$_3$ and MMe$_3$ (Ln = La, Nd; M = Al, Ga) [206]

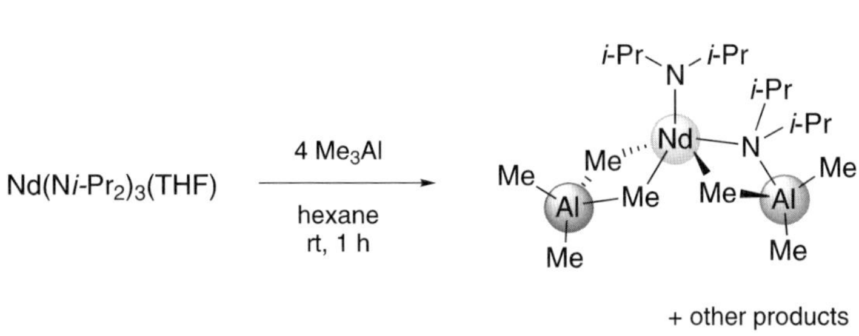

Scheme 44 Isolation of an alkylated neodymium complex with three different ligands from the reaction of neodymium diisopropylamide with TMA [207]

tion of homoleptic Ln[(μ-R)$_2$AlR$_2$]$_3$ species during the activation procedure, however, the generation of mixed-alkylated species is also feasible. The latter was evidenced by the isolation and spectroscopic and structural characterization of La[(μ-NMe$_2$)(μ-Me)GaMe$_2$]$_2$[(μ-Me)$_2$GaMe$_2$] (Scheme 43) [206], and Nd(Ni-Pr$_2$)[(μ-Ni-Pr$_2$)(μ-Me)AlMe$_2$][(μ-Me)$_2$AlMe$_2$] (Scheme 44) [207]. Interestingly, the heteroleptic neodymium diisopropylamide complex features Nd···H β-agostic interactions of the terminal diisopropylamide ligand.

6.2
Lanthanide Silylamides

6.2.1
Adduct Formation and [Amide]→[Alkyl] Exchange

Lanthanide silylamide complexes are routine precursor molecules in organo-rare-earth metal chemistry. The synthetic versatility of the Ln – N(SiMe$_3$)$_2$ moiety is well established in amine elimination reactions generating volatile HN(SiMe$_3$)$_2$ as the only byproduct (silylamide route). Complexes Ln[N(SiHMe$_2$)$_2$]$_3$(THF)$_x$ (x = 1, Ln = Sc; x = 2, Ln = Y, La – Lu) derived from a sterically less bulky amide ligand were introduced better to cope with steric restrictions in organolanthanide precatalyst design [209]. While homoleptic lanthanide bis(trimethylsilyl)amide La[N(SiMe$_3$)$_2$]$_3$ did not react

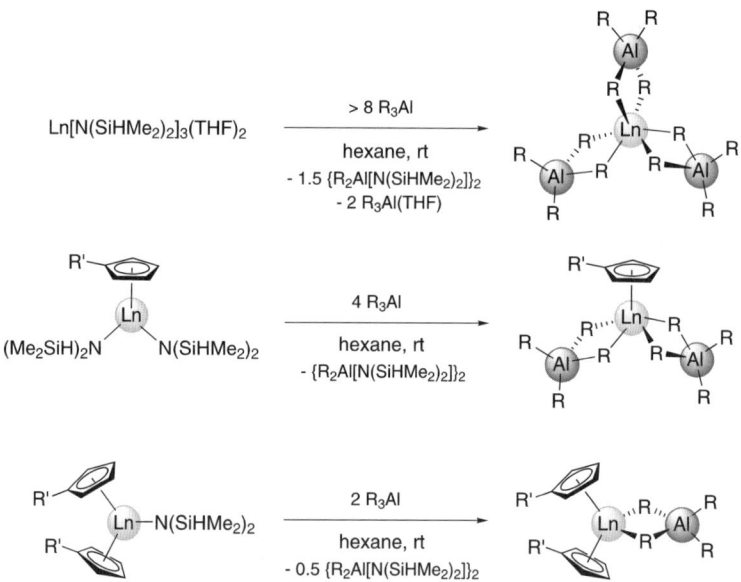

Scheme 45 Reactivity of lanthanide silylamide complexes toward trialkylaluminum R$_3$Al (R = Me, Et, i-Bu; CpR′ = substituted cyclopentadienyl ligand)

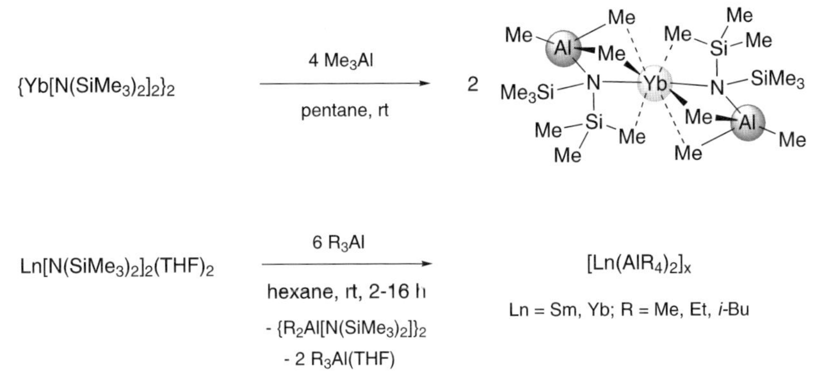

{Yb[N(SiMe$_3$)$_2$]$_2$}$_2$ $\xrightarrow{\text{4 Me}_3\text{Al}}_{\text{pentane, rt}}$ 2 [Yb complex with Me$_3$Si–N and Me–Al–Me bridges]

Ln[N(SiMe$_3$)$_2$]$_2$(THF)$_2$ $\xrightarrow[\text{hexane, rt, 2-16 h}]{\text{6 R}_3\text{Al}}$ [Ln(AlR$_4$)$_2$]$_x$

– {R$_2$Al[N(SiMe$_3$)$_2$]}$_2$
– 2 R$_3$Al(THF)

Ln = Sm, Yb; R = Me, Et, i-Bu

Scheme 46 Adduct formation and peralkylation with bis(trimethylsilyl)amides and trialkylaluminum reagents [18, 190]

with excess trimethylaluminum at ambient temperature [210], bis(dimethylsilyl)amide derivatives Ln[N(SiHMe$_2$)$_2$]$_3$(THF)$_2$ afforded peralkylated tetramethylaluminates (Scheme 45) (Fischbach et al., 2006, personal communication) [210]. Furthermore, half-lanthanidocene and lanthanidocene derivatives Cp$^{R'}$Ln[N(SiHMe$_2$)$_2$]$_2$ and Cp$^{R'}$$_2$Ln[N(SiHMe$_2$)$_2$] (Cp$^{R'}$ = substituted cyclopentadienyl ligand, e.g., C$_5$Me$_5$, C$_5$Me$_4$H, C$_5$Me$_4$SiMe$_3$, indenyl), respectively, easily undergo these [amide]→[alkyl] exchange reactions [24, 211].

Adduct formation and [amide]→[alkyl] transformation are likewise observed with divalent lanthanide amide precursors. Base-free dimeric ytterbium(II) bis(trimethylsilyl)amide Yb$_2$[N(SiMe$_3$)$_2$]$_4$ was shown to react with R$_3$Al (R = Me, Et) forming heterobimetallic adduct complexes of general composition Yb[N(SiMe$_3$)$_2$(AlR$_3$)]$_2$. The X-ray structure analysis of the methyl derivative confirmed the formation of a complex with two bridging and four semibridging Yb–Me–Al linkages (Scheme 46) [190]. Treatment of Ln[N(SiMe$_3$)$_2$]$_2$(THF)$_2$ (Ln = Sm, Yb) with excess of trialkylaluminum reagents R$_3$Al (R = Me, Et, i-Bu) led to peralkylated polymeric [Ln(AlR$_4$)$_2$]$_x$ in quantitative yields [18]. While the tetramethylaluminate derivatives are pyrophoric powders which are insoluble in aliphatic and aromatic hydrocarbons, the ethyl and isobutyl congeners are readily soluble in hexane. Perethylated [Yb(AlEt$_4$)$_2$]$_x$ forms an unusual contacted ion pair structure in the solid state exhibiting two distinct Yb environments in an intricate three-dimensional network. The aluminate ligands connect the ytterbium centers via M···(CH) "secondary" interactions combining η^1-, η^2-, and η^3-coordination modes.

6.2.2
[Amide]→[Chloride] Exchange

The sterically induced inertness of $Ln[N(SiMe_3)_2]_3$ toward trialkylaluminum reagents is abandoned in the presence of chlorinating reagents. Catalyst mixtures composed of $Nd[N(SiMe_3)_2]_3$, i-Bu_3Al, and Et_2AlCl polymerized butadiene in a highly efficient manner at elevated temperatures, yielding polymeric materials with high-cis-1,4-stereospecificities (Table 13). Bimodal molecular weight distributions suggested the formation of at least two different catalytically active species during the activation procedure. In order better to understand the diverse activation pathways within these multicomponent mixtures the reaction of $Nd[N(SiMe_3)_2]_3$ with Et_2AlCl was followed by ^1H NMR spectroscopy. The displacement of amido ligands from the neodymium center was corroborated by the formation of $Et_2Al[N(SiMe_3)_2]$ and $EtAl[N(SiMe_3)_2]_2$ [158].

Table 13 Polymerization of butadiene with ternary $Nd[N(SiMe_3)_2]_3/i$-Bu_3Al/Et_2AlCl

Catalyst ratio [a]	Yield [b] %	cis-1,4 [c] %	$trans$-1,4 [c] %	1,2 [c]
1/0/0	0	–	–	–
1/40/1	19.8	95.4	3.7	0.9
1/40/2	71.1	97.6	1.6	0.8
1/40/2.5	71.1	96.5	2.5	1.0
1/40/3	33.0	95.7	3.2	1.1
1/40/5	22.7	93.3	5.2	1.5
1/10/2	24.9	98.8	1.2	0.6
1/20/2	46.2	99.0	0.7	0.3
1/30/2	60.8	97.3	1.9	0.8

[a] General conditions: Ternary mixtures $Nd[N(SiMe_3)_2]_3/i$-Bu_3Al/Et_2AlCl in heptane, $[Nd] = 0.115$ mmol L^{-1}, [butadiene] = 1 mol L^{-1}, 15 min, $T = 70\ °C$
[b] Gravimetrically determined
[c] Measured by means of IR spectroscopy

6.2.3
[Amide]→[Hydride] Exchange

Lanthanide hydride derivatives are commonly synthesized by hydrogenolysis of lanthanide alkyl complexes [212]. In order to further exploit the thermodynamic stability of the Al – N bond diisobutylaluminum hydride (DIBAH), a common cocatalyst in diene polymerization mixtures and well-established reducing agent in organic synthesis, was used in the hydrogenol-

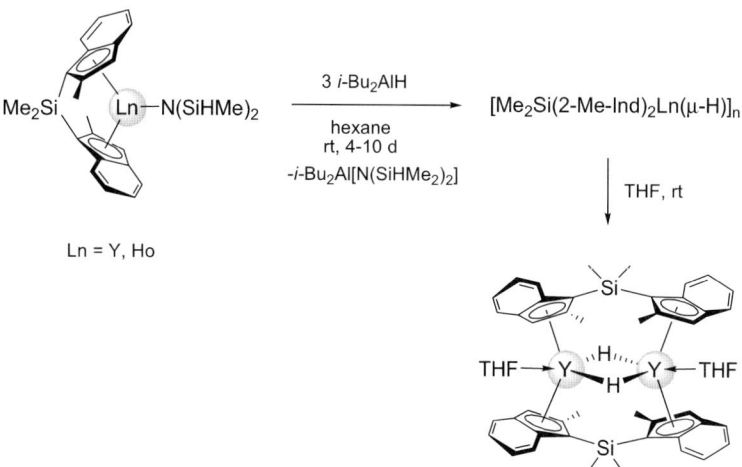

Scheme 47 DIBAH-mediated [amide]→[hydride] transformation in *ansa*-lanthanidocene(III) complexes (Ln = Y, Ho) [213]

ysis of *ansa*-lanthanidocene amide complexes. Microanalytical and spectroscopic data of the insoluble product suggested the formation of a presumably oligomeric hydride species [Me$_2$Si(2-Me-Ind)$_2$Ln(μ-H)]$_n$ (Ln = Y, Ho) via [amide]→[hydride] exchange according to Scheme 47. A mechanistic reaction pathway including the initial formation of a spectroscopically not observable DIBAH adduct Me$_2$Si(2-Me-Ind)$_2$Ln(μ-H)(μ-i-Bu)Al(i-Bu)$_2$ was presented [213].

6.2.4
Cation Formation

Butadiene polymerization initiated by a preformed cationic neodymium amide species was described by C. Boisson and coworkers [202]. Perfluoroaryl borane or trityl and ammonium borate salts, previously used to form cationic single-site group 4 metallocene catalysts for olefin polymerization [214], as well as cationic lanthanide metal complexes [215, 216] were employed as cationizing reagents in ternary mixtures, and polybutadiene was obtained in moderate to good yields (Table 14). On the basis of the successful isolation of the presumably active component [Nd{N(SiMe$_3$)$_2$}$_2$(THF)$_2$][B(C$_6$F$_5$)$_4$], a mechanistic scenario addressing the possible activation pathway and the formation of heterobimetallic Ln(III)/Al complexes was proposed (Scheme 48).

Table 14 Polymerization of butadiene with neodymium amide catalysts utilizing borate activators

Borate cocatalyst [a]	[Nd] [a] μM	Yield [b] %	cis-1,4 [c] %	trans-1,4 [c] %	M_n [d] 10^{-3}	PDI [d]
[NHMe$_2$Ph][B(C$_6$F$_5$)$_4$]	183	62	86.5	11	ins[e]	ins[e]
[Ph$_3$C][B(C$_6$F$_5$)$_4$]	270	39	85	12	171	2.8
B(C$_6$F$_5$)$_3$	154	75	84.4	13.7	274	2.2

[a] General conditions: Ternary mixtures Nd[N(SiMe$_3$)$_2$]$_3$/borate cocatalyst/i-Bu$_3$Al, polymerization in heptane, n(Nd): n(B) = 1 : 1, n(Nd): n(Al) = 1 : 10, 30 min, T = 70 °C
[b] Gravimetrically determined
[c] Measured by means of ^{13}C NMR spectroscopy
[d] Measured by means of size exclusion chromatography (SEC) against polystyrene standards
[e] Insoluble in THF

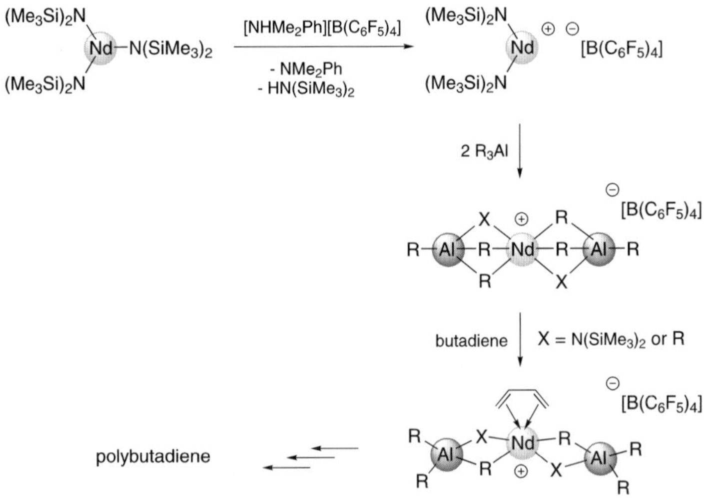

Scheme 48 Butadiene polymerization with Nd[N(SiMe$_3$)$_2$]$_3$/[NHMe$_2$Ph][B(C$_6$F$_5$)$_4$]/R$_3$Al and proposed activation mechanism [202]

6.3
Lanthanide Anilides

Compared to the extensive use of phenolate ligands and the isolation and structural characterization of a broad variety of differently alkylated heterobimetallic Ln(III)/Al complexes, only two examples of anilide→AlR^3 adduct formation and [anilide]→[alkyl] exchange are reported in the literature. There exist no reports on the use of simple anilide ligands in diene polymerization.

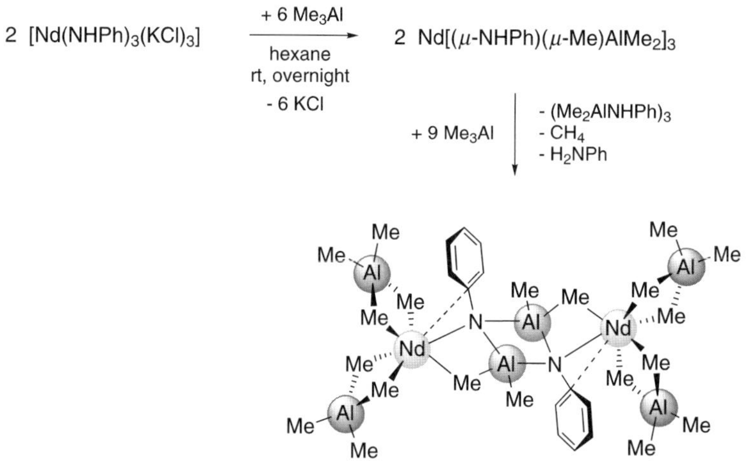

Scheme 49 Formation of a heterobimetallic Nd(III)/Al complex from Nd(NHPh)$_3$(KCl)$_3$ and Me$_3$Al [217]

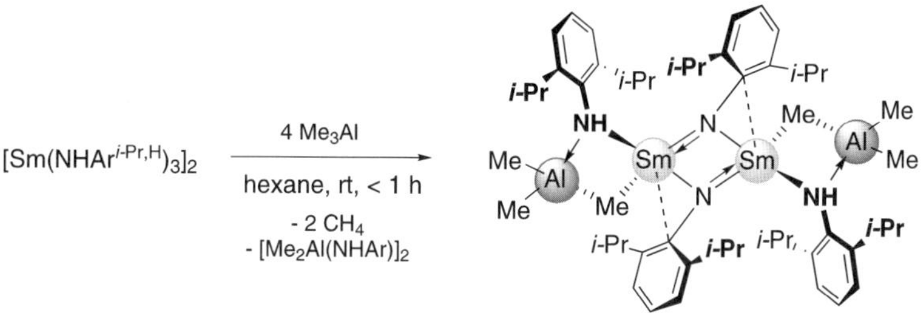

Scheme 50 Formation of a "μ_2-imido" functionality located between two samarium metal centers [218]

Treatment of a suspension of Nd(NHPh)$_3$(KCl)$_3$ with trimethylaluminum in hexane yielded the heteroleptic heterobimetallic cluster {Nd[(μ_2-Me)$_2$AlMe$_2$]$_2$(μ_3-NPh)(μ_2-Me)AlMe}$_2$ in low yield (Fig. 29). The formation of the cyclic byproduct (Me$_2$AlNHPh)$_3$ was proven by means of NMR spectroscopy and X-ray crystallography [217]. The molecular structure of the Nd$_2$Al$_6$ dimer revealed the presence of a doubly deprotonated "imido" ligand bridging each a neodymium center and two aluminum atoms.

Dimeric π-arene-bridged samarium anilide [Sm(NHAr$^{i\text{-Pr,H}}$)$_3$]$_2$ was employed in the isolation and characterization of a dimeric amido-imido compound. Addition of 4 equivalents of TMA led to (1) cleavage of one of the Ln–NHPh bonds, (2) deprotonation, and (3) adduct formation. Density functional theory calculations, carried out in order better to understand the nature of the imido bonding, revealed the presence of Sm–N π-bonding interactions involving the samarium 5d orbitals [218].

6.4
Other Systems

Organoaluminum-promoted alkylation/activation gained also entry into non-cyclopentadienyl or postmetallocene organolanthanide chemistry [219]. For example, an efficient neodymium catalyst precursor was prepared via pretreatment of $NdCl_3(THF)_3$ with MeLi at 0 °C followed by the reaction with 2,6-diiminopyridine. Upon activation with MMAO butadiene was polymerized with good activities in a predominantly *cis*-1,4-fashion (95–97% *cis*) (Scheme 51) [220]. Similarly, structurally characterized hexa-1,5-diene-1,6-diamide neodymium complexes were prepared from $NdBr_3(THF)_{3.5}$ and the appropriate dilithium reagent and subsequently activated by MMAO to polymerize butadiene. Activities and selectivities were comparable to highly efficient carboxylate-based catalyst mixtures [221].

Structurally characterized heterobimetallic Ln/Al complexes bearing donor-functionalized dianionic diamido ligands of the types $[NNN]^{2-}$ and $[NON]^{2-}$ were obtained from [amide]→[alkyl] (Scheme 52) or [aluminate]→[amide] interchange reactions (Scheme 53).

Accordingly, postlanthanidocene tetramethylaluminate complexes $[NNN]Ln[(\mu\text{-Me})_2AlMe_2]$ of the smallest rare-earth metal centers scandium and lutetium were accessible in good yields, whereas the formation of an aluminum-free terminal methyl complex was suggested for the "larger" yttrium metal center (Estler et al., 2006, personal communication) [222]. Noteworthy, this [amide]→[alkyl] transformation is selective with respect to the terminal alkylamido ligand. The reaction of homoleptic lanthanide tris(tetramethylaluminate) complexes $Ln(AlMe_4)_3$ with a "THF"-functionalized $H_2[NON]$ ligand precursor took place according to an alkane elimination reaction (tetramethylaluminate route, Sect. 7.2.2) [223]. As shown in Scheme 53, postlanthanidocene tetramethylaluminate trimethylaluminum adducts were also obtained for the largest rare-earth metal center lan-

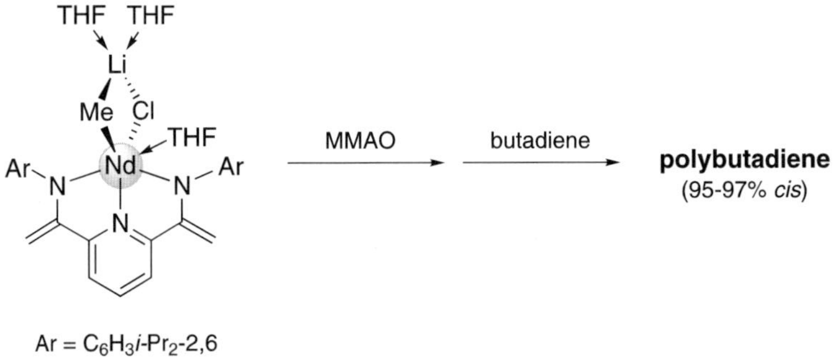

Scheme 51 Stereospecific butadiene polymerization based on a MMAO-activated neodymium methyl complex supported by a dianionic modification of neutral 2,6-diiminopyridine [190]

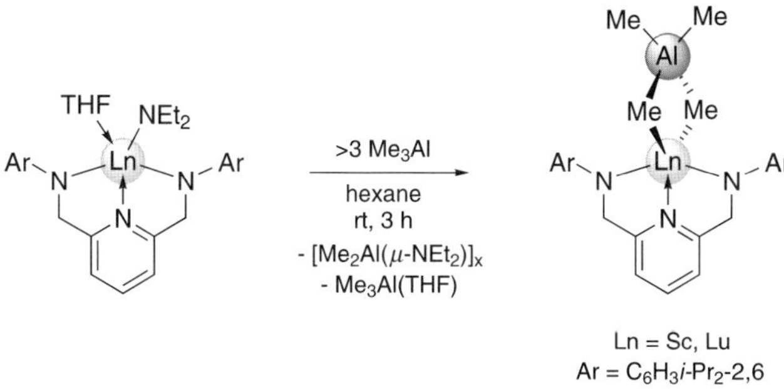

Scheme 52 Generation of terminal tetramethylaluminate complexes via selective [amide]→[alkyl] interchange [222]

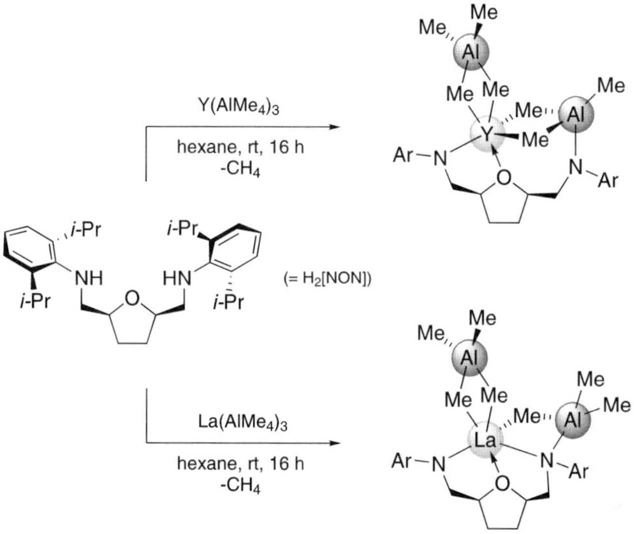

Scheme 53 Generation of heteroleptic tetramethylaluminate complexes via [aluminate]→[amide] interchange [223]

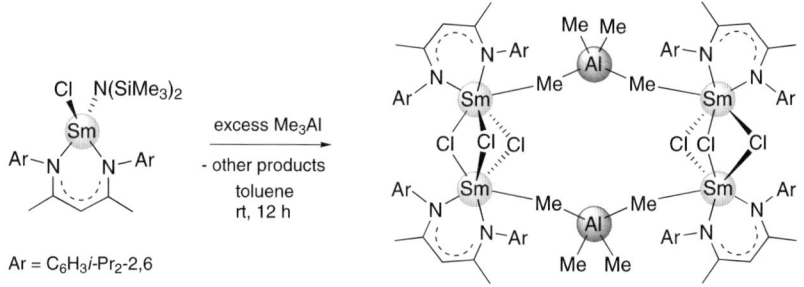

Scheme 54 Formation of a β-diketiminato-derived postsamarocene complex with μ_2-η^1:η^1 aluminate coordination [224]

thanum. X-ray crystallography confirmed unusual organoaluminum coordination modes as evidenced by the formation of an atypically bent heterobimetallic [Ln(μ-Me)$_2$AlMe$_2$] moiety within the lanthanum complex (Fig. 26) and a novel [Ln(III)(μ-Me)$_2$AlMe(NR$_2$)] unit for the yttrium derivative (Fig. 27) [223].

Finally, a samarium aluminate tetramer featuring β-diketiminato ligands was obtained in low yield (20%) by J. Arnold et al. (Scheme 54) [224]. Once again this example corroborates the feasibility of selective [amide]→[alkyl] transformations in N-ligand-based postlanthanidocene chemistry, previously observed for octaethylporphyrin yttrium complexes by C.J. Schaverien [225, 226].

7
Lanthanide Alkyls

7.1
Lanthanide Allyl Complexes

Lanthanide allyl complexes have been discussed as true active species in diene polymerization catalysis since the 1960s. In 1981, A. Mazzei reported that anionic Li[Nd(η^3-C$_3$H$_5$)$_4$] · 1.5dioxane [227] acts as a moderately active single-component catalyst, producing predominantly *trans*-1,4-polydienes (84%) with large amounts of 1,2-vinyl connectivities (15%). This was rationalized by the cleavage of allyllithium and its initiator abilities [228]. The "allyl theory" was impressively substantiated by the most valuable work of R. Taube addressing structure-reactivity issues of diene polymerization based on various structurally evidenced homo- and heteroleptic allyl complexes (Table 15). Although organoaluminum reagents, such as MAO, and Me$_2$AlCl/Me$_3$Al mixtures, were used to obtain highest catalytic activities and *cis*-stereospecificities, heterobimetallic Ln/Al species were not considered as major intermediates. This section briefly summarizes allyl-based diene polymerization mixtures utilizing organoaluminum activation. Mechanistic investigations of single-component allyl catalysts are not addressed.

Highly active *cis*-stereoselective butadiene polymerization catalysts were obtained upon activation of tris(allyl)neodymium complex Nd(η^3-C$_3$H$_5$)$_3$ · dioxane with alkylaluminoxanes, e.g., MAO, HIBAO (turnover frequencies as high as 15 000 mol butadiene (mol Nd h)$^{-1}$) [152, 239]. When activated with 30–50 equivalents of MAO heteroleptic bis- and mono(allyl) neodymium chlorides Nd(η^3-C$_3$H$_5$)$_2$Cl · 1.5THF and Nd(η^3-C$_3$H$_5$)Cl$_2$ · 2THF, respectively, catalyzed the *cis*-1,4-polymerization of butadiene in toluene or heptane with even higher activities (ca. 575 000 mol butadiene (mol Nd h)$^{-1}$) [153]. A dicationic monobutenyl neodymium(III) complex, originating from an anion transfer reaction to the Lewis acidic methylaluminoxane, was discussed as

Table 15 Anionic, cationic, and neutral lanthanide allyl complexes reported in diene polymerization

Allyl complexes	Refs.
Anionic	
Li[Ln(η^3-C$_3$H$_5$)$_4$]·1.5dioxane (Ln = La, Nd, Sm)	[229–231]
Li[Cp^{R}La(η^3-C$_3$H$_5$)$_3$] · 2dioxane (Cp^R = C$_5$H$_5$, C$_5$Me$_5$, C$_9$H$_7$, C$_{13}$H$_9$)	[232]
Li[Cp^{R}Nd(η^3-C$_3$H$_5$)$_3$] · 2dioxane (Cp^R = C$_5$H$_5$)	[230]
Li[Cp^{R}Nd(η^3-C$_3$H$_5$)$_3$] · 3DME (Cp^R = C$_5$Me$_5$)	[230]
Li[(Me$_2$CC$_5$H$_4$)$_2$Ln(η^3-C$_3$H$_5$)$_2$] · DME (Ln = Nd, Sm)	[231, 233–238]
cationic	
[Nd(η^3-C$_3$H$_5$)Cl · 5THF][B(C$_6$H$_5$)$_4$] · THF	[241]
neutral	
La(η^3-C$_3$H$_5$)$_3$·1.5dioxane	[239]
La(η^3-C$_3$H$_5$)$_3$ · D (D = DME, TMEDA, 2 HMPT)	[240]
Nd(η^3-C$_3$H$_5$)$_3$·dioxane	[152, 239, 241]
Nd(η^3-C$_3$H$_5$)$_3$	[242]
Nd(η^3-C$_3$H$_5$)$_2$Cl·dioxane	[243]
Nd(η^3–C$_3$H$_5$)$_2$Cl · 1.5THF	[153]
Nd(η^3-C$_3$H$_5$)Cl$_2$ · 2THF	[153]
La(η^3-C$_3$H$_5$)$_2$X · 2THF (X = Cl, Br, I)	[244]
(C$_5$Me$_5$)Nd(η^3-C$_3$H$_5$)$_2$· dioxane	[241]
YCl[1,3-C$_3$H$_3$(SiMe$_3$)$_2$]$_2$	[245]
LaCl[1,3-C$_3$H$_3$(SiMe$_3$)$_2$]$_2$ · THF	[245]
NdI$_2$[1,3-C$_3$H$_3$(SiMe$_3$)$_2$] · 1.25THF	[245]

catalytically active species (Scheme 55). It may be stabilized through coordinative interaction with the intricate counter anion as well as the growing polymer polybutadiene chain [152, 153].

A similar mechanistic scenario was presented for the formation of a catalytically active η^3-butenyl neodymium(III) complex in the ternary system Nd(C$_3$H$_4$R)$_3$/2Me$_2$AlCl/30Me$_3$Al [243]. Mixtures of "preformed" polybutadienyl derivative Nd(C$_3$H$_4$R)$_3$, generated in situ from the reaction of solvent-free [Nd(η^3-C$_3$H$_5$)$_3$]$_n$ with approximately 15 equivalents of butadiene, and dialkylaluminum reagents were proposed to give heterobridging chloroaluminate ligands stabilizing the (formally dicationic) neodymium center (Scheme 56).

Recently, the first example of a structurally proven [allyl] → [alkyl] exchange reaction was reported. The addition of excess trimethylaluminum to the cyclopentadienyl-functionalized yttrocene monoallyl complex [C$_5$Me$_4$(SiMe$_2$CH$_2$CH=CH$_2$)]$_2$Y(η^3-C$_3$H$_5$) quantitatively yielded the dimeric alumi-

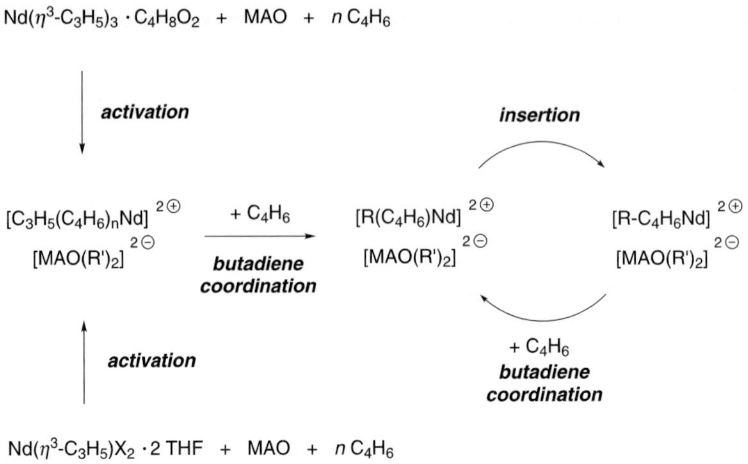
Nd(η3-C3H5)3 · C4H8O2 + MAO + n C4H6
activation
insertion
[C3H5(C4H6)nNd] 2⊕
[MAO(R')2] 2⊖
+ C4H6
butadiene coordination
[R(C4H6)Nd] 2⊕
[MAO(R')2] 2⊖
[R-C4H6Nd] 2⊕
[MAO(R')2] 2⊖
activation
+ C4H6
butadiene coordination
Nd(η3-C3H5)X2 · 2 THF + MAO + n C4H6

Scheme 55 Proposed activation and butadiene polymerization for MAO-activated $Nd(\eta^3\text{-}C_3H_5)_3 \cdot$ dioxane and $Nd(\eta^3\text{-}C_3H_5)X_2 \cdot$ THF catalysts (adopted from [152, 153]). $R\text{-}C_4H_6 =$ polybutadienyl residue (growing polymer chain). $R' =$ abstracted allyl or halide X

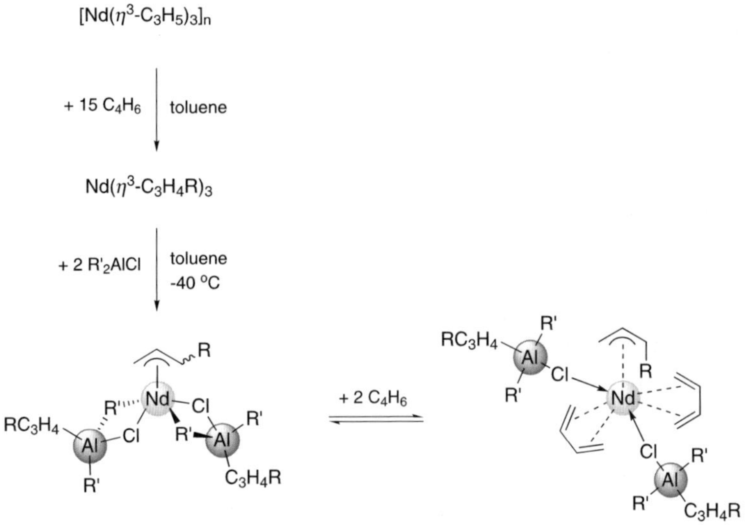
[Nd(η3-C3H5)3]n
+ 15 C4H6
toluene
Nd(η3-C3H4R)3
+ 2 R'2AlCl
toluene
-40 °C
R
Nd
Cl
R'
RC3H4
Al
Cl
R'
Al
C3H4R
+ 2 C4H6
RC3H4
Al
Cl
R
Nd
R'
C3H4R

Scheme 56 Proposed activation scenario in ternary $Nd(C_3H_4R)_3/2R'_2AlCl/30Me_3Al$ catalyst mixtures (adopted from [243]). R = growing polymer chain; R′ = Me, Et

nate shown in Scheme 57, which was X-ray crystallographically characterized [246]. A similar allyl exchange reaction between lanthanide bis(allyl) complexes $LnCl[1,3\text{-}C_3H_3(SiMe_3)_2]_2(THF)_x$ (Ln = La, $x = 1$; Ln = Y, $x = 0$)

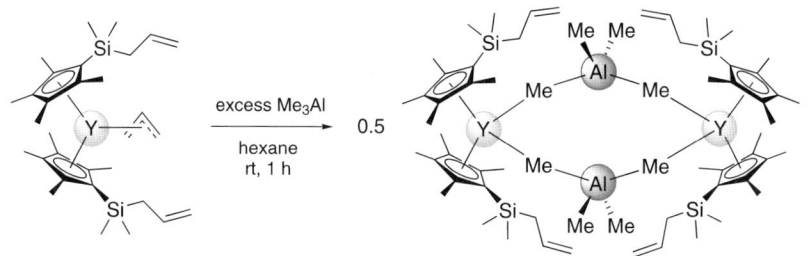

Scheme 57 Structurally evidenced [allyl]→[alkyl] transformation in yttrocene chemistry [246]

and tri*iso*butylaluminum had been previously proposed on the basis of NMR-spectroscopic investigations [245].

7.2
Lanthanide Tetraalkylaluminates

In the meantime, tetraalkylaluminate moieties feature a ubiquitous structural motif in organo-rare-earth metal chemistry. This is in contrast to group 4 organometallic chemistry where M-AlR$_4$ bonding is considered as extremely labile [25] being only documented by the X-ray structure analysis of [Ti(N*t*-Bu)(Me$_3$[9]aneN$_3$)(μ-Me)$_2$AlMe$_2$][B(C$_6$F$_5$)$_4$] (Me$_3$[9]aneN$_3$ = 1,4,7-trimethyltriazacyclononane) [26]. Several examples of structurally distinct heteroleptic Ln/Al heterobimetallic complexes bearing the anionic [AlR$_4$]$^-$ unit (R = Me, Et, *i*-Bu) were reported in the literature and/or presented in the previous sections. As a rule higher activities in diene polymerization were observed for a higher degree of metal alkylation [148]. Homoleptic lanthanide(III) tris(tetraalkylaluminates) Ln(AlR$_4$)$_3$ are preferentially accessible from high-yield R$_3$Al-mediated [amide] → [alkyl] transformations, and the methyl derivatives Ln(AlMe$_4$)$_3$ were isolated and fully characterized for differently sized lanthanide metal centers (Ln = Y, La, Ce, Pr, Nd, Sm, Gd, Lu) (Scheme 58, Fig. 5) (Fischbach et al., 2006, personal communication) [149–151]. The isolation of the ethyl analogues proved to be difficult and so far, only La(AlEt$_4$)$_3$ was successfully crystallized from the alkylated byproducts (Scheme 59) [150]. Homoleptic Ln(AlMe$_4$)$_3$ complexes were first reported in 1995 [149] and are currently employed as versatile synthetic precursors for the generation of a variety of heterobimetallic Ln/Al complexes according to alkane elimination reactions [148, 150, 151, 223, 247].

In the course of X-ray crystallographic investigations different types of [AlMe$_4$]$^-$ coordination modes (**a**)–(**f**) were found (Chart 6). Among these, terminal (**a**) and bridging η^1:η^1-coordinated ligands (**e**) seem to be favored as evidenced for homoleptic Ln(AlMe$_4$)$_3$, heteroleptic non-metallocene mono- and bis(aluminates), R$_2$Ln(AlMe$_4$) and RLn(AlMe$_4$)$_2$, as well as lan-

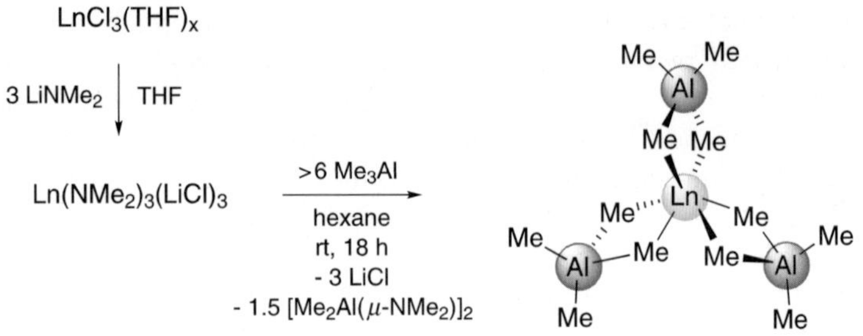

LnCl$_3$(THF)$_x$

3 LiNMe$_2$ $\downarrow$ THF

Ln(NMe$_2$)$_3$(LiCl)$_3$ $\xrightarrow[\substack{\text{hexane} \\ \text{rt, 18 h} \\ -\text{3 LiCl} \\ -\text{1.5 [Me}_2\text{Al}(\mu\text{-NMe}_2)]_2}]{>6 \text{ Me}_3\text{Al}}$

Scheme 58 Synthesis of homoleptic lanthanide tris(tetramethylaluminate) complexes Ln[(μ-Me)$_2$AlMe$_2$]$_3$ (Ln = Y, La, Ce, Pr, Nd, Sm, Gd, Lu)

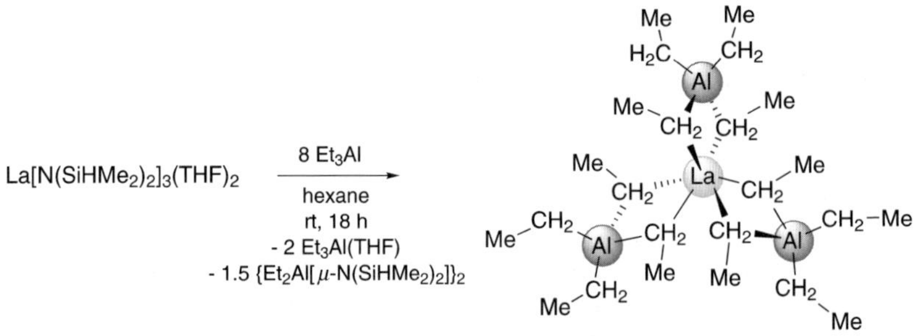

La[N(SiHMe$_2$)$_2$]$_3$(THF)$_2$ $\xrightarrow[\substack{\text{hexane} \\ \text{rt, 18 h} \\ -\text{2 Et}_3\text{Al(THF)} \\ -\text{1.5 \{Et}_2\text{Al}[\mu\text{-N(SiHMe}_2)_2]\}_2}]{8 \text{ Et}_3\text{Al}}$

Scheme 59 Synthesis of homoleptic lanthanum tris(tetraethylaluminate) La[(μ-Et)$_2$AlEt$_2$]$_3$ from silylamide precursor La[N(SiHMe$_2$)$_2$]$_3$(THF)$_2$ and Et$_3$Al [150]

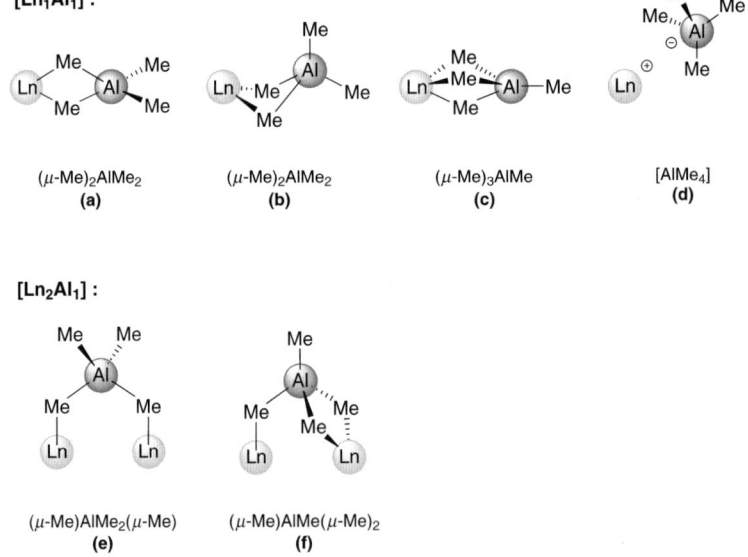

Chart 6 Structurally characterized coordination modes of the [AlMe$_4$]-fragment in organo-rare-earth metal chemistry

thanidocene complexes. However, in the presence of sterically highly unsaturated rare-earth metal centers additional bonding modes were observed. Atypically bent [Ln(μ-Me)$_2$Al] moieties (b) with elongated Ln-(μ-Me) distances and shortened Ln···Al contacts are adopted in half-lanthanidocene bis(tetramethylaluminate) complexes (C$_5$Me$_5$)Ln(AlMe$_4$)$_2$, the extent of distortion from [LnMe$_2$Al] coplanarity increasing with increasing size of the metal center (La > Nd > Y > Lu) (Fig. 6) [211, 247]. Terminal (c) and bridging (μ-Me)AlMe(μ-Me)$_2$-coordinated aluminate ligands (f) appeared in alkylated polynuclear chloride clusters (Sect. 7.2.3) [248]. A rare example of a structurally authenticated non-coordinating [AlMe$_4$]$^-$ unit (d) was found in fluorenyl lanthanide(II) complexes by H. Yasuda and Y. Kai [249].

7.2.1
Lappert's Concept of Donor-Induced Aluminate Cleavage: Reversibility and Adduct-Formation

Conventionally, terminal lanthanidocene(III) alkyl complexes are synthesized from lanthanidocene(III) chloride precursors and alkyllithium reagents [212, 250–254]. An organoaluminum-based strategy toward lanthanidocene(III) alkyls was reported as early as 1978 by M.F. Lappert and coworkers [10]. Dimeric (μ-Me)$_2$-bridged lanthanidocene complexes [(C$_5$H$_5$)$_2$Ln(μ-Me)]$_2$ were obtained in high yields (ca. 80%) from the reaction of equimolar quantities of tetramethylaluminate complexes (C$_5$H$_5$)$_2$Ln(AlMe$_4$) (Ln = Y, Dy, Ho, Er, Tm, Yb) and the Lewis base pyridine (= donor-induced aluminate cleavage). Since scandium rather than aluminum is the more electrophilic metal center, the reaction of (C$_5$H$_5$)$_2$Sc(AlMe$_4$) with Lewis base molecules (e.g., pyridine or THF) did not produce donor-free methyl complexes but donor adducts of general type (C$_5$H$_5$)$_2$ScMe(Do) (Scheme 60).

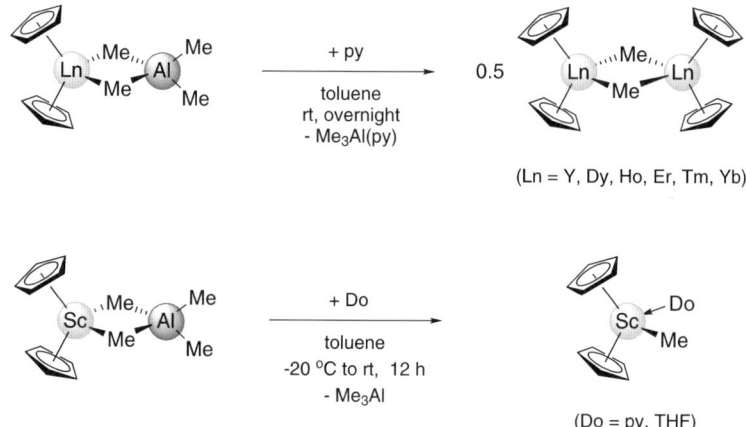

Scheme 60 Donor(Do)-induced aluminate cleavage

Alternatively, lanthanidocene tetraalkylaluminate complexes have been synthesized from divalent $(C_5Me_5)_2Sm(THF)_x$ ($x = 0, 2$) and excess of R_3Al (R = Me, Et) [108, 109] or from solvent-free ate complexes. As reported by P.L. Watson et al. in 1982 [16], the reaction of $[(C_5Me_5)_2LnMe_2]Li$ (Ln = Yb, Lu) with excess of trimethylaluminum gave tetramethylaluminate complexes $(C_5Me_5)_2Ln(AlMe_4)$ and $LiAlMe_4$ as an insoluble byproduct.

The reversibility of the tetraalkylaluminate cleavage reaction was another important detail of this early work by Lappert and was later on exploited for the synthesis of mixed-alkylated aluminate complexes. Accordingly, C_2-symmetric *ansa*-bridged yttrocene(III) alkyl complexes were reacted with a slight excess of different trialkylaluminum reagents to (re-)generate the initial homo-alkylated complexes or fully characterized mixed-alkylated species (Scheme 61) [24]. An X-ray structure analysis unequivocally confirmed the formation of complex rac-$[Me_2Si$-$(2$-Me-$C_9H_5)_2]Y[(\mu$-$Me)(\mu$-$Et)AlEt_2]$

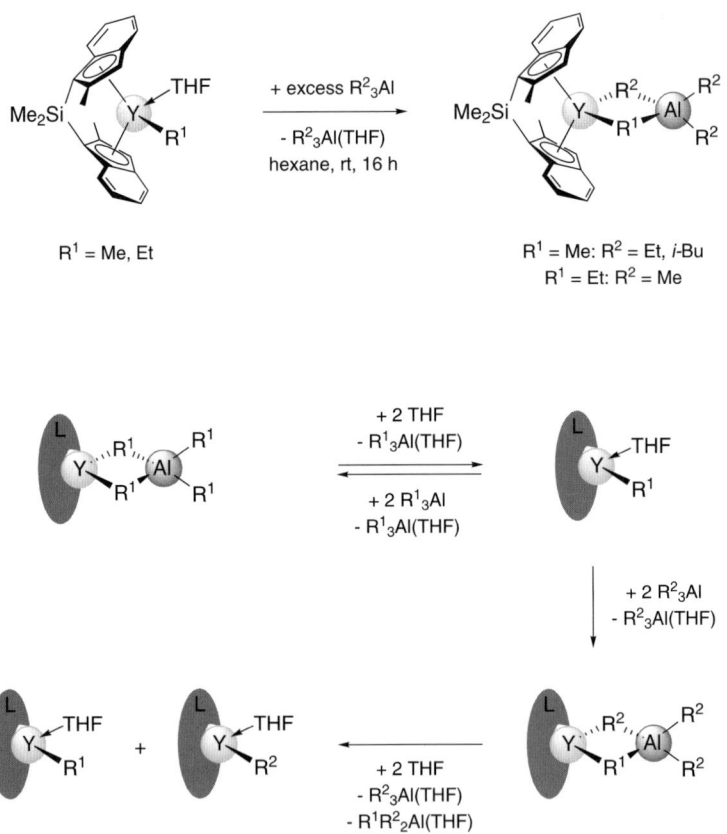

Scheme 61 Synthesis of mixed-alkylated yttrocene(III) tetraalkylaluminate complexes (*top*) and reversibility phenomena in aluminate chemistry (*bottom*) [24]

with bridging methyl and ethyl groups (Scheme 61). Donor-induced cleavage of the heterobridged alkylaluminate complexes, performed on a NMR-preparative scale by adding d_8-THF, revealed the intrinsic coordination capability of the differently sized alkyl moieties. Not surprisingly, methyl-ethyl-bridged derivatives gave a 1 : 1-mixture of terminal methyl and ethyl complexes, respectively. However, the preferred formation of a LnMe(THF) over a Ln(i-Bu)(THF) yttrocene complex (6 : 1-ratio) from a methyl-*iso*butyl-bridged Ln/Al precursor suggests that the bulkier alkyl group, which mimics a polymer chain, is preferentially transferred to the aluminum alkyl. Such studies reinforce that high concentrations of trialkylaluminum considerably affect both olefin pre-coordination and polymerization. Furthermore, due to their markedly decreased coordination capability, organoaluminum compounds carrying bulkier, branched alkyl groups comparatively increase the polymerization activity.

Lappert's concept of donor-induced aluminate cleavage was also successfully applied for the generation of solvent-free "LnMe$_3$" as well as half-lanthanidocene and lanthanidocene methyl derivatives carrying the bulky C$_5$Me$_5$ ligand (Scheme 62) [10, 12, 151].

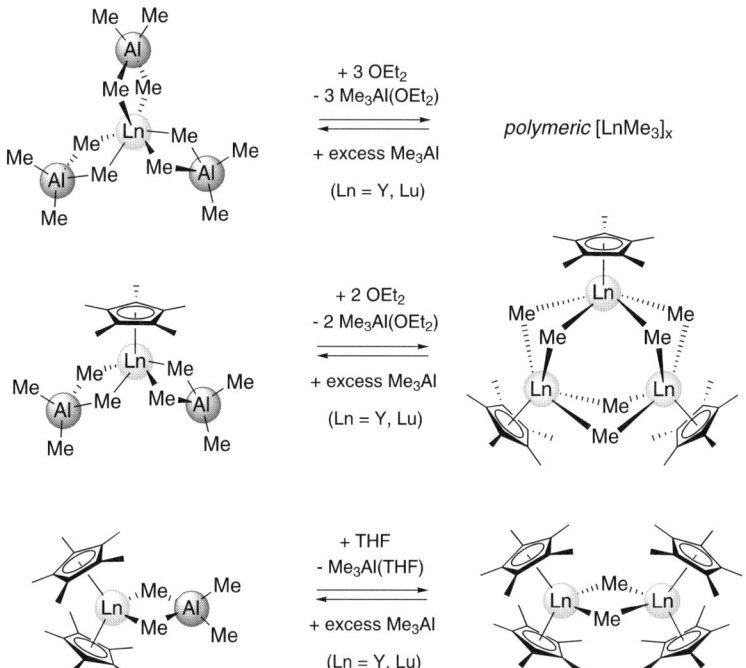

Scheme 62 Donor-induced aluminate cleavage of lanthanide(III) tetramethylaluminates as a general approach to unsolvated hydrocarbyl complexes [10, 12, 151]

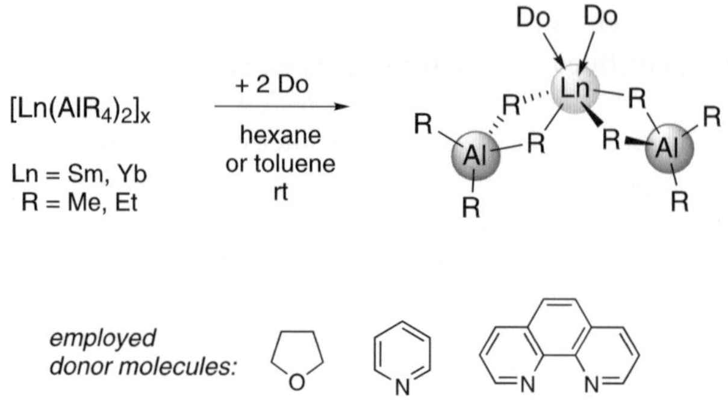

Scheme 63 Synthesis of donor molecule adducts of lanthanide(II) bis(tetraalkylaluminate) complexes (Ln = Sm, Yb; R = Me, Et) [255, 256]

As evidenced for many lanthanidocene tetramethylaluminate complexes, aluminate cleavage in the presence of excess of Do molecules leads to LnMe(Do) species. This reaction behavior was also observed for $(C_5Me_5)_2$Sm$[(\mu\text{-Me})_2AlMe_2]$, however, the tetraethylaluminate $(C_5Me_5)_2$Sm$[(\mu\text{-Et})$ AlEt$_2$ $(\mu\text{-Et})]$ $_2$Sm$(C_5Me_5)_2$ gave mono(THF) adduct $(C_5Me_5)_2$Sm$[(\mu\text{-}\eta^2\text{-}$Et)AlEt$_3$](THF) [115]. Moreover, unlike homoleptic Ln(AlMe$_4$)$_3$ and various heteroleptic tetraalkylaluminate derivatives of the trivalent rare-earth metal centers, "aluminate" cleavage does not occur at Ln(II) metal centers. Instead, the interaction of polymeric [Ln(AlR$_4$)$_2$]$_x$ with donor molecules (Do) accomplished discrete monolanthanide donor adduct complexes [Ln(AlR$_4$)$_2$](Do)$_x$ (Ln = Sm, Yb; Do = THF, pyridine ($x = 2$); phenanthroline ($x = 1$); Scheme 63) in good yields with Do of varying bond strength and bonding mode (Fig. 7) [255, 256].

This different reactivity of heterobimetallic homoleptic Ln^{III}Al^{III}$_3$Me$_{12}$ and Ln^{II}Al^{III}$_2$R$_8$ (R = Me, Et) toward Lewis base molecules clearly reflects a different Ln–C bonding. While the former display true aluminate complexes Ln[AlMe$_4$]$_3$ like Li[AlEt$_4$] and Mg[AlMe$_4$]$_2$ [257, 258], divalent derivatives are better described as lanthanidate complexes [AlEt$_2$]$_2$[LnEt$_4$(Do)$_x$] similar to [Li(Do)$_x$]$_3$[Ln^{III}Me$_6$] [259]. This implicates that the Ln–C bonding nature cannot be rationalized on the basis of an electronegativity scale E_N of the metal centers involved (E_N scale according to Pauling: Li = 1.0, Ln^{III} = 1.1 – 1.3, Al^{III} = 1.6) [260, 261]. Because of the dependency of E_N on the oxidation state of the metal center the E_N value for Ln^{II} centers should be < 1.1 and therefore favor aluminate bonding. Also, the Lewis acidity criterion (Al^{III} > Ln^{III} >> Ln^{II}) commonly considered as the driving force for AlR$_3$Do separation seems to be not applicable. It rather has to be the increased covalent Ln^{II}-ligand bonding which controls such easily performed Lewis base addition reactions.

7.2.2
Alkane Elimination: The Tetramethylaluminate Route

The above-mentioned donor-induced cleavage reactions reveal that tetraalkylaluminate complexes can behave as metal alkyls in "disguise". This bonding feature is in accordance with various alkane elimination reactions observed for homoleptic and heteroleptic tetramethylaluminate complexes.

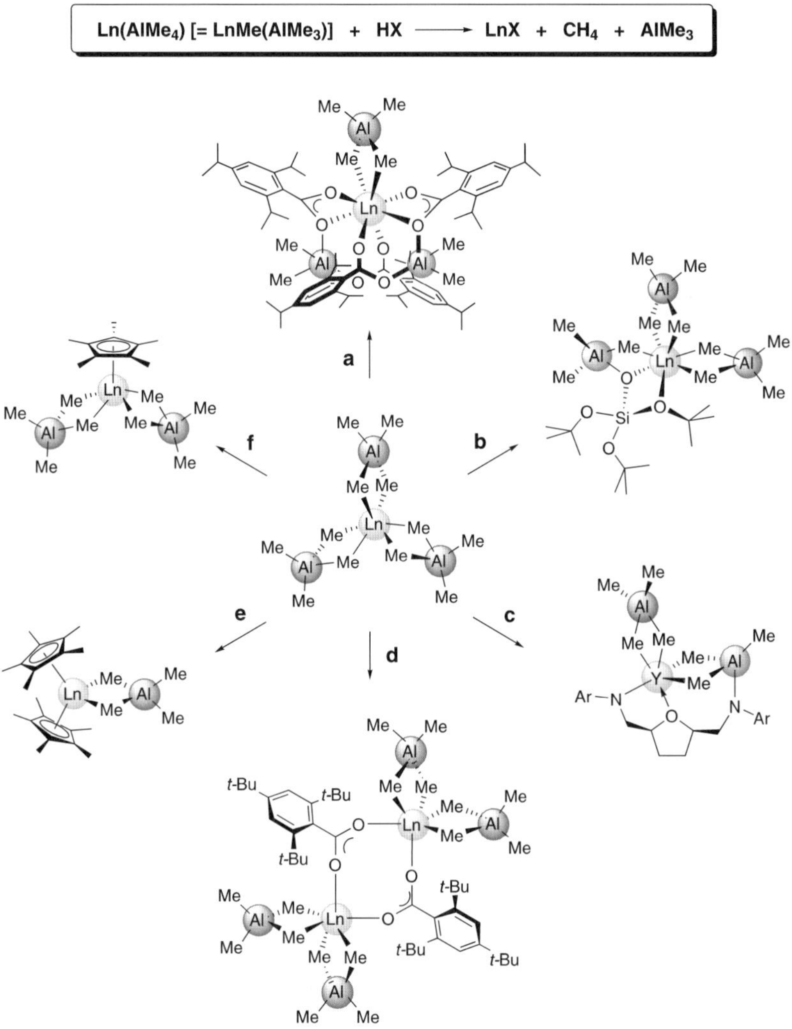

Scheme 64 Alkane elimination in lanthanide(III) tetramethylaluminate chemistry: the tetramethylaluminate route. (**a**): + 4HO$_2$CC$_6$H$_2$(*i*-Pr)$_3$, – CH$_4$, alkylated byproducts; (**b**): + HOSi(O*t*-Bu)$_3$, – CH$_4$, – Me$_3$Al; (**c**): + H$_2$[NON], – 2CH$_4$, – Me$_3$Al; (**d**): + HO$_2$CC$_6$H$_2$(*t*-Bu)$_3$, – CH$_4$, – Me$_3$Al; (**e**): + 2C$_5$Me$_5$H, – 2CH$_4$, – 2Me$_3$Al; (**f**): + C$_5$Me$_5$H, – CH$_4$, – Me$_3$Al

Scheme 64 summarizes examples of [aluminate]→[X] transformation reactions (tetramethylaluminate route), which were accomplished for diverse Brønsted acidic substrates HX.

7.2.3
[Alkyl]→[Chloride] Exchange

Preformation of homoleptic lanthanide tris(tetramethylaluminate) complexes Ln(AlMe$_4$)$_3$ (Ln = Y, La, Ce, Pr, Nd, Sm, Gd) and Et$_2$AlCl generated extremely active and highly cis-1,4-stereoselective binary diene polymerization initiators (Fischbach et al., 2006, personal communication) [150]. Optimal performance in isoprene polymerization was observed for most of the lanthanide metal centers for n(Cl):n(Ln) ratios of 2:1 (Table 16). Only the samarium

Table 16 Polymerization of isoprene with binary Ln(AlMe$_4$)$_3$/Et$_2$AlCl catalysts

Ln [a]	$n_{Cl} : n_{Ln}$ [b]	Yield [c] %	cis-1,4 [d] %	M_n [e] 10^{-3}	M_w [e] 10^{-3}	PDI [e]
Y	1:1	5	67.1	54	226	4.21
Y	2:1	97	75.9	101	400	3.95
Y	3:1	2	97.3	–	–	–
La	1:1	92	> 99	128	546	4.25
La	2:1	99	> 99	184	600	3.26
La	3:1	< 1	> 99	–	–	–
Ce	1:1	> 99	> 99	160	386	2.41
Ce	2:1	> 99	> 99	152	469	3.08
Ce	3:1	13	> 99	66	241	3.66
Pr	1:1	> 99	> 99	386	732	1.90
Pr	2:1	> 99	> 99	320	735	2.30
Pr	3:1	> 99	> 99	345	704	2.02
Nd	1:1	> 99	> 99	228	788	3.45
Nd	2:1	> 99	> 99	117	326	2.78
Nd	3:1	> 99	> 99	113	329	2.92
Gd	1:1	> 99	> 99	278	937	3.33
Gd	2:1	> 99	> 99	146	377	2.58
Gd	3:1	> 99	> 99	195	486	2.49

[a] General conditions: 8 mL hexane, 0.02 mmol precatalyst, 0.02–0.06 mmol Et$_2$AlCl (1–3 equiv), 20 mmol isoprene, 24 h, T = 40 °C
[b] Catalyst preformation 30 min at ambient temperature
[c] Gravimetrically determined
[d] Measured by means of ^{13}C NMR spectroscopy
[e] Measured by means of size exclusion chromatography (SEC) against polystyrene standards

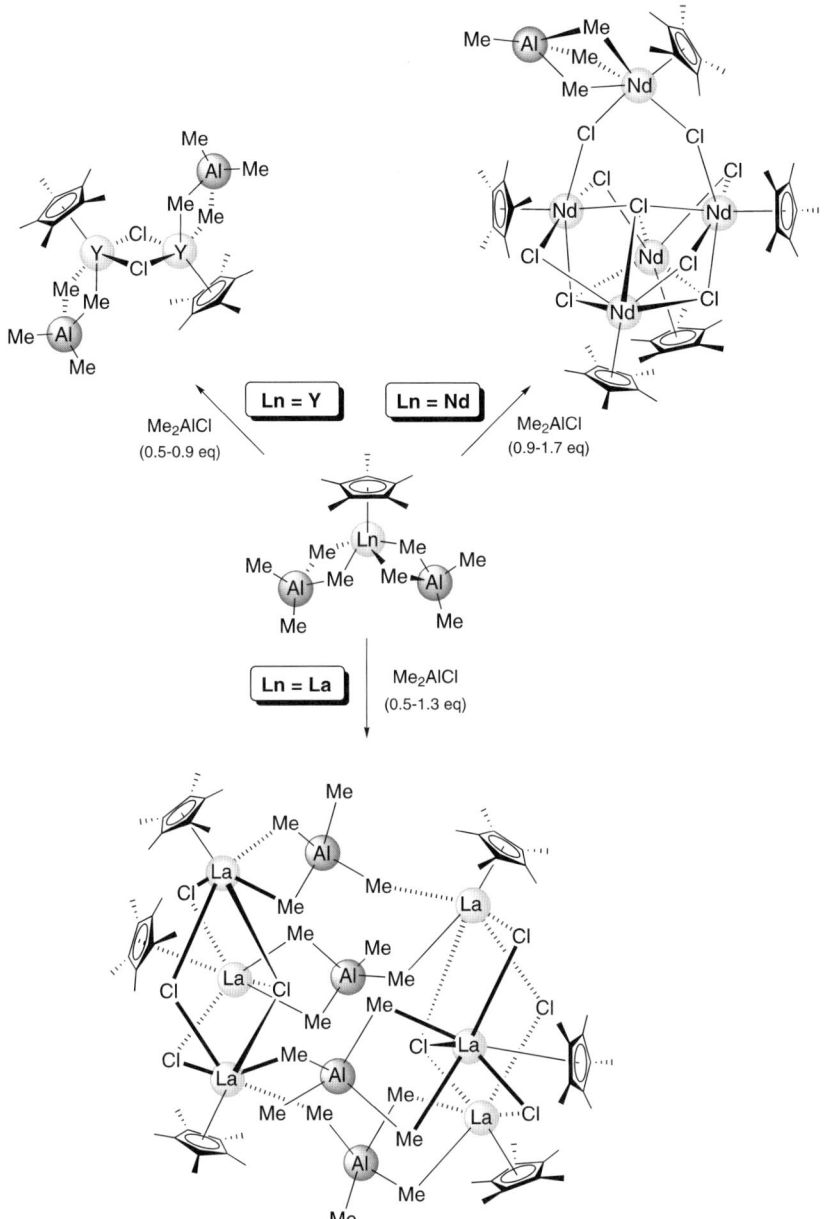

Scheme 65 [Alkyl]→[chloride] interchange reactions in half-lanthanidocene chemistry [248]

derivative did not produce larger amounts of polyisoprene, probably due to the formation of low-valent samarium species (color change!) upon interaction with the cocatalyst [150].

NMR spectroscopic studies suggested the formation of a variety of differently alkylated compounds within the catalyst mixture. These findings were in accordance with (i) chloride transfer to an alkylated rare-earth metal center being the crucial activation step and (ii) neodymium being the most active rare-earth metal center ("neodymium effect") [150]. Direct structural evidence of an [alkyl]→[chloride] interchange reaction was obtained from half-lanthanidocene model compounds. Alkylated [Y_2Al_2], [Nd_5Al] (Fig. 12), and [La_6Al_4] cluster molecules (Fig. 11) were isolated from binary (C_5Me_5)Ln(AlMe$_4$)$_2$/Me$_2$AlCl mixtures (Scheme 65) [248]. These structurally characterized heterobimetallic clusters feature rare examples of alkali metal-free organolanthanide complexes with both alkyl and chloride ligands [115, 116, 224, 262]. The wide-open half-lanthanidocene coordination sphere facilitates new coordination modes (c) and (f) of the tetramethylaluminate ligand (Chart 6, Scheme 65) involving extremely long Ln – C bond distances. Apparently, subtle changes in the rare-earth metal size considerably affect the [aluminate]→[chloride] exchange reaction and coordination behavior of the [AlMe$_4$] moiety. Here, one can speculate about additional evidence of the "obscure neodymium effect". Note that both binary Ln(AlMe$_4$)$_3$/Me$_2$AlCl and commercially employed ternary Ln(carboxylate)$_3$/i-Bu$_2$AlH/Et$_3$Al$_2$Cl$_3$ "Ziegler Mischkatalysatoren" initially produce a fine precipitate which re-dissolves upon addition of monomer [134, 150]. Clearly, such an activation scenario suggests cluster formation being part of the initiating steps.

7.2.4
Cation Formation

The formation of catalytically active dicationic metal centers was originally proposed by R. Taube and coworkers for MAO-activated allyl-based lanthanide catalysts [152, 153]. Several examples of lanthanidocene/alkylaluminoxane systems are mentioned in the literature, using tri- and divalent rare-earth metal centers, as well as different types of cationizing cocatalysts (e.g., MAO, MMAO) [119, 120, 263–268]. Novel cationic complexes [(C_5Me_5)$_2$Ln][B(C_6F_5)$_4$] (Ln = Pr, Nd, Gd) were synthesized from lanthanidocene tetramethylaluminate complexes [(C_5Me_5)$_2$Ln(AlMe$_4$)]$_2$ and fully characterized (Scheme 66) [264, 265]. X-ray structure analysis confirmed the formation of cationic metal centers and "bridging" borate anions. Furthermore, highly efficient initiators for the cis-1,4-stereospecific polymerization of butadiene were achieved in the presence of i-Bu$_3$Al (Table 17). Once again, catalytic activity and stereoregularity were strongly dependent on the rare-earth metal center, with best results obtained for

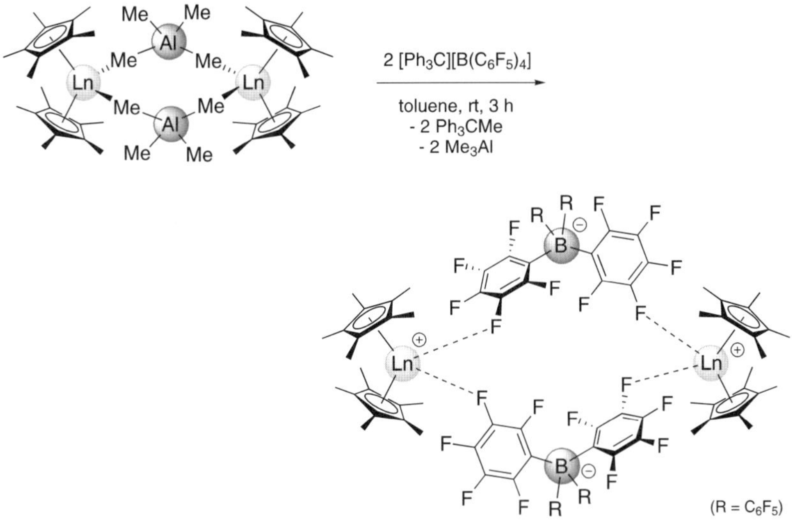

Scheme 66 Formation of cationic lanthanidocene(III) metal centers in borate complexes {[(C$_5$Me$_5$)$_2$Ln][B(C$_6$F$_5$)$_4$]}$_2$ (Ln = Pr, Nd, Gd) [262]

Table 17 Polymerization of butadiene with [(C$_5$Me$_5$)$_2$Ln][B(C$_6$F$_5$)$_4$] activated by i-Bu$_3$Al (adopted from [264, 265])

Ln	T °C	Time min	Yield %	cis-1,4 [b] %	trans-1,4 [b] %	M_n [c] 10^{-3}	PDI [c]
Pr	50	300	95	90.2	6.8	76	1.65
Nd	50	25	96	91.3	6.3	130	1.37
Gd	50	3	100	97.5	1.0	245	1.73
Gd	−20	30	93	99.6	0.0	405	1.41
Gd	−78	12 h	54	>99.9	0.0	301	1.45

[a] General conditions: Polymerization in toluene, [butadiene]$_0$ = 0.90 M (10 mmol), [Ln]$_0$ = 1.80 × 10^{-3} M (2.0 × 10^{-5} mol), n(Al):n(Ln) = 5
[b] Determined by means of ^{1}H and ^{13}C NMR spectroscopy in CDCl$_3$
[c] Measured by means of size exclusion chromatography (SEC) against polystyrene standards

gadolinium. Even at −78 °C polybutadiene was obtained in 54% yield after 12 h. A "perfect" cis-1,4-microstructure (> 99.9%) was reported under these conditions [264, 265].

Structural evidence of the generation of mono- and dicationic alkyl species was reported by J. Okuda and coworkers [155, 269]. Monocationic yttrium(III) dimethyl and dicationic yttrium(III) methyl complexes, [YMe$_2$(THF)$_5$][BPh$_4$] and [YMe(THF)$_6$][BPh$_4$]$_2$, respectively, were synthe-

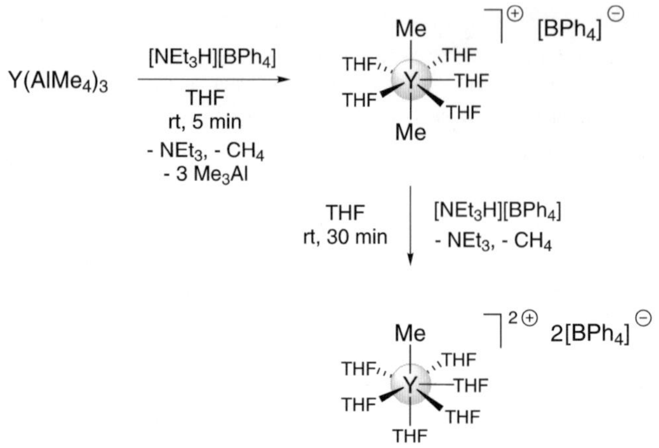

Scheme 67 Synthesis of mono- and dicationic yttrium methyl complexes from Y(AlMe$_4$)$_3$ and [NEt$_3$H][BPh$_4$] [269]

Table 18 Polymerization of butadiene and isoprene with Y(AlMe$_4$)$_3$ activated by [NMe$_2$PhH][B(C$_6$F$_5$)$_4$] (adopted from [155]) [a]

$n_B : n_{Ln}$	Cocatalyst [d]	Time min	Conv. %	Selectivity [e] (cis:trans:vinyl)	M_n [f] 10^{-3}	PDI [f]
Butadiene polymerization [b]						
1:1		30	< 5	–	–	–
1:1		240	73	–	186	4.4
1:1	i-Bu$_3$Al	840	100	90:8:2	50	2.6
2:1		30	26	95:3:2	140	2.0
2:1		240	93	–	312	2.6
2:1	i-Bu$_3$Al	30	18	95:3:2	61	1.7
2:1	i-Bu$_3$Al	840	100	97:2:1	100	2.1
Isoprene polymerization [c]						
1:1		10	56	60:26:14	117	1.6
1:1	i-Bu$_3$Al	10	86	89:0:11	315	2.1
2:1		10	65	67:13:20	260	1.5
2:1	i-Bu$_3$Al	10	78	90:0:10	133	2.9

[a] General conditions: 5 mmol Y, $T = 25\,°C$
[b] $V_{butadiene} = 7.5$ mL (14 wt% in toluene), $V_{total} = 30$ mL
[c] $V_{isoprene} = 1$ mL, $V_{total} = 30$ mL
[d] i–Bu$_3$Al: 0.5 mmol
[e] Determined by means of ^{1}H and ^{13}C NMR spectroscopy
[f] Measured by means of size exclusion chromatography (SEC) against polystyrene standards

sized from homoleptic Y(AlMe$_4$)$_3$ according to Scheme 67. Similar products were obtained from hexamethyl lanthanide complexes Li$_3$LnMe$_6$(THF) (Ln = Sc, Lu, Yb, Ho, Nd) [155]. Although no catalytic activity in diene polymerization was observed with the X-ray crystallographically characterized tetraphenylborate complexes, activation of the initial tetramethylaluminate by [PhNHMe$_2$][B(C$_6$F$_5$)$_4$] gave polymeric materials with high-*cis*-1,4-stereospecificities (Table 18). In butadiene polymerization, higher activities and higher molecular weights were observed for the dicationic species. Moreover, the addition of *i*-Bu$_3$Al prevented the formation of cross-linked polydienes [155].

Similarly, monocationic yttrium bis(alkyl) complexes with non-coordinated tetraalkylaluminate counteranions, [Y(CH$_2$SiMe$_3$)$_2$(THF)$_4$]$^+$[Al(CH$_2$SiMe$_3$)$_4$]$^-$, were isolated from the reaction of Y(CH$_2$SiMe$_3$)$_3$(THF)$_2$ with Al(CH$_2$SiMe$_3$)$_3$ in 75% yield. Ethylene was polymerized after the addition of a 5-fold excess of [PhNHMe$_2$][B(C$_6$F$_5$)$_4$]. Note that a structurally authenticated fluorenyl ytterbium(II) cation [(η^5-C$_{13}$H$_9$)Yb(THF)$_4$]$^+$ with a non-coordinated tetramethylaluminate fragment [AlMe$_4$]$^-$ has been previously described by H. Yasuda and Y. Kai [249].

7.3
Other Systems

A rare example of isospecific 3,4-polymerization of isoprene mediated by a constrained-geometry rare-earth metal initiator was reported by Z. Hou [270]. Binuclear silyl-linked cyclopentadienyl phosphido lanthanide dialkyl complexes were synthesized in good yields and activated with an equimolar amount of [Ph$_3$C][B(C$_6$F$_5$)$_4$] (Scheme 68). Cationic alkyl species were proposed as intermediates and an activation scenario was presented based on DFT calculations [270].

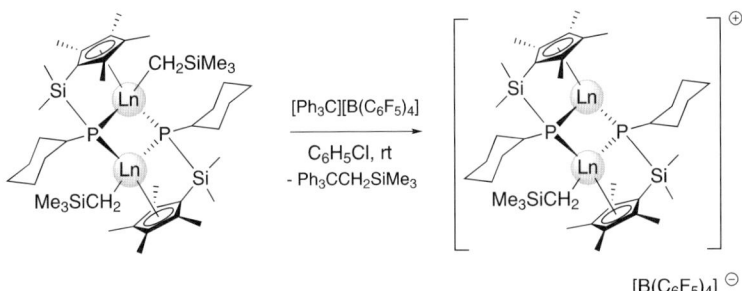

Scheme 68 Proposed initial cation formation during the activation reaction of a half-lanthanidocene phosphide complex with [Ph$_3$C][B(C$_6$F$_5$)$_4$] [270]

8
Lanthanide Hydrides

The formation of catalytically active Ln – H initiator bonds upon treatment of rare-earth metal precursors (carboxylates, alkoxides, silylamides, alkyls) with direct (DIBAH) or indirect (Et$_3$Al, i-Bu$_3$Al: via β-H elimination at elevated temperature) organoaluminum hydride transfer reagents seems to be a viable procedure. Subsequent [H]→[allyl] transformation and generation of the supposedly catalytically active allyl species is a well-established reaction pathway in organolanthanide chemistry [271, 272]. Indeed, DIBAH is routinely used as a preferred cocatalyst species in multicomponent Ziegler catalysts [41, 49, 156]. However, only a few reports exist on the reactivity of discrete rare-earth metal complexes toward organoaluminum hydrides. For example, on the basis of NMR spectroscopic investigations a DIBAH-mediated [carboxylate]→[alkoxide] reduction was postulated to occur in commercially applied neodymium(lanthanum)-based ternary diene polymerization initiators [125]. Moreover, the reaction of Me$_2$Si(2-Me-Ind)$_2$Ln[N(SiHMe$_2$)$_2$] with DIBAH resulted in a structurally evidenced [silylamide]→[hydride] interchange (see Sect. 6.2.3) [213].

Catalyst mixtures with discrete rare-earth metal components featuring preformed Ln – H bonds have not been employed in diene polymerization. Therefore, in this section the interaction of purely inorganic aluminohydrides (alanes) with organolanthanide compounds will be only briefly addressed, as well as the utilization of borohydrides in the fabrication of polydienes.

8.1
Lanthanidocene Aluminohydrides

The X-ray structural characterization of a series of mono-, di-, and poly-lanthanidocene aluminohydride and alane complexes by B.M. Bulychev and G.L. Soloveichik revealed a variety of coordination modes of the [AlH$_4$]$^-$, [AlH$_3$] and [AlH$_2$]$^+$ moieties [273–281]. Routinely, these complexes were synthesized by the reaction of a lanthanidocene chloride with LiAlH$_4$ or AlH$_3$. The nuclearity and AlH$_x$ coordination mode depend on the presence of additional Lewis bases, such as THF, diethyl ether, or triethylamine (Chart 7). Diene polymerization based on lanthanidocene aluminohydrides has not been reported.

8.2
Lanthanide Borohydrides

Because of the low solubilities of lanthanide halides in aliphatic and aromatic solvents catalytic activities of the initially reported binary systems LnCl$_3$/R$_3$Al were rather low. Pseudohalogen borohydride ligands [BH$_4$]$^-$, suc-

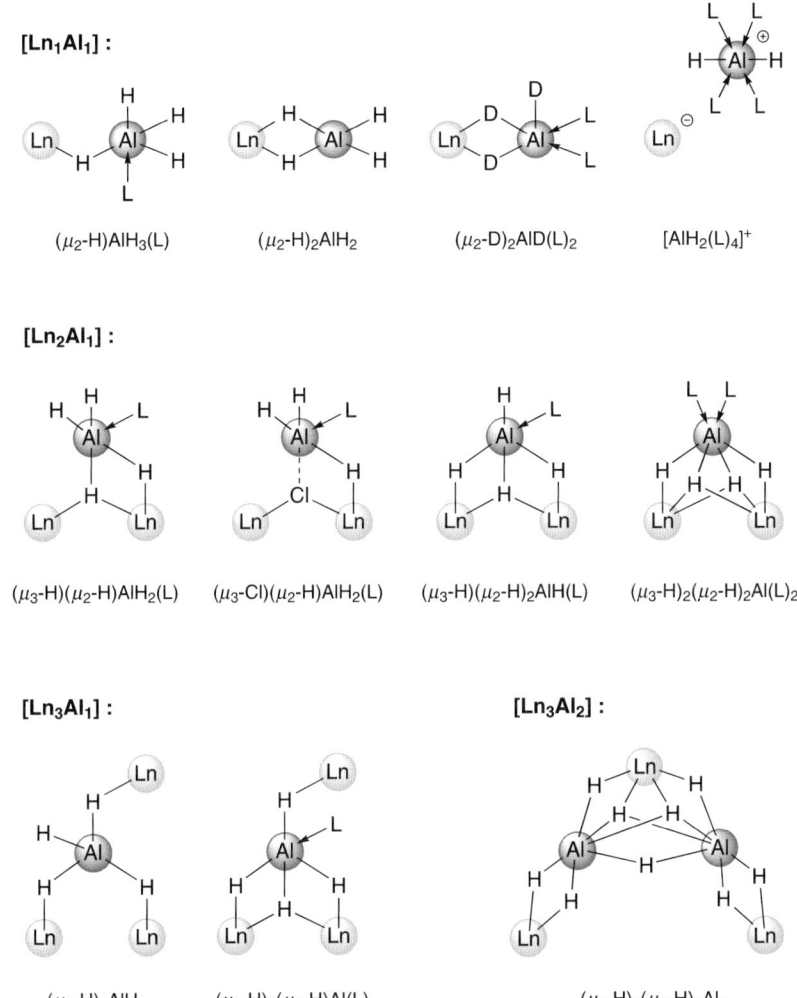

Chart 7 Aluminohydride and alane coordination modes in lanthanidocene complexes

cessfully employed as alternative synthetic precursors in organolanthanide chemistry [282–284], led to more soluble and more active precatalysts for the controlled polymerization of 1,3-dienes. Neodymium tris(borohydride), $Nd(BH_4)_3(THF)_3$, efficiently polymerized isoprene upon activation with Et_3Al or Bu_2Mg [285]. However, no activity was observed in the presence of MAO, allylMgBr, Et_2AlCl, and i-Bu_3Al cocatalysts. Furthermore, lanthanidocene and half-lanthanidocene derivatives were prepared and successfully employed for the fabrication of polydienes (Tables 19 and 20) [284, 286, 287]. On the basis of proton NMR spectroscopic studies and the well-documented ability of the borohydride group to act as a bridging ligand, the

Table 19 Polymerization of isoprene (IP) with binary $Nd(BH_4)_3(THF)_3/Et_3Al$ and $Nd(BH_4)_3(THF)_3/Bu_2Mg$ catalyst mixtures (adopted from [285]) [a]

Cocatalyst (equiv)	[IP]/[Nd]	Time h	Yield %	cis-1,4 %	trans-1,4 %	M_n[b] 10^{-3}	PDI[b]
$Et_3Al(5)$	135	20	90	61.5	34.1	8.2	1.67
$Et_3Al(30)$	400	20	16	53.1	42.7	5.5	1.92
$Et_3Al(15)$	400	90	100	39.3	58.3	10.0	3.00
$Et_3Al(50)$	200	90	59	49.2	46.4	2.8[c]	1.91
$Bu_2Mg(1)$	600	2.75	69	3.2	95.1	27.6	1.57
$Bu_2Mg(1)$	1 000	2.75	63	3.1	95.3	43.0	1.79
$Bu_2Mg(2)$	600	2.75	95	1.8	96.2	23.6	1.14
$Bu_2Mg(3)$	1 000	20	91	2.4	95.5	15.0	1.50
$Bu_2Mg(0)$	200	24	0	–	–	–	–

[a] General conditions: Polymerization in toluene (0.5 ml), $V_{isoprene} = 1$ mL, $T = 50\,°C$. Microstructure determined on the basis of ^1H NMR integration in $CDCl_3$
[b] Measured by means of size exclusion chromatography (SEC) against polystyrene standards
[c] Bimodal shape

Table 20 Polymerization of isoprene (IP) with Bu_2Mg activated lanthanide borohydride catalysts (adopted from [286]) [a,b]

Catalytic system	[IP]/[Nd]	Activity $(g\,mol_{Nd}^{-1})\,h^{-1}$	trans-1,4 %	M_n[c] 10^{-3}	PDI[c]
1 / 1.0 Bu_2Mg	600	15600	95.1	27.6	1.57
2 / 1.2 Bu_2Mg	1180	37300	98.1	50.5	1.16
2 / 0.9 Bu_2Mg	120	2900	98.5	9.5	1.15
3 / 1.2 Bu_2Mg	160	700	98.2	9.7	1.22
4 / 1.0 Bu_2Mg	600	1100	98.4	15.4	2.01
5 / 1.5 BuLi	1000	12000	95.0	90.0	1.50
6 / 1.0 Bu_2Mg	460	340	89.5	–	–

[a] General conditions: Polymerization in toluene, 2 h 45 min, $T = 50\,°C$
[b] Precatalysts: $Nd(BH_4)_3(THF)_3$ (1), $(C_5Me_4Pr)Nd(BH_4)_2(THF)_2$ (2), $(C_5Me_5)Nd(BH_4)_2(THF)_2$ (3), $\{(C_5Me_4Pr)[C(p\text{-tol})NMe]_2Nd(BH_4)\}_2$ (4), $(C_5Hi\text{-}Pr_4)Sm(BH_4)_2(THF)$ (5), $NdCl_3(THF)_3$ (6)
[c] Measured by means of size exclusion chromatography (SEC) against polystyrene standards

formation of $Nd(\mu\text{-}BH_4)(\mu\text{-}R)MgR$ moieties in alkylmagnesium-activated catalyst mixtures was discussed. Proposed molecular structures of the active heterobimetallic species are presented in Scheme 69 [286].

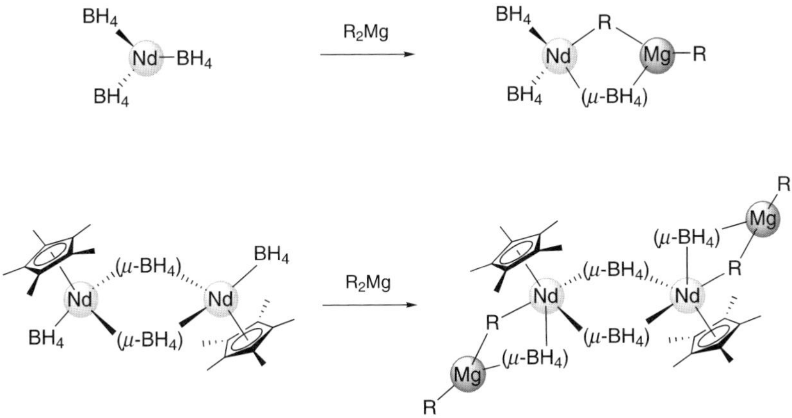

Scheme 69 Proposed formation of active Nd/Mg heterobimetallic species from $Nd(BH_4)_3(THF)_3$ and $(C_5Me_5)Nd(BH_4)_2(THF)_2$. Solvent molecules are omitted for clarity (adopted from [286])

9
Heterogenized Lanthanide Complexes

Solution grafting has been the predominant approach for the immobilization of rare-earth metal precatalyst components [288]. The identification of the catalytically active surface species, commonly formed upon interaction with organoaluminum compounds, is difficult and assisted by molecular model complexes. Several types of support materials including magnesium chloride [289], silica [290], and organic (co-)polymers [291, 292], were examined both in the gas-phase and the slurry polymerization of 1,3-dienes.

9.1
Organic (Co)Polymer Supported Complexes

Recyclable polymer-bonded lanthanide diene polymerization catalysts were first utilized by D.E. Bergbreiter et al. [293]. Since lanthanide carboxylates were found to be useful catalyst precursors, and several types of insoluble carboxylated polymers are commercially available, initial work focused on polymer-supported carboxylate salts of cerium and neodymium. According to Scheme 70, carboxylated polystyrene (cross-linked with 2% divinylbenzene) and carboxylated polyethylene were employed as support materials. Upon activation with i-Bu$_3$Al and Et$_2$AlCl the stereospecific (slurry) polymerization of butadiene was achieved (benzene solution) while no significant difference between the two types of catalysts was observed. Recycling of the catalysts was performed under anhydrous, oxygen-free conditions. No significant loss of activity and selectivity was

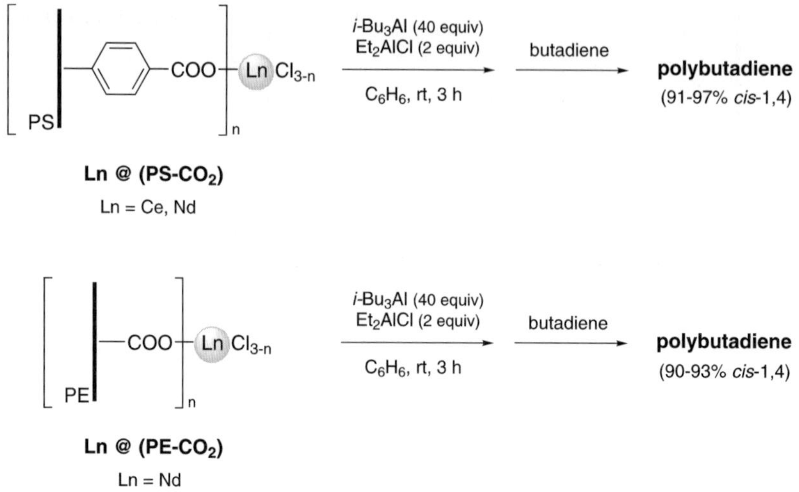

Scheme 70 Polymer-bonded lanthanide-based diene polymerization catalysts. PS = DVB-cross-linked polystyrene (2% divinylbenzene), PE = polyethylene (adopted from [293])

reported, however, subsequent cycles required additional cocatalyst [293]. Other polymeric support materials which have been employed comprise poly(styrene-acrylic acid), poly(ethylene-acrylic acid), polystyrene-*graft*-poly(acrylic acid), polyethylene-*graft*-poly(acrylic acid), polypropylene-*graft*-poly(acrylic acid), and poly(styrene-acrylamide) [292, 294–299].

9.2
Silica-Supported Complexes

Modified and unmodified silica materials also display versatile supports for heterogeneous diene polymerization catalysts. Accordingly, gas-phase polymerization of butadiene with SiO_2-supported rare-earth metal catalysts was the focus of several laboratory scale studies [300–305]. In a recent study a variety of Lewis acidic metal chlorides MCl_x ($MCl_x = BCl_3$, $TiCl_4$, $ZrCl_4$, $HfCl_4$, $SnCl_4$, $SbCl_5$) were used in order to modify the silica surface (Silica I 332, Grace Davison, pore volume 1.55 cm^3 g^{-1}, surface area ≈ 320 m^2 g^{-1}) [306]. As shown in Scheme 71, complex (η^6-C_6H_5Me)Nd(AlCl$_4$)$_3$ and Et$_3$Al were immobilized in a two-step consecutive procedure. High-*cis*-1,4-polybutadiene was obtained with all of the supported [M/Al/Nd] catalysts (> 99% *cis*), with highest activities observed for the BCl$_3$-modified silica surfaces (1188 g$_{polymer}$ g$_{catalyst}^{-1}$ h^{-1}).

The heterogenization of MAO-activated Nd(η^3-C_3H_5)$_3$ · dioxane on MAO-functionalized SiO_2 was reported by T. Rühmer et al. [307]. In situ DRIFT (= diffuse reflectance infrared Fourier transform) spectroscopy and TPRS (= temperature-programmed reaction spectroscopy) were employed

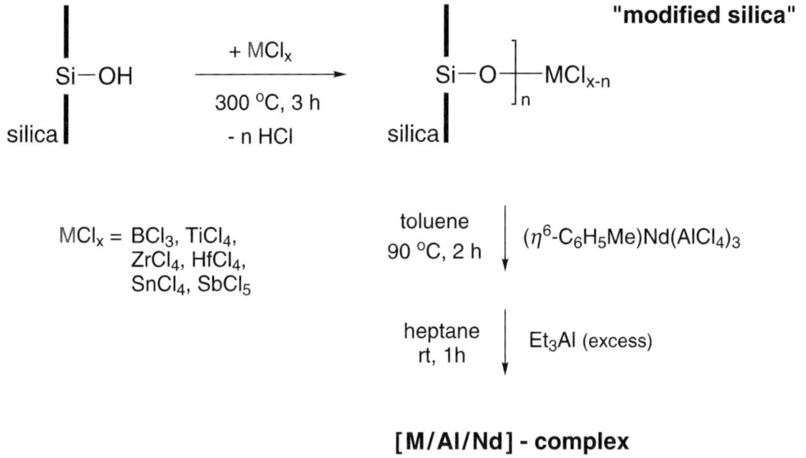

Scheme 71

Si—OH (silica) + MCl$_x$ $\xrightarrow[\text{- n HCl}]{300\,°C,\ 3\,h}$ Si—O$\overline{}$MCl$_{x-n}$ (silica) "modified silica"

MCl$_x$ = BCl$_3$, TiCl$_4$, ZrCl$_4$, HfCl$_4$, SnCl$_4$, SbCl$_5$

$\xrightarrow[\text{90 °C, 2 h}]{\text{toluene}}$ (η^6-C$_6$H$_5$Me)Nd(AlCl$_4$)$_3$

$\xrightarrow[\text{rt, 1 h}]{\text{heptane}}$ Et$_3$Al (excess)

[M/Al/Nd] - complex

Scheme 71 Preparation of heterogeneous neodymium-based diene polymerization catalysts on modified silica (adopted from [306])

to study the interaction of butadiene with the supported allyl lanthanide complexes [308, 309]. The investigation of "adducts" of Nd(C$_3$H$_5$)$_3$ with various aluminum trialkyls suggested that a repeating butadiene insertion, i.e., polymer formation, is only feasible with sterically demanding alkyl groups [309].

Rare-earth metal alkyl species heterogenized on silica Vulkasil S (BAYER AG, 175 ± 20 m^2 g^{-1}) were proposed by G. Fink/M. Bochmann and coworkers for the butadiene homo-, and ethylene/butadiene copolymerization [310]. Accordingly, active Ln – C bonds formed via the interaction of surface-bonded Ln – N(SiMe$_3$)$_2$ moieties with tri*iso*butylaluminum in the course of the catalyst activation. Polybutadienes with *cis*-1,4-stereospecificities between 60.2 and 91.2% were obtained in moderate yields for different lanthanide metal centers (Ln = Sc, Y, La, Nd, Sm, Gd, Dy) [310]. The polymerization conditions were (1) suspension of catalysts in light petroleum/condensed butadiene, (2) sealed tubes, and (3) 50 °C. Surprisingly, all of the different lanthanide metal centers achieved very similar polymerization results with turnover numbers in the range of 800–1500 mol BD (mol metal)$^{-1}$ h^{-1} and neodymium "among the worst" in terms of both activity and stereoselectivity.

Recently, periodic mesoporous silica of the M41S and SBA families were introduced as versatile supports for organometallic species [288, 311]. In a preliminary study, cubic MCM-48 featuring a three-dimensional mesopore system was applied to generate heterogeneous rare-earth metal-based diene polymerization catalysts [150]. Two synthesis protocols were described which differ in the order of neodymium source Nd(AlMe$_4$)$_3$ and cocatalyst Et$_2$AlCl addition (see Scheme 72). The different organometallic-inorganic hybrid materials were characterized by means of FTIR spectroscopy, elemental analysis,

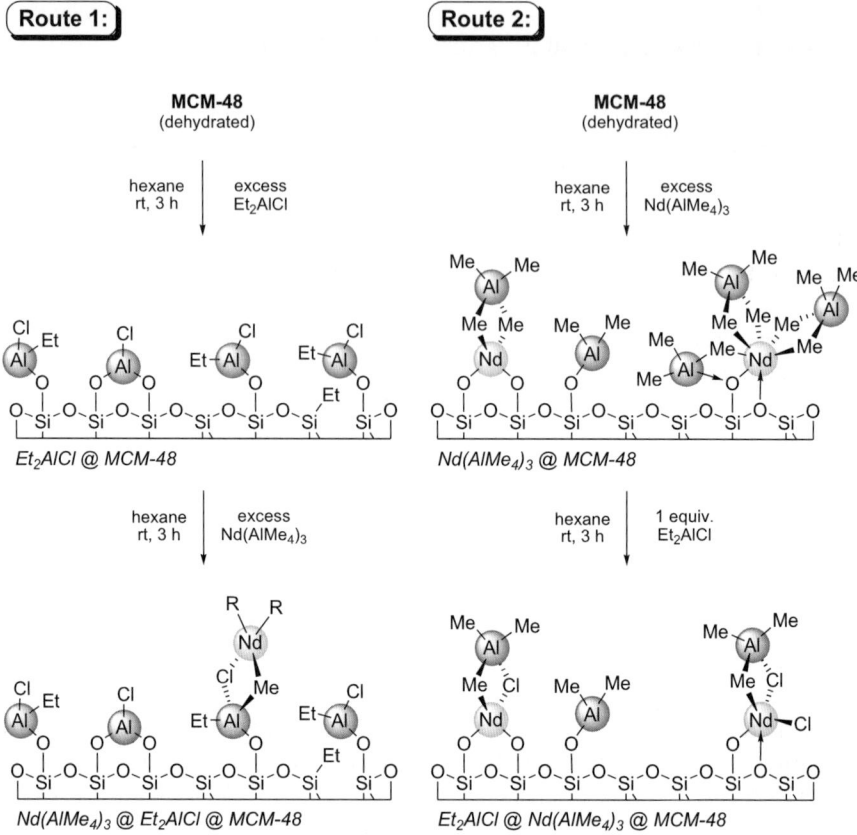

Scheme 72 Proposed surface species of hybrid materials after the immobilization of Nd(AlMe$_4$)$_3$ and Et$_2$AlCl on dehydrated MCM-48. R = Me or (AlMe$_4$) [142, 150]

and nitrogen physisorption. The proposed surface species were modeled with hexane soluble lanthanide alkyl and siloxide complexes [150, 200]. All of the neodymium-*grafted* materials performed as efficient single-component catalysts in the slurry polymerization of isoprene (conditions: hexane, 40 °C). Polymer analysis revealed high-*cis*-1,4-stereospecificities (> 99% *cis*), with higher molecular weights ($M_w \approx 10^6$ g mol^{-1}) and significantly smaller polydispersities (PDI = 1.33–1.88) as compared to the homogeneous binary Nd(AlMe$_4$)$_3$/Et$_2$AlCl catalysts (Sect. 7.2.3). Furthermore, the order of the grafting sequence seemed to have minor implications for the catalyst performance. The narrow molecular weight distributions were attributed to the absence of any organoaluminum cocatalyst dissociation/reassociation processes at the heterogenized active neodymium centers [150].

10
Structural Features of Ln/Al Heterobimetallic Moieties

This paragraph summarizes important details of alkylaluminate bonding and coordination, which emerged from solution NMR and X-ray crystallographic studies.

10.1
Solution Structure: Site Mobility in Tetraalkylaluminate Complexes

In general, homo- and heterobridged alkylaluminate complexes display enhanced solubility in aliphatic and aromatic solvents. The majority of isolated rare-earth metal derivatives has been subjected to detailed solution NMR studies. A common feature of the alkyl ligands is their high mobility, particularly for larger sized and/or sterically unsaturated Ln centers. Such high alkyl group exchangeability is anticipated to crucially affect the polymerization performance of Ziegler–Natta-type catalysts with respect to, for example, catalyst resting states (dormant species) and polymer chain transfer.

10.1.1
Monomer–Dimer Equilibrium

As evidenced by X-ray structure analysis, bis(pentamethylcyclopentadienyl) lanthanide(III) tetramethylaluminates preferentially crystallize as dimeric complexes, with [AlMe$_4$]$^-$ groups linking two [(C$_5$Me$_5$)$_2$Ln]$^+$ units in a "Ln(μ-Me)AlMe$_2$(μ-Me)Ln" fashion [$\mu,\eta^1{:}\eta^1$ coordination mode] [Ln = Y [110], Sm [109], La (Dietrich et al., 2006, personal communication)]. The unambiguous structural evidence of this unique alkylaluminate coordination, featuring two linear "Ln(μ-Me)Al" linkages in the solid state, facilitated the interpretation of the corresponding solution NMR spectra. Both monomeric [(C$_5$Me$_5$)$_2$Ln][(μ-Me)$_2$AlMe$_2$] and dimeric [(C$_5$Me$_5$)$_2$Ln][(μ-Me)AlMe$_2$(μ-Me)]$_2$[Ln(C$_5$Me$_5$)$_2$] are observable by NMR spectroscopy at

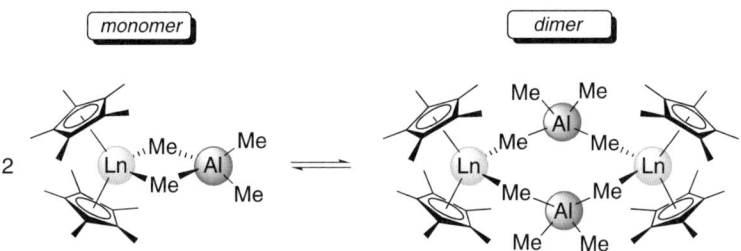

Scheme 73 Monomer–dimer equilibrium observed for lanthanidocene tetramethylaluminate complexes

Table 21 Thermodynamic data for the monomer–dimer equilibrium shown in Scheme 73

$(C_5Me_5)_2Ln$	ΔG^0_{298K} kJ mol^{-1}	ΔH^0 kJ mol^{-1}	ΔS^0 J K^{-1} mol^{-1}	Refs.
$(C_5Me_5)_2Y$	−6.4(9)	−43.6(9)	−125(13)	[110]
$(C_5Me_5)_2Sm$	−7(2)	−36(2)	−100(9)	[109]
$(C_5Me_5)_2Lu$	0.5(9)	−33.8(9)	−115(13)	[110]

ambient temperature. Monomeric complexes with terminal η^2-coordinated tetramethylaluminate ligands are observed at "higher temperatures", or in the presence of sterically more demanding cyclopentadienyl ligands. Thermodynamic data for the monomer–dimer equilibrium (Scheme 73) were reported for the yttrium, samarium, and lutetium derivatives (see Table 21). Note that this intrinsic dimer formation is also observed in the presence of non-cyclopentadienyl ancillary ligand sets [224].

10.1.2
Alkyl Exchange in Terminal Tetramethylaluminates

Alkyl group site interchange reactions in "terminal" η^2-coordinated lanthanidocene tetramethylaluminate complexes featuring unsubstituted cyclopentadienyl ligands were first reported by the Lappert group [9]. Proton and carbon NMR spectroscopy at 40 °C revealed a fluxional behavior of the alkyl ligands in $(C_5H_5)_2Y[(\mu\text{-Me})_2AlMe_2]$ and $(C_5H_5)_2Y[(\mu\text{-Et})_2AlEt_2]$. However, low temperature NMR experiments gave well-resolved spectra for both complexes, with separated signals for bridging and terminal alkyl groups. Because of $^2J_{HY}$ coupling involving the ^{89}Y nuclei (spin = 1/2), the bridging alkyl groups were observed as doublet (R = Me) and eight-line multiplet (R = Et), respectively [9]. In contrast, both methyl and ethyl derivatives of the smallest rare-earth metal center scandium did not show any fluxional behavior at ambient temperature, with two distinct alkyl environments for bridging and terminal alkyl groups [9]. In the following, VT-NMR experiments were performed for a variety of terminal tetramethylaluminate complexes in order to distinguish between the two feasible exchange mechanisms (associative, dissociative) (Scheme 74).

As a representative example, a region of the VT-NMR spectra of the heterobimetallic Ln/Al carboxylate complex $Y[(O_2CC_6H_2i\text{-Pr}_3)_2(AlMe_2)]_2[(\mu\text{-Me})_2AlMe_2]$ is shown in Fig. 3 (cf., Sect. 3.4.2.1). The activation parameters $\Delta H^{\ddagger}$ and $\Delta S^{\ddagger}$ for the exchange of bridging and terminal methyl groups (Table 22) are consistent with a strongly bonded aluminate ligand and a dissociative methyl group exchange [148], as reported earlier for Me_6Al_2 [312]. In contrast, an associative methyl group exchange

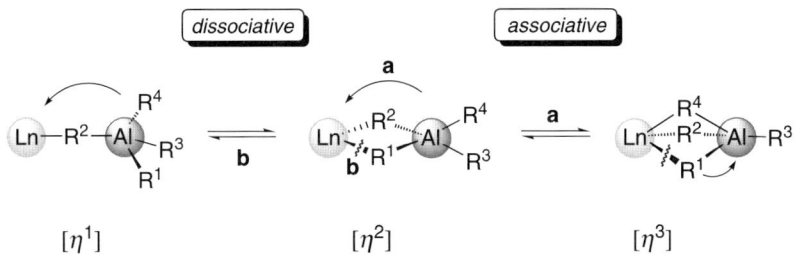

Scheme 74 Dissociative and associative alkyl site exchange pathways in tetraalkylaluminate complexes [148]

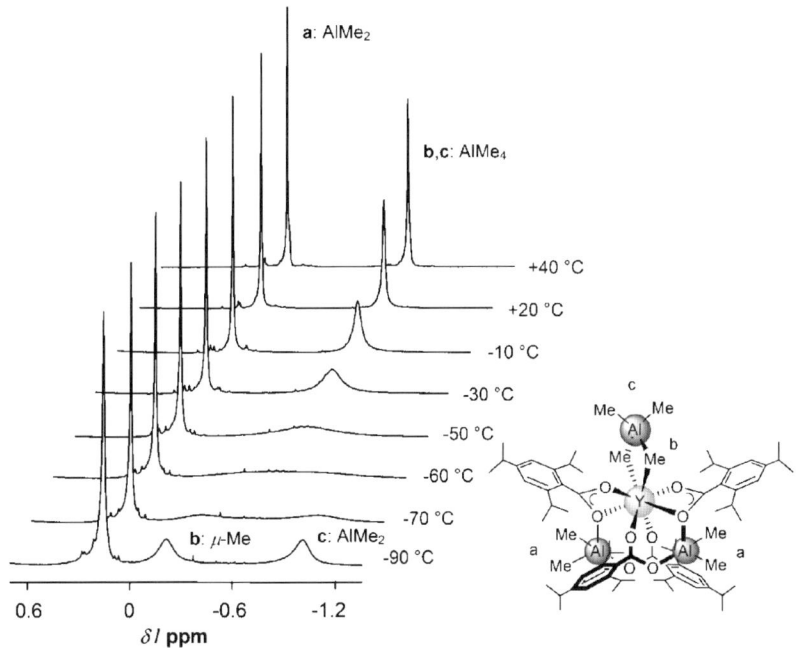

Fig. 3 Region of the VT-NMR spectra (400 MHz, d_8-toluene) of the heterobimetallic Y/Al carboxylate complex Y[(O$_2$CC$_6$H$_2$i-Pr$_3$)$_2$(AlMe$_2$)$_2$]$_2$[(μ-Me)$_2$AlMe$_2$] [147]

was observed for the lanthanum derivative, as proposed earlier for half-lanthanidocene bis(tetramethylaluminate) complexes and C_2-symmetric *ansa*-yttrocene complex Me$_2$Si(2-MeBenzInd)$_2$Y[(μ-Me)$_2$AlMe$_2$] [211, 212]. Similar metal size dependent distinct exchange mechanisms were observed in postlanthanidocene chemistry [222].

While the [η^2]-coordination mode of the tetraalkylaluminate [AlR$_4^-$] ligand is routinely observed in the solid state, the two extreme coordination modes [η^1] and [η^3] have been structurally evidenced only for

Table 22 Thermodynamic data for the exchange of bridging and terminal methyl groups in Ln/Al heterobimetallic complexes

Complex	T_c K^a	$\Delta G^{\ddagger}_{298K}$ [b] kJ mol^{-1}	$\Delta H^{\ddagger}$ [b] kJ mol^{-1}	$\Delta S^{\ddagger}$ [b] J K^{-1} mol^{-1}	Refs.
[1] Y(AlMe$_4$)	263	53(3)	73(4)	66(3)	[148]
[1] La(AlMe$_4$)	213	45(2)	28(2)	−58(3)	[148]
[2] Sc(AlMe$_4$)	348[c]	73(1)	109(1)	122(1)	[222]
[2] Lu(AlMe$_4$)	213	50(2)	34(1)	−56(4)	[222]
[3] Y(AlMe$_4$)	–	63.0	24.3	−130	[204]

[a] Decoalescence temperature
[b] Extracted from line-shape analysis; uncertainties mainly based on temperature errors
[c] Coincides with decomposition point

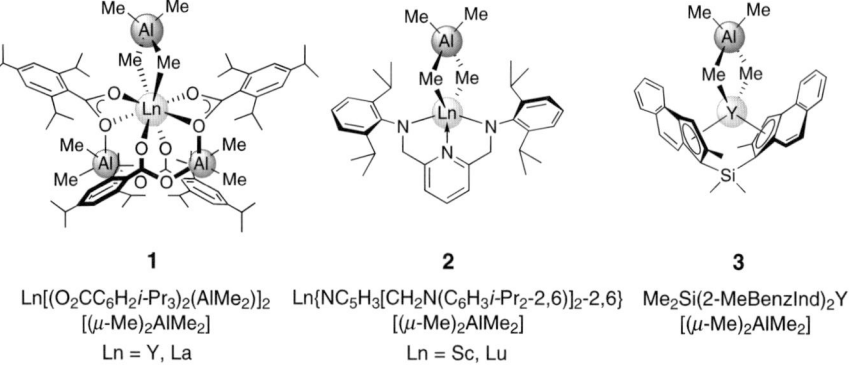

1
Ln[(O$_2$CC$_6$H$_2$i-Pr$_3$)$_2$(AlMe$_2$)]$_2$
[(μ-Me)$_2$AlMe$_2$]
Ln = Y, La

2
Ln{NC$_5$H$_3$[CH$_2$N(C$_6$H$_3$i-Pr$_2$-2,6)]$_2$-2,6}
[(μ-Me)$_2$AlMe$_2$]
Ln = Sc, Lu

3
Me$_2$Si(2-MeBenzInd)$_2$Y
[(μ-Me)$_2$AlMe$_2$]

Chart 8 Heteroleptic tetramethylaluminate complexes examined via VT-NMR spectroscopy

a few examples. Such either sterically oversaturated or sterically highly unsaturated rare-earth metal centers were found in (C$_5$Me$_5$)$_2$Sm(THF)(μ-Et)AlEt$_3$ [115], [Al$_3$Nd$_6$(μ-Cl)$_6$(μ_3-Cl)$_6$(μ-C$_2$H$_5$)$_9$(C$_2$H$_5$)$_5$(Oi-Pr)]$_2$ [170, 313], (C$_5$Me$_5$)$_5$Nd$_5$Cl$_9$(AlMe$_4$) [248], (C$_5$Me$_5$)$_6$La$_6$Cl$_8$(AlMe$_4$)$_4$ [248], and [Yb(AlEt$_4$)$_2$]$_n$ [18]. For the latter homoleptic Yb(II) complex, DFT calculations at the BPW91/I level of theory are supportive of a highly fluxional nature of the aluminate coordination [ΔE ($\eta^2 \rightarrow \eta^3$) = −8 kcal mol^{-1}] and, hence, consistent with the VT-NMR spectroscopic investigations [18]. Accordingly, a complex with two η^3-coordinated tetraethylaluminate ligands appears to be the most stable conformation: six bridging methylene groups surround the Yb center in a distorted antiprismatic fashion.

10.2
Alkyl Exchange in Trimethylaluminum Adduct Complexes

The high mobility of the methyl ligands of Ln/Al organometallics in solution was also evidenced for "heterobridged" trimethylaluminum adduct complexes (Scheme 75). As for the "homobridged" tetramethylaluminate complexes, this fluxional process may be suppressed at lower temperature and/or in the presence of sterically demanding ligand environments and smaller rare-earth metal centers.

For example, complexes $Ln(OC_6H_3 i\text{-}Pr_2\text{-}2,6)_3(AlMe_3)_2$ (Ln = La, Sm), featuring two $[(\mu\text{-}OAr^{i\text{-}Pr,H})(\mu\text{-}Me)AlMe_2]$ ligands, did not show any decoalescence of the alkylaluminum signals in the proton NMR spectra in the [20 to −90]°C temperature range [186, 187]. However, the TMA methyl resonance of the smaller yttrium derivative separated at −80°C into bridging and

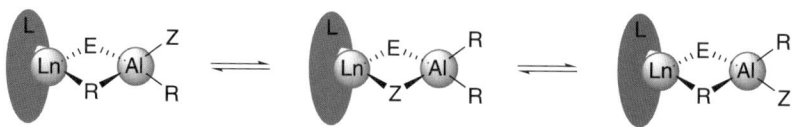

Scheme 75 Alkyl exchange in trialkylaluminum R_3Al adduct complexes (Z = R)

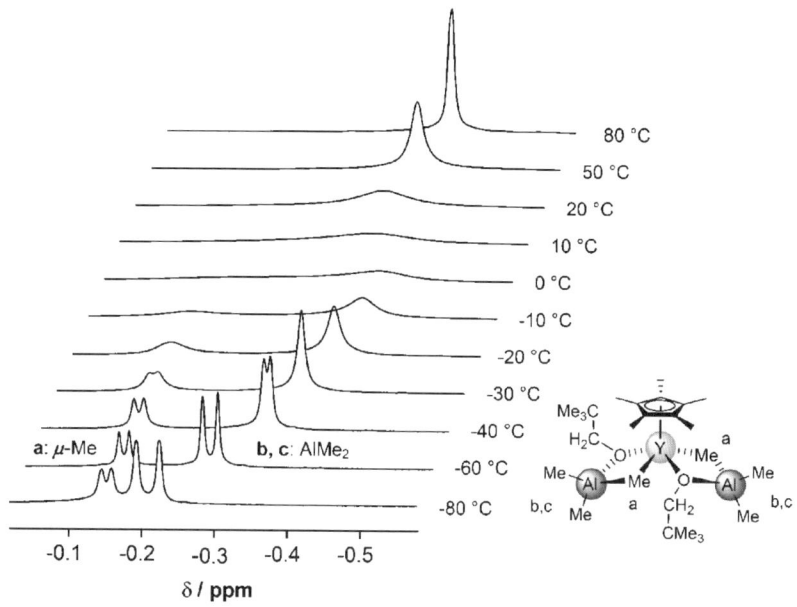

Fig. 4 Alkyl region of VT-NMR spectra (400 MHz, d_8-toluene) of the heterobimetallic Y/Al alkoxide complex $(C_5Me_5)Y(OCH_2CMe_3)_2(AlMe_3)_2$ [142]

terminal groups, indicating the "freezing out" of bridging and terminal ligands [188]. Similarly, half-lanthanidocene complexes $(C_5Me_5)Ln[(\mu\text{-}OCH_2t\text{-}Bu)(\mu\text{-}Me)AlMe_2]_2$ (Ln = Y, Lu) showed a fast methyl group exchange at ambient temperature and separated peaks for bridging and terminal methyl groups at lower temperatures (Fig. 4) [142].

10.3
Solid-State Structures

Most of the knowledge about aluminate and alkylaluminum coordination stems from X-ray crystallographic studies. The basic idea of this section is to compile a "rare-earth metal aluminate library" categorizing this meanwhile comprehensive class of heterobimetallic compounds. Main classification criteria are the type of homo- and heterobridging aluminate ligand (tetra-, tri-, di-, and monoalkylaluminum complexes), the type of "co-ligand" (cyclopentadienyl, carboxylate, alkoxide, siloxide, amide), and the Ln center oxidation state. In addition, related Ln/Al heterobimetallic alkoxide complexes ("nonalkylaluminum" complexes) are surveyed. Emphasis is not put on wordy structure discussions but on the different coordination modes (charts) and important structural parameters in tabular form. An arbitrary collection of molecular structure drawings complements this structural report.

In addition to the numerous X-ray structure analyses, there exists one neutron diffraction study on $Nd(AlMe_4)_3 \cdot 1/2Al_2Me_6$, which was performed at 100 K and gave the first detailed look at a bridging methyl group with a trigonal-bipyramidal carbon atom [208]. The Nd–H distances were found to range from 2.58(2) to (2.75 ± 2) Å, being quite long and corresponding to weaker-than-covalent forces. The "tipping" (or tilting) of the CH_3 groups toward the large Lewis acidic Nd center $(Nd-C-H_{a,b}(80.3\pm4)°$, $NdC-H_c(173.8\pm8)°)$ was interpreted as a $C-H\cdots Ln$ multi- or polyagostic interaction.

A topological analysis of the total electron density of $Y(AlR_4)_3$ (R = Me, Et) using Bader's "Atoms in Molecules" (AIM) approach was carried out in order to gain a better insight into the electronic structure of the "routine" η^2-bonding of aluminate ligands [18, 314]. The contour map of the Laplacian of the electron density for $Y(AlEt_4)_3$ clearly revealed the hypervalent character of the bridging C atoms (CN = 5). Furthermore, the Y–C–Al bond paths appear curved inwards, confirming the electron-deficient character of the $[YC_2Al]$ fragments. From this study it was also concluded that the aluminate bonding cannot be simply described as contact ion pairs Y^{3+}/AlR_4^- because the bridging Al–C bonds seem to be weakened at the expense of significant $Y\cdots C$ interactions, i.e., the occurrence of enhanced charge delocalization within the "Y–C–Al" fragment. A significant degree of covalency was further evidenced by a marked polarization of the outer shell of the core of the Y atom.

10.3.1
Tetraalkylaluminates

Table 23 Structural parameters of heterobimetallic Ln/Al complexes featuring [AlR$_4$] or [AlR$_3$R'] moieties (tetraalkylaluminate complexes)

	Mode (CN)	Ln···Al Å	Ln – C$_{Al}$ Å	Refs.
"homoleptic" aluminates				
Y(AlMe$_4$)$_3$	A1 (6)	3.067–3.068	2.505(7)–2.514(8)	[149]
Pr(AlMe$_4$)$_3$	A1 (6)	3.167(2)–3.187(1)	2.589(3)–2.616(3)	d
Nd(AlMe$_4$)$_3$	A1 (6)	3.150–3.170	2.563(14)–2.609(14)	[149]
Lu(AlMe$_4$)$_3$	A1 (6)	3.006(1)–3.018(1)	2.455(2)–2.471(2)	e
Yb^{II}Al$_2$Et$_8$	A1/A5 (6)	2.809(2)/ 3.188(3)–3.218(2)	2.744(6)–2.824(5)/ η^2: 2.649(5)–2.700(7), η^1: 2778(6)–2.818(5)	[18]
Yb^{II}Al$_2$Me$_8$(phen)	A1 (6)	3.126(2), 3.149(2)	2.600(5)–2.621(6)	[255]
Yb^{II}Al$_2$Et$_8$(THF)$_2$	A1 (6)	3.2139(5)	2.652(2), 2.673(2)	[256]
Sm^{II}Al$_2$Et$_8$(THF)$_2$	A1 (6)	3.324(8)	2.765(3), 2.783(2)	[256]
lanthanidocenes				
(C$_5$H$_5$)$_2$Y(AlMe$_4$)	A1 (8)	3.056	2.601, 2.565	[107]
(C$_5$H$_5$)$_2$Yb(AlMe$_4$)	A1 (6)	3.014(6)	2.609(23), 2.562(18)	[9]
(C$_5$Me$_5$)$_2$Y(AlEt$_4$)	A1 (8)	3.172	2.607(5), 2.622(4)	[315]
(C$_5$Me$_5$)$_2$Sm(AlEt$_4$)	A1 (8)	3.248	2.639(4), 2.6794	[108]
(C$_5$Me$_5$)$_2$Sm(THF)(AlEt$_4$)	A3 (8)	–	2.757(3)$_\alpha$, 2.938(4)$_\beta$	[115]
[Me$_2$Si(C$_5$H$_3$t-Bu)$_2$]Sc(AlMe$_4$)	A1 (8)	2.932	2.425, 2.490	f
(C$_{12}$H$_3$Me$_4$i-Pr$_2$t-Bu)Sc(AlMe$_4$)	A1 (8)	2.918	2.414, 2.442	g
[Me$_2$Si(2-Me-C$_9$H$_5$)$_2$]Y(AlMeEt$_3$)	A1 (8)	3.045(1)	2.557(3), 2.561(3)	[24]
[Me$_2$Si(2-Me-C$_9$H$_5$)$_2$]Y(AlEt$_4$)	A1 (8)	3.040	2.536, 2.548	[315]
[Me$_2$Si(2-Me-C$_{13}$H$_7$)$_2$]Y(AlMe$_4$)	A1 (8)	3.069(1)	2.540(6)	[204]
[(C$_5$Me$_5$)$_2$Y(μ-AlMe$_4$)]$_2$	A5 (8)	–	2.65(2), 2.67(2)	[110]
[(C$_5$Me$_5$)$_2$La(μ-AlMe$_4$)]$_2$	A5 (8)	–	2.862(3), 2.872(3)	e
[(C$_5$Me$_5$)$_2$Sm(μ-AlMe$_4$)]$_2$	A5 (8)	–	2.743(16), 2.750(16)	[109]
[(C$_5$Me$_4$Et)$_2$Sm(μ-AlMe$_4$)]$_2$	A5 (8)	–	2.720(6), 2.732(6)	[316]
{[(C$_5$Me$_4$)SiMe$_2$(CH$_2$CH=CH$_2$)]$_2$Sm(μ-AlMe$_4$)}$_2$	A5 (8)	–	2.702(6), 2.703(6)	[246]
{(C$_5$Me$_5$)$_2$Sm[μ-AlMe$_3$(C$_5$Me$_5$)]}$_2$	A5 (8)	–	2.733–2.897	[317]

Table 23 (continued)

	Mode (CN)	Ln···Al Å	Ln – C$_{Al}$ Å	Refs.
half-lanthanidocenes				
(C$_5$Me$_5$)Y(AlMe$_4$)$_2$	A1 (7)	2.9257(7)/3.1112(7)	2.644(3), 2.655(3)/ 2.549(3), 2.556(3)	[211]
(C$_5$Me$_5$)La(AlMe$_4$)$_2$	A1 (7)	3.0141(9)/3.2687(9)	2.794(2), 2.802(4)/ 2694(3), 2.707(3)	[247]
(C$_5$Me$_5$)Nd(AlMe$_4$)$_2$	A1 (7)	2.959(1)/3.200(1)	2.745(3), 2.745(3)/ 2.629(3), 2.644(3)	[247]
(C$_5$Me$_5$)Lu(AlMe$_4$)$_2$	A1 (7)	2.914(2)/3.061(2)	2.597(3), 2.572(3)/ 2.501(3), 2.501(9)	[211]
[(C$_5$Me$_5$)$_2$Y(AlMe$_4$)(μ-Cl)]$_2$	A1 (7)	3.087(2)	2.539(4), 2.530(4)	[248]
(C$_5$Me$_5$)$_6$La$_6$Cl$_8$(AlMe$_4$)$_4$	A4 (8)	3.318(2), 3.353(2)	η^2: 2.765(4)–2.819(4), η^1: 2.949(3), 2.949(3)	[248]
(C$_5$Me$_5$)$_5$Nd$_5$Cl$_9$(AlMe$_4$)	A2 (8)	2.920(5)	2.779(16)–2.879(18)	[248]
carboxylates				
[La(AlMe$_4$)$_2$ (μ-O$_2$CC$_6$H$_2$$t$-Bu$_3$)]$_2$	A1 (6)	3.255(1), 3.255(1)	2.679(3)–2.694(3)	[148]
[Nd(AlMe$_4$)$_2$ (μ-O$_2$CC$_6$H$_2$$t$-Bu$_3$)]$_2$	A1 (6)	3.190(1), 3.178(1)	2.614(2)–2.626(3)	[148]
alk(aryl)oxides, siloxides				
C$_{62}$H$_{154}$Al$_6$Cl$_{26}$Nd$_{12}$O$_2$	A2 (7)	2.829	2.585–2.769	[313]
Y(OC$_6$H$_3$$t$-Bu$_2$-2,6)$_2$(AlMe$_4$)	A1 (4)	3.067(1)	2.494(3), 2.500(4)	[188]
Lu(OC$_6$H$_3$$t$-Bu$_2$-2,6)$_2$(AlMe$_4$)	A1 (4)	3.000(1)	2.435(4), 2.425(4)	[188]
amides				
[NNN]Lu(AlMe$_4$) [a]	A1 (5)	3.018(1)	2.435(4), 2.435(4)	[222]
other complexes				
{[NN]$_2$Sm$_2$Cl$_3$(μ-AlMe$_4$)}$_2$ [b]	A5 (6)	–	2.720(6)–2.759(6)	[224]
Sm^{II}[(9-Me$_3$Si-Flu)(AlMe$_3$)]$_2$ [c]	A1 (8)	–	2.856(8)	[249]
Sm^{II}[(9-Me$_3$Si-Flu)(AlEt$_3$)]$_2$ [c]	A1 (8)	–	2.94(3), 2.96(3)	[249]
Yb^{II}[(9-Me$_3$Si-Flu)(AlMe$_3$)]$_2$ [c]	A2 (8)	3.05(1)	2.672, 2.782	[249]

[a] [NNN] = NC$_5$H$_3$(CH$_2$NC$_6$H$_3$$i$-Pr$_2$-2,6)$_2$-2,6.
[b] [NN] = CH[C(CH$_3$)NC$_6$H$_3$$i$-Pr$_2$-2,6]$_2$-1,2.
[c] Me$_3$Si-fluorenyl ligand is considered as pseudo-alkyl substituent R
[d] Fischbach et al., 2006, personal communication
[e] Dietrich et al., 2006, personal communication.
[f] Day et al., 2006, personal communication.
[g] Day et al., 2006, personal communication.

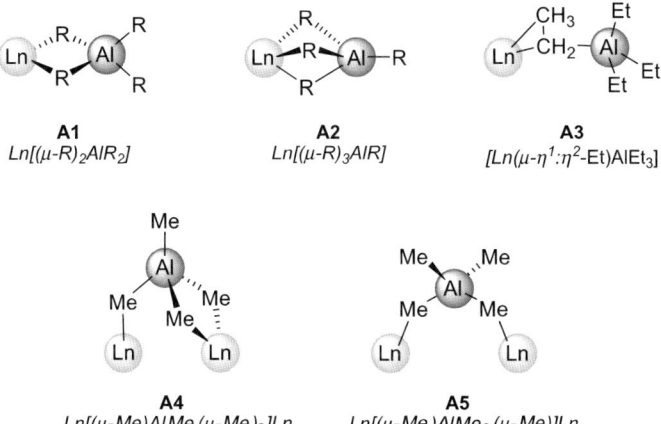

Chart 9 Coordination modes in lanthanide tetraalkylaluminate complexes

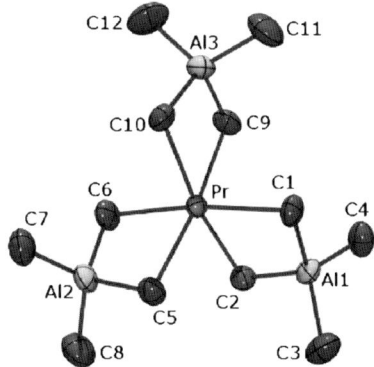

Fig. 5 Solid-state structure of Pr(AlMe$_4$)$_3$ (motif **A1**)

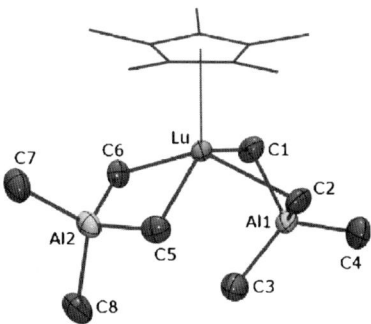

Fig. 6 Solid-state structure of (C$_5$Me$_5$)Lu(AlMe$_4$)$_2$ (motif **A1**) [211]

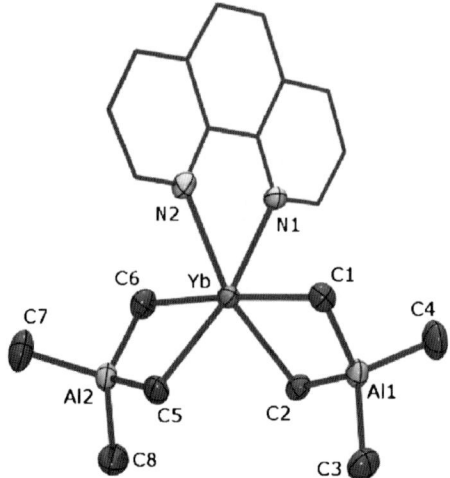

Fig. 7 Solid-state structure of Yb(AlMe$_4$)$_2$(phen) (motif **A1**) [255]

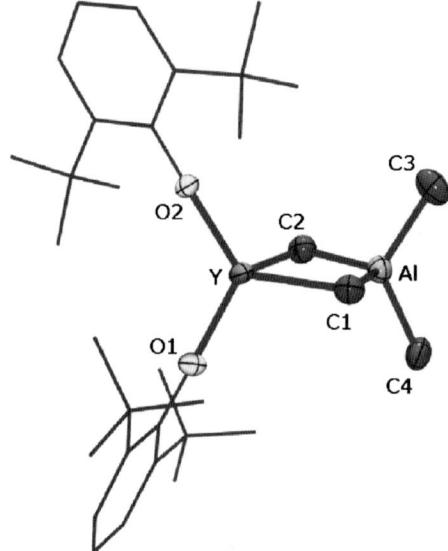

Fig. 8 Solid-state structure of Y(OC$_6$H$_3$t-Bu$_2$-2,6)$_2$(AlMe$_4$) (motif **A1**) [188]

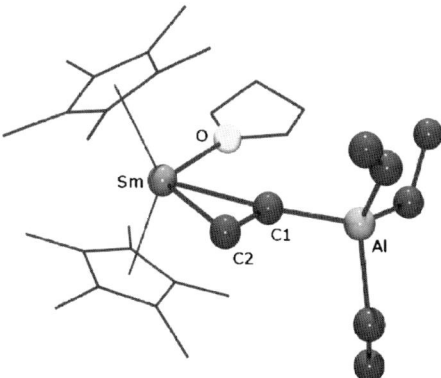

Fig. 9 Solid-state structure of $(C_5Me_5)_2Sm(THF)(AlEt_4)$ (motif **A3**) [115]

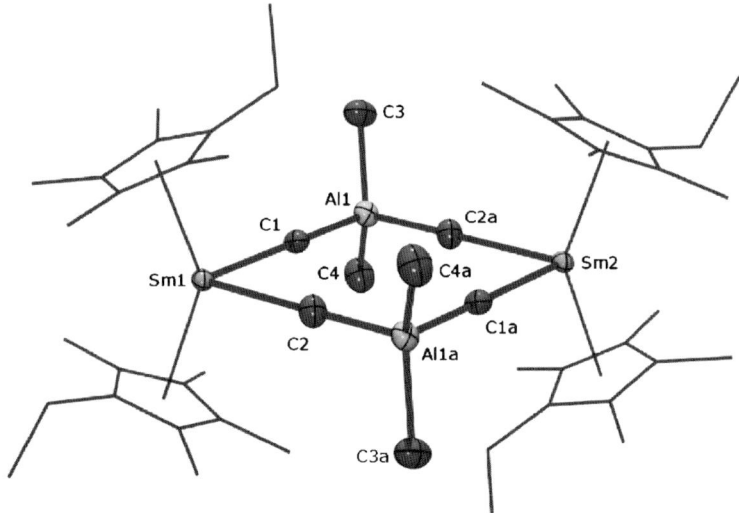

Fig. 10 Solid-state structure of $[(C_5Me_4Et)_2Sm(\mu\text{-}AlMe_4)]_2$ (motif **A5**) [316]

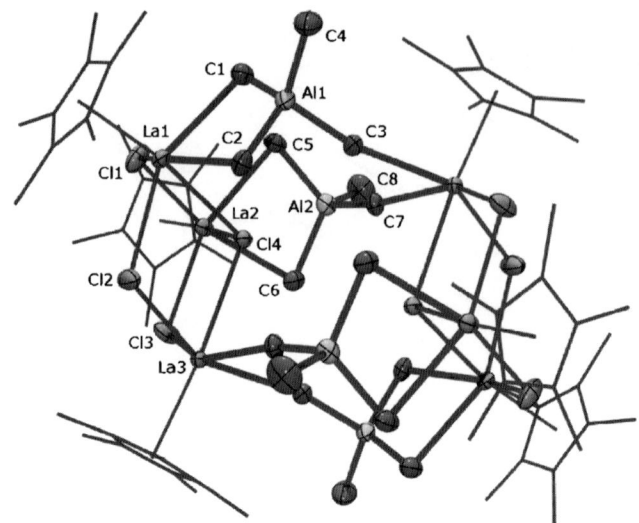

Fig. 11 Solid-state structure of $(C_5Me_5)_6La_6Cl_8(AlMe_4)_4$ (motif **A4**) [248]

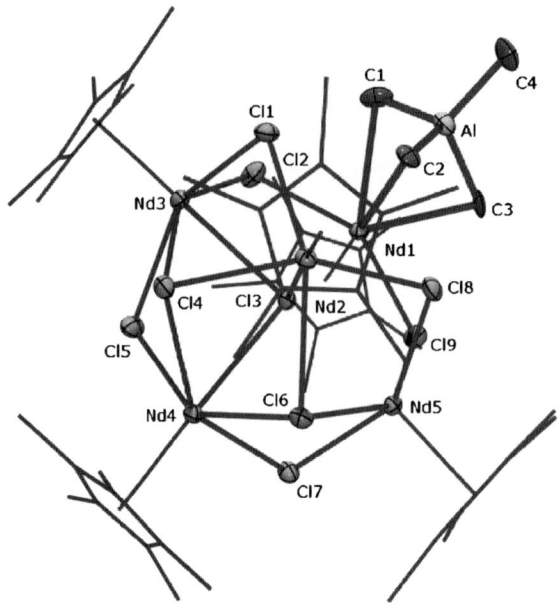

Fig. 12 Solid-state structure of $(C_5Me_5)_5Nd_5Cl_9(AlMe_4)$ (**A2**) [248]

10.3.2
Trialkylaluminum Complexes

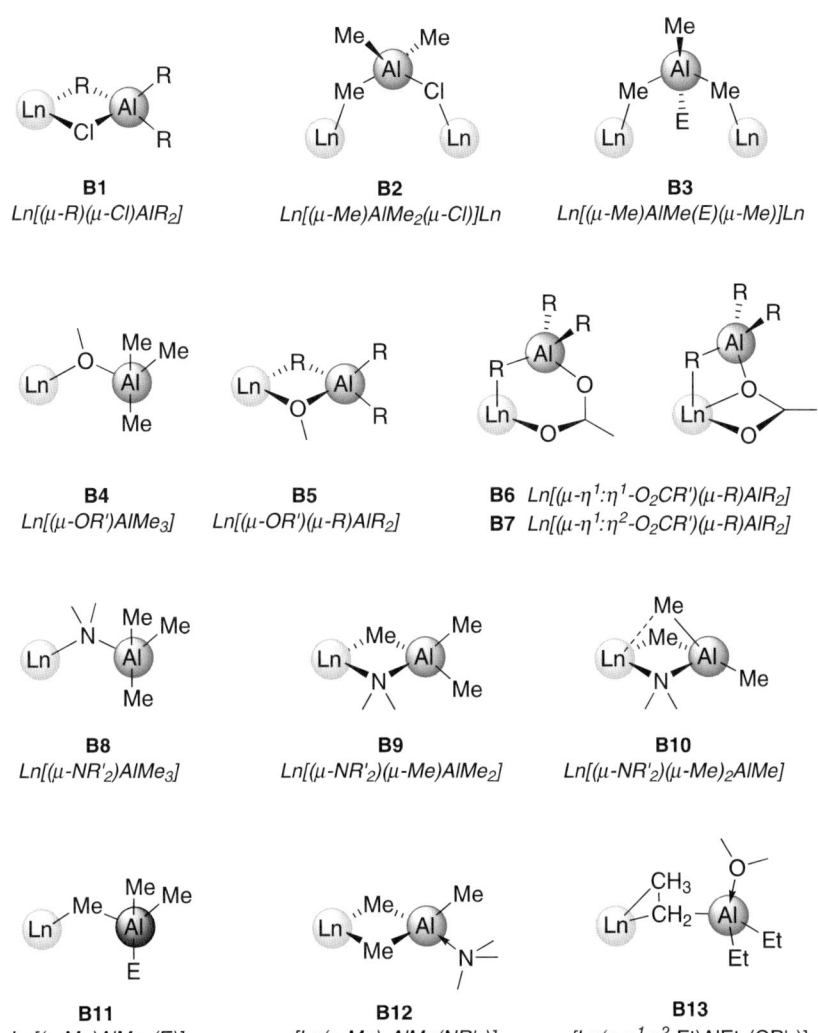

Chart 10 Coordination modes in lanthanide trialkylaluminum complexes

Table 24 Structural parameters of heterobimetallic Ln/Al complexes featuring [R_3Al] moieties

	Mode (CN)	Ln···Al Å	Ln – C_{Al} Å	Refs.
lanthanidocenes				
(C_5Me_5)$_2$Y[(μ-Cl)(μ-Et)AlEt$_2$]	**B1** (8)	3.525	2.721(5)	[115]
(C_5Me_5)$_2$Y[(μ-Cl)(μ-*i*-Bu)Al(*i*-Bu)$_2$]	**B1** (8)	3.628	2.790(2)	[115]
[(C_5Me_5)$_2$Sm(μ-Cl)AlMe$_2$(μ-Me)]$_2$	**B2** (8)	–	2.745(3)	[116]
[(C_5Me_5)$_2$Sm(μ-Cl)AlEt$_2$(μ-Et)]$_2$[a]	**B2** (9)	–	2.749(3)$_\alpha$, 2.976(3)$_\beta$	[116]
(C_5Me_5)$_2$Sm[(μ-Cl)(μ-*i*-Bu)Al(*i*-Bu)$_2$]	**B1** (8)	3.659	2.836(3)	[116]
[(C_5Me_5)$_2$Yb(AlMe$_3$)$_2$(S-*p*-Tol)]$_2$	**B11** (8)	–	2.670(7), 2.667(6)	[183]
(C_5Me_5)$_2$Yb^{II}(AlEt$_3$)(THF)	**B13** (8)	–	2.854(18)$_\alpha$, 2.939(21)$_\beta$	[144]
(C_5Me_5)$_2$Sm[(O_2CPh)(Al*i*-Bu$_3$)]	**B7** (9)/ **B6** (8)	3.732/ –	3.209(3)/2.962(3)	[143]
half-lanthanidocenes				
($C_5H_4SiMe_3$)Y(O*t*-Bu)$_2$(AlMe$_3$)$_2$	**B5** (7)	3.094(2), 3.160(2)	2.577(7), 2.562(6)	[180]
(C_5Me_5)Y(OCH$_2$*t*-Bu)$_2$(AlMe$_3$)$_2$	**B5** (7)	3.166(1), 3.176(1)	2.584(3), 2.576(4)	[142][b]
(C_5Me_5)Lu(OCH$_2$*t*-Bu)$_2$(AlMe$_3$)$_2$	**B5** (7)	3.119(2), 3.123(2)	2.540(4), 2.533(4)	[142][b]
alk(aryl)oxides, siloxides				
Y(O*t*-Bu)$_3$(AlMe$_3$)$_3$	**B5** (6)	3.195(7)	2.688(28)	[178]
Nd(O*t*-Bu)$_3$(AlMe$_3$)$_3$	**B5** (6)	3.30(1)	2.784(11)	[179]
Y(O*t*-Bu)$_3$(AlMe$_3$)$_2$(THF)	**B5** (6)	3.244(3), 3.200(3)	2.735(6), 2.668(1)	[178]
Y(OC$_6$H$_3$*i*-Pr$_2$-2,6)$_3$(AlMe$_3$)$_2$	**B5** (5)	3.192(1), 3.187(1)	2.544(2), 2.541(2)	[188]
La(OC$_6$H$_3$*i*-Pr$_2$-2,6)$_3$(AlMe$_3$)$_2$	**B5** (5)	3.410(3), 3.367(3)	2.801(5), 2.759(5)	[187]
Nd(OC$_6$H$_3$*i*-Pr$_2$-2,6)$_3$(AlMe$_3$)$_2$	**B5** (5)	3.295(3), 3.312(3)	2.652(4), 2.681(6)	[142]
Sm(OC$_6$H$_3$*i*-Pr$_2$-2,6)$_3$(AlMe$_3$)$_2$	**B5** (5)	3.284(3), 3.273(3)	2.632(5), 2.620(5)	[186]
Sm(OC$_6$H$_3$*i*-Pr$_2$-2,6)$_3$(AlEt$_3$)$_2$	**B5** (5)	3.293(2), 3.286(2)	2.649(4), 2.627(4)	[187]

Table 24 (continued)

	Mode (CN)	Ln···Al Å	Ln–C$_{Al}$ Å	Refs.
amides				
Nd(NMe$_2$)$_3$(AlMe$_3$)$_3$	**B9**/ **B8** (6)	3.274(2)	2.708(5)	[206]
[Sm(NC$_6$H$_3$*i*-Pr$_2$)(NHC$_6$H$_3$*i*-Pr$_2$)(AlMe$_3$)]$_2$	**B9** (6)	3.29(2)	2.65(2)	[218]
Yb^{II}[N(SiMe$_3$)$_2$]$_2$(AlMe$_3$)$_2$	**B10** (6)	2.926, 2.963	2.788(2), 2.756(2), 3.042(2), 3.202(3)	[190]
other complexes				
(C$_5$Me$_5$)$_4$Y$_4$(CH)$_2$Me$_2$(AlMe$_3$)$_4$	**B3** (7)	3.075(1)– 3.086(1)	2.640(3)–2.700(3)	[318]

^{a} Bridging Et coordinates (μ-η^1:η^2)
^{b} Fischbach et al., 2006, personal communication

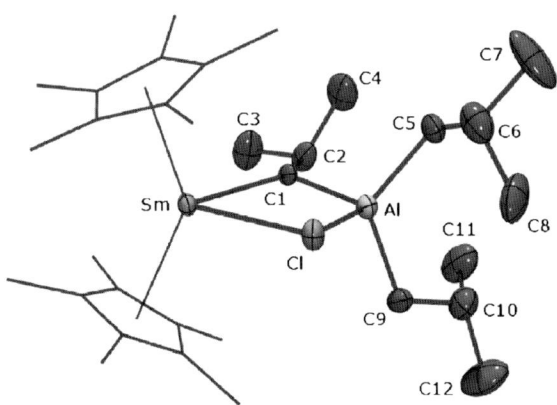

Fig. 13 Solid-state structure of (C$_5$Me$_5$)$_2$Sm[(μ-Cl)(μ-*i*-Bu)Al(*i*-Bu)$_2$] (motif **B1**) [116]

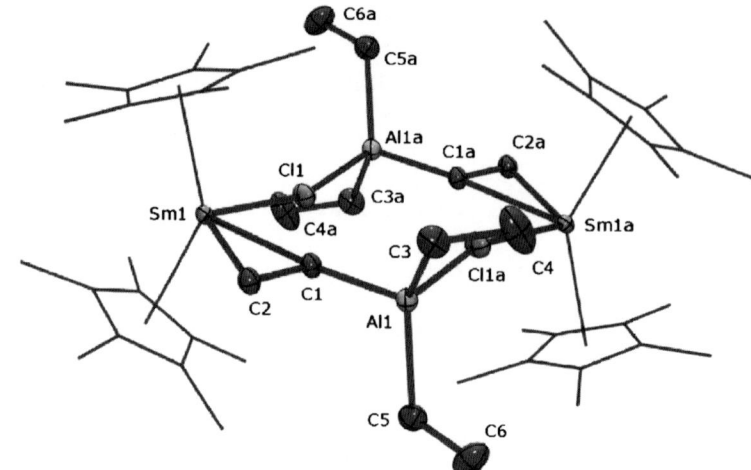

Fig. 14 Solid-state structures of [(C$_5$Me$_5$)$_2$Sm(μ-Cl)AlEt$_2$(μ-Et)]$_2$ (motif **B2**) [116]

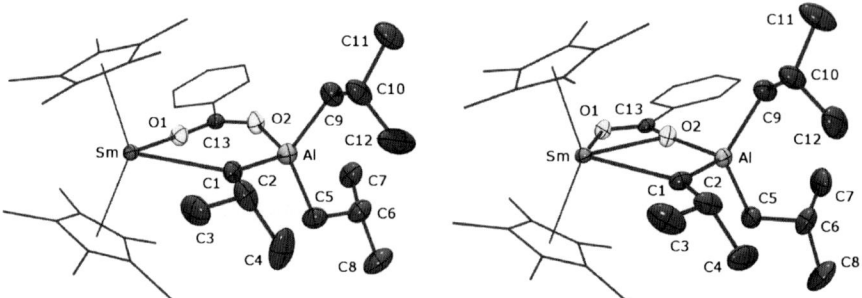

Fig. 15 Solid-state structures of (C$_5$Me$_5$)$_2$Sm(O$_2$CPh)(i-Bu$_3$Al) (motifs **B6** (*left*) and **B7** (*right*)) [143]

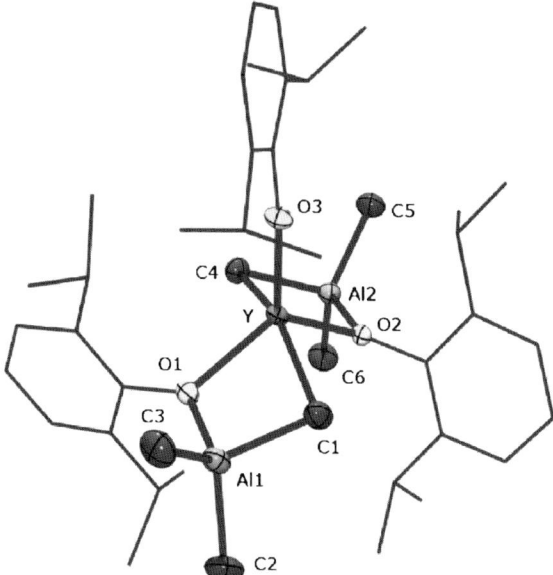

Fig. 16 Solid-state structure of Y(OC$_6$H$_3$*t*-Bu$_2$-2,6)$_3$(AlMe$_3$)$_2$ (motif **B5**) [188]

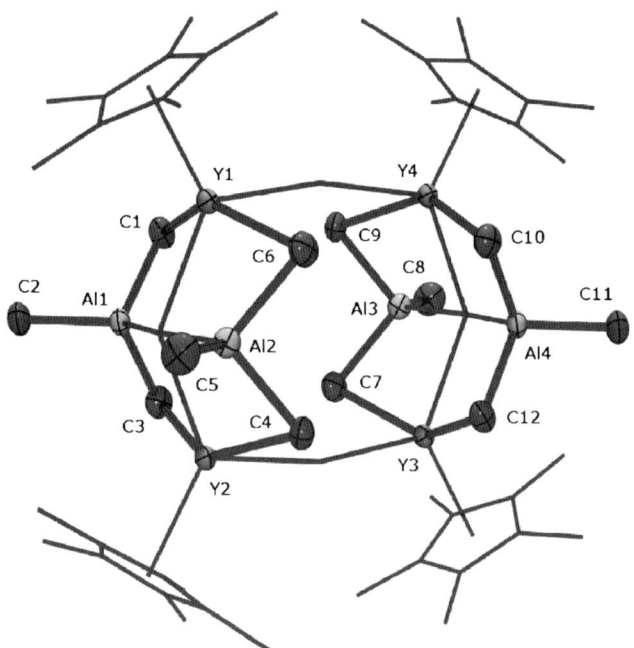

Fig. 17 Solid-state structure of (C$_5$Me$_5$)$_4$Y$_4$(CH)$_2$Me$_2$(AlMe$_3$)$_4$ (motif **B3**) [318]

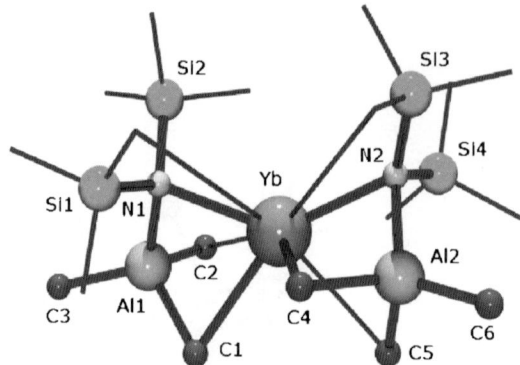

Fig. 18 Solid-state structure of Yb[N(SiMe$_3$)$_2$]$_2$(AlMe$_3$)$_2$ (motif **B10**) [190]

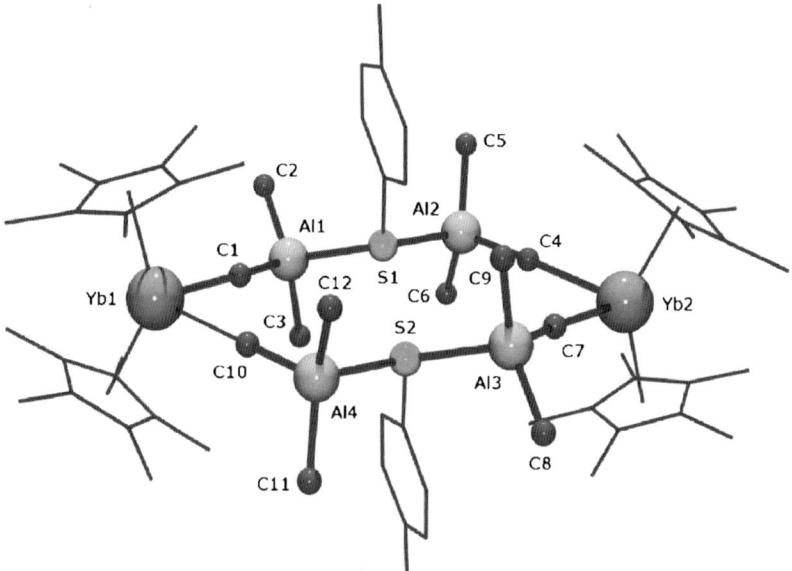

Fig. 19 Solid-state structure of [(C$_5$Me$_5$)$_2$Yb(AlMe$_3$)$_2$(S-p-Tol)]$_2$ (motif **B11**) [183]

10.3.3
Dialkylaluminum Complexes

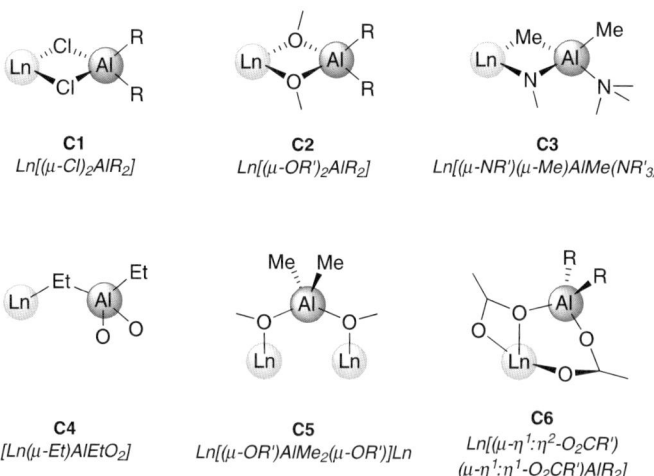

Chart 11 Coordination modes in lanthanide dialkylaluminum complexes

Table 25 Structural parameters of heterobimetallic Ln/Al complexes featuring [R_2Al] moieties

	Mode (CN)	Ln···Al Å	Ln – C_{Al} Å	Refs.
lanthanidocenes				
(C_5Me_5)$_2$Y[(μ-Cl)$_2$AlMe$_2$]	C1 (8)	3.720(1)	–	[115]
(C_5Me_5)$_2$Y[(μ-Cl)$_2$AlEt$_2$]	C1 (8)	3.691(1)	–	[115]
(C_5Me_5)$_2$Y[(μ-Cl)$_2$Al(*i*-Bu)$_2$]	C1 (8)	3.717(1)	–	[115]
(C_5Me_5)$_2$Sm[(μ-Cl)$_2$AlMe$_2$]	C1 (8)	3.791(1)	–	[115]
(C_5Me_5)$_2$Sm[(μ-Cl)$_2$AlEt$_2$]	C1 (8)	3.765(1)	–	[115]
(C_5Me_5)$_2$Sm[(μ-Cl)$_2$Al(*i*-Bu)$_2$]	C1 (8)	3.772(1)	–	[115]
carboxylates				
[Pr(CF$_3$CO$_2$)$_2$(CF$_3$CHO$_2$)(AlEt$_2$)(THF)$_2$]$_2$	C6 (8)	–	–	[319]
[Eu(CF$_3$CO$_2$)$_2$(CF$_3$CHO$_2$)(AlEt$_2$)(THF)$_2$]$_2$	C6 (8)	–	–	[132]
[Nd(CF$_3$CO$_2$)$_2$(CF$_3$CHO$_2$)(Al*i*-Bu$_2$)(THF)$_2$]$_2$	C6 (8)	–	–	[132]
alk(aryl)oxides				
Nd[(OC$_6$H$_4$Me)$_2$AlMe$_2$]$_3$	C2 (6)	3.363(6)	–	[185]
Nd[(OC$_6$H$_3$Me$_2$)$_2$AlEt$_2$](OC$_6$H$_3$Me$_2$)$_2$(THF)$_2$	C2 (6)	3.411(6)	–	[184]
Yb[(OC$_6$H$_3$Me$_2$)$_2$AlMe$_2$](OC$_6$H$_3$Me$_2$)$_2$(THF)$_2$	C2 (6)	3.298(7)	–	[184]
Sm[(O*t*-Bu)$_2$Al*i*-Bu$_2$](O*t*-Bu)$_2$(THF)$_2$	C2 (6)	3.36(1)	–	[320]

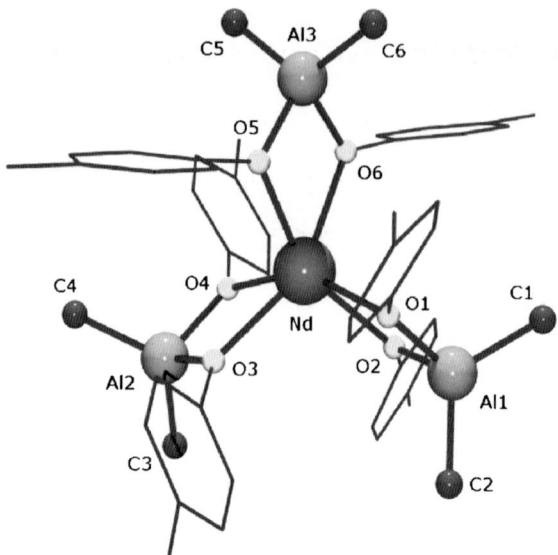

Fig. 20 Solid-state structure of $Nd[(OC_6H_4Me-4)_2AlMe_2]_3$ (motif **C2**) [185]

10.3.4
Monoalkylaluminum and "Non-Alkylaluminum" Complexes

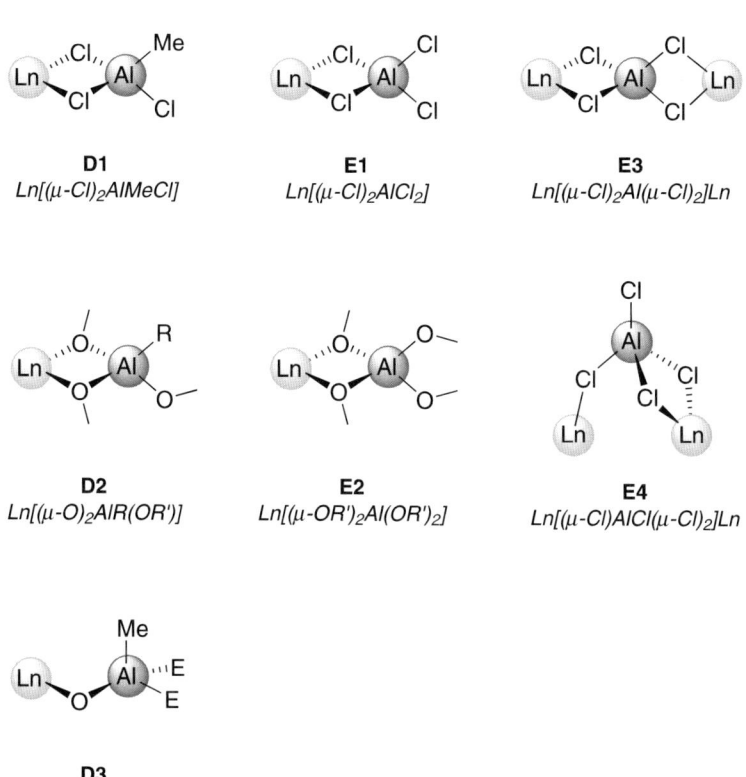

Chart 12 Coordination modes in heterobimetallic Ln/Al complexes featuring [AlE$_3$R] or [AlE$_4$] moieties (E = Cl, O)

Table 26 Structural parameters of heterobimetallic Ln/Al complexes featuring [AlE$_3$R] moieties (E = Cl, O)

	Mode (CN)	Ln···Al Å	Ln – C$_{Al}$ Å	Ref.
(η^6-C$_6$H$_5$Me)Nd[(μ-Cl)$_2$AlMeCl]$_3$	D1 (9)	3.749–3.839	–	[105]
{Sm[(Oi-Pr)$_2$Ali-Bu(Oi-Pr)](Oi-Pr)$_2$(HOi-Pr)}$_2$	D2 (6)	3.295	–	[320]
(C$_5$H$_5$)$_2$Yb(THF)(μ-O)AlMe[NN] [a]	D3 (8)	–	–	[321]
(C$_5$H$_5$)$_2$Er(THF)(μ-O)AlMe[NN] [a]	D3 (8)	–	–	[321]
(C$_5$H$_5$)$_2$Dy(THF)(μ-O)AlMe[NN] [a]	D3 (8)	–	–	[321]
(C$_5$H$_5$)$_2$Yb(μ-O)AlMe[NN] [a]	D3 (7)	–	–	[321]
(C$_5$H$_5$)$_3$Dy(μ-OH)AlMe[NN] [a]	D3 (10)	–	–	[321]

[a] [NN] = CH[C(CH$_3$)NC$_6$H$_3$$i$-Pr$_2$-2,6]$_2$-1,2.

Table 27 Structural parameters of heterobimetallic Ln/Al complexes featuring [AlE$_4$] moieties (E = Cl, Br, O)

	Mode (CN)	Ln···Al Å	Ln – C$_{Al}$ Å	Ref.
tetrahalogenoaluminates (Ln^{III})				
[Tb(AlCl$_4$)$_3$]$_x$	E1/E3 (8)	–	–	[322]
[Dy(AlCl$_4$)$_3$]$_x$	E1/E3 (8)	–	–	[323]
[Ho(AlCl$_4$)$_3$]$_x$	E1/E3 (8)	–	–	[324]
(η^6-C$_6$H$_4$Me$_2$-1,3)Pr(AlCl$_4$)$_3$	E1 (9)	3.766–3.817	–	[325]
(η^6-C$_6$H$_6$)Nd(AlCl$_4$)$_3$	E1 (9)	3.680–3.782	–	[99]
(η^6-C$_6$H$_5$Me)Nd(AlCl$_4$)$_3$	E1 (9)	3.715–3.804	–	[326]
(η^6-C$_6$H$_3$Me$_3$)Nd(AlCl$_4$)$_3$	E1 (9)	3.715–3.809	–	[327]
(η^6-C$_6$H$_5$Me)Gd(AlBr$_4$)$_3$	E1 (9)	3.896–3.994	–	[100]
(η^6-C$_6$H$_6$)Sm(AlCl$_4$)$_3$	E1 (9)	3.660–3.766	–	[99]
(η^6-C$_6$H$_5$Me)Sm(AlCl$_4$)$_3$	E1 (9)	3.697–3.783	–	[328]
(η^6-C$_6$H$_4$Me$_2$-1,3)Sm(AlCl$_4$)$_3$	E1 (9)	3.663–3.783	–	[98]
(η^6-C$_6$Me$_6$)Sm(AlCl$_4$)$_3$	E1 (9)	3.710–3.795	–	[96, 97]
(η^6-C$_6$H$_5$Me)Yb(AlCl$_4$)$_3$	E1 (9)	3.628–3.750	–	[100]
(η^6-C$_6$Me$_6$)Yb(AlCl$_4$)$_3$	E1 (9)	3.677–3.752	–	[101]
(C$_5$Me$_5$)$_2$Yb(AlCl$_4$)	E1 (8)	3.65	–	[113]
tetrahalogenoaluminates (Ln^{II})				
[Sm(AlCl$_4$)$_2$]$_x$	–	–	–	[329]
(η^6-C$_6$H$_6$)Sm(AlCl$_4$)$_2$	E4 (9)	3.835, 3.846	–	[330]
(η^6-C$_6$H$_3$Me$_3$)Sm(AlCl$_4$)$_2$	E1/E4 (9)	3.835/3.815, 3.869	–	[330]

Table 27 (continued)

	Mode (CN)	Ln···Al Å	Ln–C$_{Al}$ Å	Ref.
Sm[(η^6-C$_6$H$_5$Me)Sm(AlCl$_4$)$_3$]$_2$	E4/E3 (8); E4/E3 (9)	–/3.892; 3.745, 3.822/3.816	–; –	[329]
Na[(η^6-C$_6$H$_6$)Sm(AlCl$_4$)$_3$]	E1 (9)	3.785–3.860	–	[330]
Na[(η^6-C$_6$H$_3$Me$_3$)Sm(AlCl$_4$)$_3$]	E1 (9)	3.762–3.859	–	[330]
[(η^6-C$_6$H$_6$)Eu^{II}(AlCl$_4$)$_2$]$_4$	E1/E3 (9)	3.787/3.888, 3.902	–	[331]
(η^6-C$_6$H$_6$)Yb^{II}(AlCl$_4$)$_2$	E1/E3 (9)	3.661/3.766, 3.803	–	[332]
(η^6-C$_6$H$_3$Me$_3$)Yb(AlCl$_4$)$_2$	E1/E4 (8)	3.682/3.618	–	[332]
Na[(η^6-C$_6$H$_6$)$_2$Yb$_2$(AlCl$_4$)$_6$]	E1/E3 (9)	3.640, 3.708/3.828	–	[332]
tetraalkoxyaluminates				
Eu[Al(Oi-Pr)$_4$]$_3$	E2 (6)	3.228–3.252	–	[333]
Er[Al(Oi-Pr)$_4$]$_3$	E2 (6)		–	[334]
{Y[Al(Oi-Pr)$_4$](Oi-Pr)$_2$(HOi-Pr)}$_2$	E2 (6)	3.227	–	[335]
{Er[Al(Oi-Pr)$_4$](Oi-Pr)$_2$(HOi-Pr}$_2$	E2 (6)	3.204(1)	–	[336]
Nd[Al(Oi-Pr)$_4$]$_3$	E2 (6)	3.258–3.283	–	[337]
Nd[Al(Oi-Pr)$_4$]$_3$(HOi-Pr)	E2 (7)	3.284–3.347	–	[338]
{Nd[Al(Oi-Pr)$_4$](Oi-Pr)$_2$(HOi-Pr)}$_2$	E2 (6)	3.310	–	[339]
Sm(OC$_6$H$_3$$i$-Pr$_2$)$_3$[Al$_2$(O$t$-Bu)$_6$]	E2 (5)	3.406	–	[320]
{Pr[Al(Oi-Pr)$_4$]$_2$(HOi-Pr)(μ-Cl)}$_2$	E2 (7)	3.293, 3.306	–	[340]

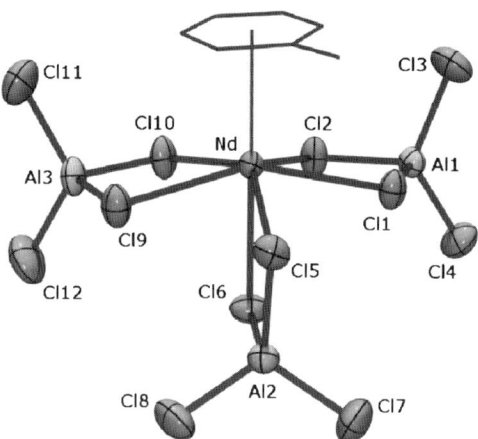

Fig. 21 Solid-state structure of (η^6-C$_6$H$_5$Me)Nd(AlCl$_4$)$_3$ (motif E1) [326]

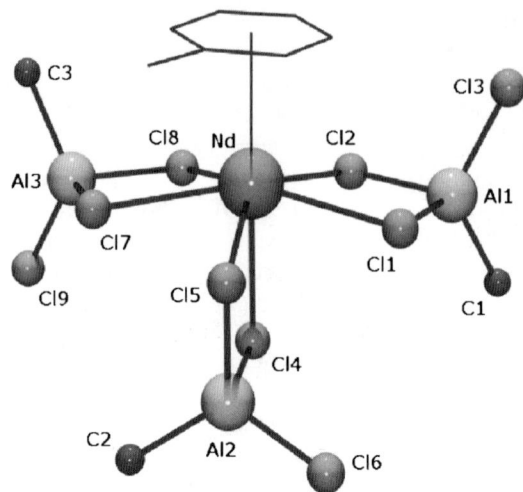

Fig. 22 Solid-state structure of $(\eta^6\text{-}C_6H_5Me)Nd(AlCl_3Me)_3$ (motif **D1**) [105]

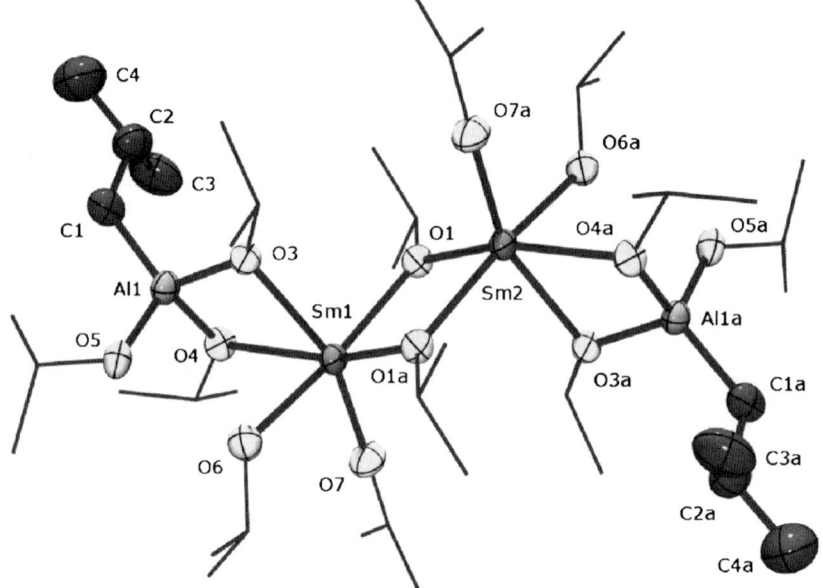

Fig. 23 Solid-state structure of $\{Sm[(Oi\text{-}Pr)_2Al(i\text{-}Bu)(Oi\text{-}Pr)](Oi\text{-}Pr)_2(HOi\text{-}Pr)\}_2$ (motif **D2**) [320]

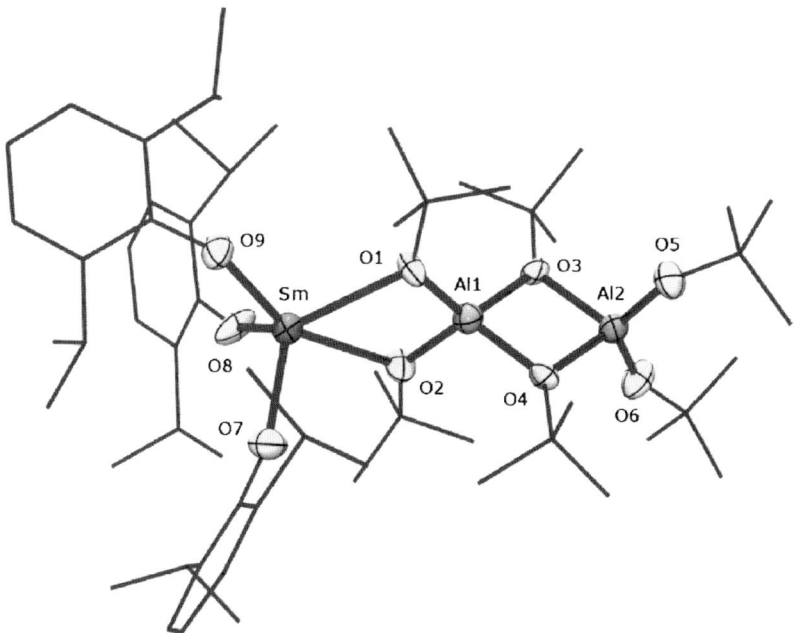

Fig. 24 Solid-state structure of $Sm(OC_6H_3i\text{-}Pr_2)_3[Al_2(Ot\text{-}Bu)_6]$ (motif **E2**) [320]

10.3.5
"Mixed-Alkylated" Complexes

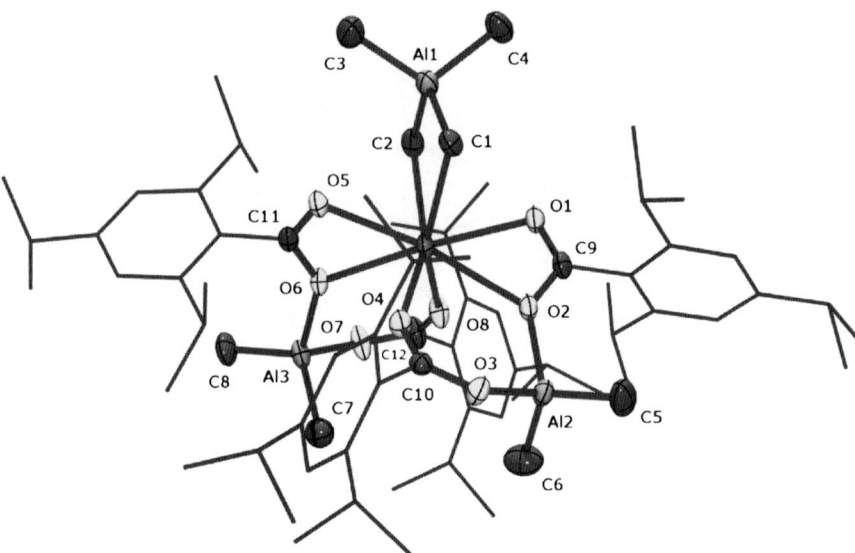

Fig. 25 Solid-state structure of Nd[(O$_2$CC$_6$H$_2$$i$-Pr$_3$)$_2$(AlMe$_2$)]$_2$(AlMe$_4$) (motifs **A1** and **C6**) [147]

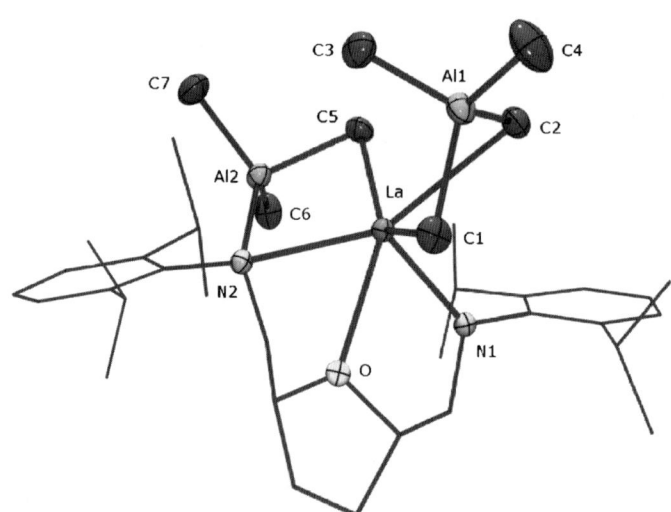

Fig. 26 Solid-state structure of [NON]La(AlMe$_4$)(AlMe$_3$) (motifs **A1** and **B9**) [223]

Table 28 Structural parameters of heterobimetallic Ln/Al complexes featuring *different* alkylated aluminum centers ("mixed-alkylated complexes")

	Mode (CN)	Ln···Al Å	Ln–C$_{Al}$ Å	Refs.
lanthanidocenes				
[(C$_5$Me$_5$)$_2$Sm(Al$_2$Et$_5$O)]$_2$	B13/C4 (9)	–	2.686(5)$_\alpha$, 3.014(4)$_\beta$/3.035(4)	[116]
carboxylates				
Nd[(O$_2$CC$_6$H$_2$i-Pr$_3$)$_2$(AlMe$_2$)]$_2$(AlMe$_4$)	A1/C6 (8)	3.199	2.650(2), 2.659(2)	[147]
alk(aryl)oxides, siloxides				
La(AlMe$_4$)$_2$[OSi(Ot-Bu)$_3$(AlMe$_3$)]	B5/A1 (7)	3.347(2)/3.310(2), 3.288(2)	2.798(3)/2.668(5)–2.800(4)	[150]
Pr(AlMe$_4$)$_2$[OSi(Ot-Bu)$_3$(AlMe$_3$)]	B5/A1 (7)	3.301(1)/3.263(2), 3.235(2)	2.754(2)/2.618(2)–2.792(2)	b
amides, imides				
Nd(Ni-Pr$_2$)(AlMe$_4$)[(Ni-Pr$_2$)(AlMe$_3$)]	A1/B9 (5)	3.216(1)/3.284(1)	2.639(3), 2.659(3)/2.718(2)	[207]
[NON]Y(AlMe$_4$)(AlMe$_3$)[a]	A1/B12 (6)	3.097(1)/3.111(1)	2.573(3), 2.516(3)/2.607(3), 2.584(3)	[223]
[NON]La(AlMe$_4$)(AlMe$_3$)[a]	A1/B9 (6)	3.066(1)/3.394(1)	2.920(2), 2.780(2)/2.696(2)	[223]
{Nd[AlMe$_4$]$_2$[(NPh)–(AlMe$_2$)]}$_2$	A1/C3 (7)	3.236(7), 3.185(7)/3.276(7)	2.641(6), 2.707(7), 2.612(6), 2.616(7)/2.859(7)	[217]
other complexes				
[Eu^{II}(MeOCH$_2$CH$_2$O)$_2$(AlMe$_2$)(AlMe$_3$)]$_2$	B4/C5 (6)	–	–	[191]

[a] [NON] = OC$_4$H$_6$(CH$_2$NC$_6$H$_3$i-Pr$_2$-2,6)$_2$-2,5.
[b] Fischbach et al., 2006, personal communication

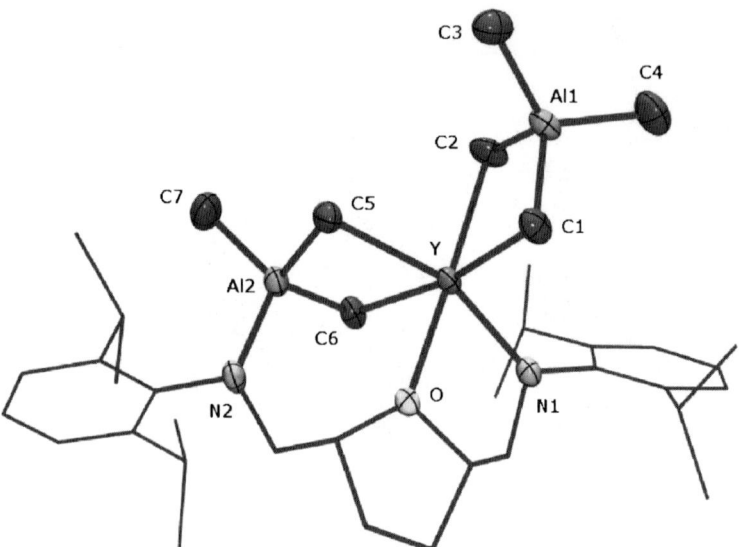

Fig. 27 Solid-state structure of [NON]Y(AlMe$_4$)(AlMe$_3$) (motifs **A1** and **B12**) [223]

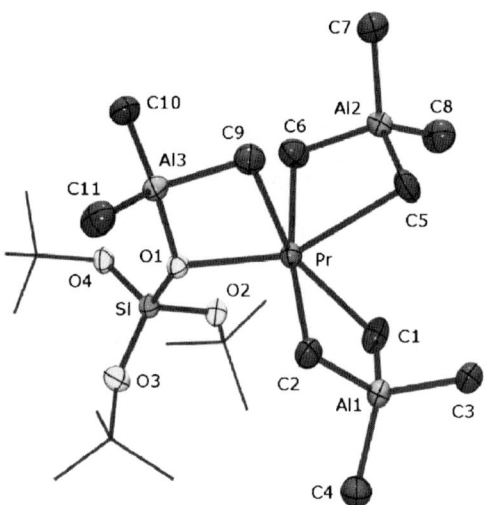

Fig. 28 Solid-state structure of Pr(AlMe$_4$)$_2$[OSi(Ot-Bu)$_3$(AlMe$_3$)] (motifs **A1** and **B7**)

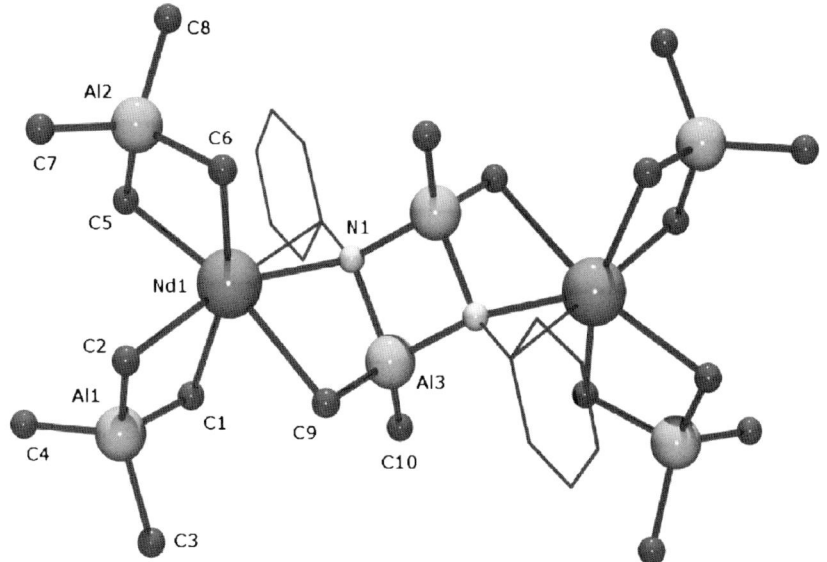

Fig. 29 Solid-state structure of {Nd[AlMe$_4$]$_2$[(NPh)(AlMe$_2$)]}$_2$ (motifs **A1** and **C3**) [217]

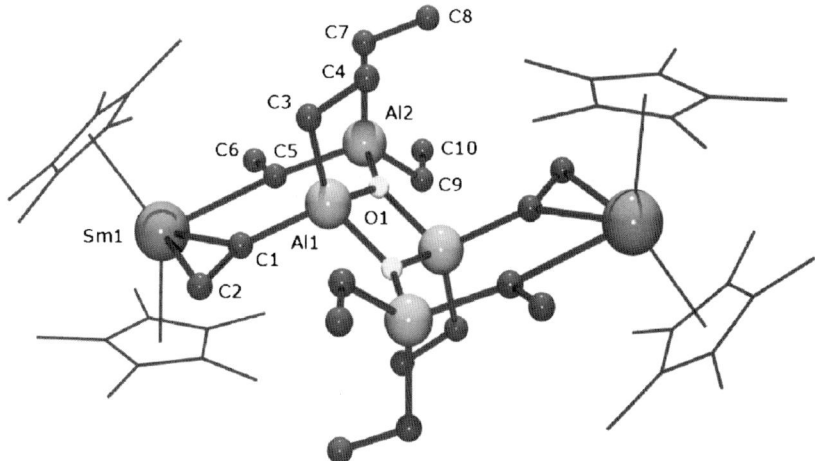

Fig. 30 Solid-state structure of [(C$_5$Me$_5$)$_2$Sm(Al$_2$Et$_5$O)]$_2$ (motifs **B13** and **C4**) [116]

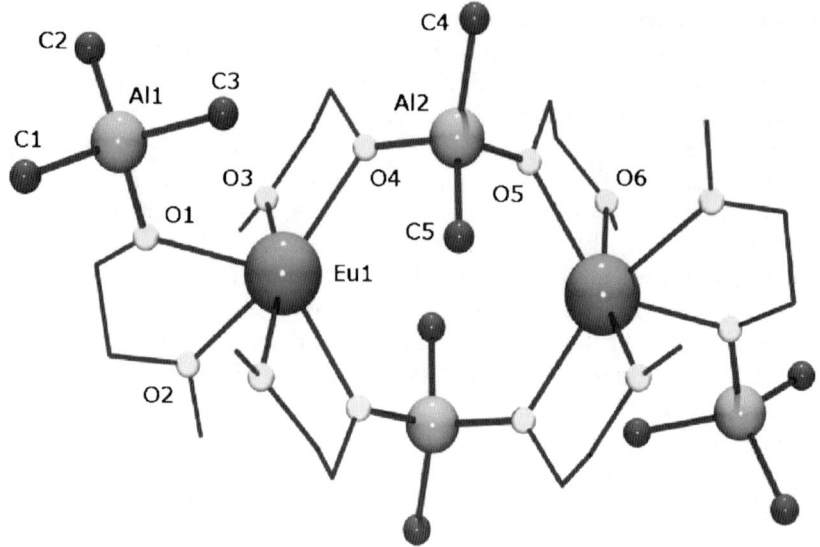

Fig. 31 Solid-state structure of [Eu^{II}(MeOCH$_2$CH$_2$O)$_2$(AlMe$_2$)(AlMe$_3$)]$_2$ (motifs **B4** and **C5**) [191]

11
Conclusion and Perspective

The past five years have witnessed the synthesis and structural elucidation of a rapidly increasing number of Ln/Al organometallics. The reactivity of heterobimetallic moieties of type [Ln(μ-alkyl)$_x$Al(alkyl)$_y$] ($x + y = 4$) and [Ln(μ-X)(μ-alkyl)Al(alkyl)$_2$] (= [Ln(X){Al(alkyl)$_3$}]) document a predominantly ionic Ln(III) – C(alkyl) bonding motif featuring moderate thermodynamic stability and enhanced kinetic lability. The detailed investigation of such Ln – [AlR$_4$] aluminate bonding has not only enriched the *Lanthanide Ziegler–Natta Model*, which emulates important steps of the transition-metal catalyzed α-olefin polymerization. The intrinsic aluminate bonding has been also exploited in organometallic synthesis sequences, i.e., the *tetramethylaluminate route*, which utilizes such "methyl complexes in disguise" according to alkane elimination reactions.

Crucially, discrete Ln/Al organometallics were unambiguously identified as intermediates of the commercially applied neodymium-based diene polymerization and subsequently employed in *binary* initiator mixtures. Particularly for the industrially relevant O-only bonded carboxylate- and alk(aryl)oxide rare-earth metal components, the use of pre-alkylated Ln derivatives developed into valuable structure-reactivity relationships partially uncovering the blackbox, which is provided by *ternary Ziegler Mischkatalysatoren*. Accordingly, rare-earth metal centers provide a unique stereoelectronic environment for studying the reactivity of simple aluminum alkyls (Me$_3$Al [TMA], Et$_3$Al [TEA], i-Bu$_3$Al) in *Ziegler*-type catalysts, revealing fundamental reactivity pattern (including unprecedented alkylation capability) and bonding features. Variable temperature NMR studies and X-ray crystallographic analyses give precise insight into the Ln metal-size dependent dynamic behavior of the [AlR$_4$] ligand and the coordination principles of homoleptic and heteroleptic complexes. Furthermore, the enhanced kinetic lability of the Ln – [AlR$_4$] aluminate bonding, particularly in the presence of large rare-earth metal centers, and the existence of extensive Al-to-Ln and Ln-to-Al ligand transfer sequences corroborate the ease of polymer chain transfer via Ln – Al heterobimetallic complexes in *Ziegler*-type catalysts. This is realized in a modified allyl insertion mechanism of the diene polymerization as shown in Scheme 76.

Comprehensive studies of isoprene polymerization revealed that prealkylated Ln – Al heterobimetallic carboxylate and alk(aryl)oxide complexes are highly efficient precatalyst components in *binary* Ziegler-type catalysts. After activation with Et$_2$AlCl, the catalyst systems confirmed classic features of rare-earth metal-based catalysts such as high *cis*-1,4 polymerization (> 99%), neodymium as the most active metal center (*neodymium effect*), and a favorable $n_{Cl}:n_{LN}$ ratio of ~ 2. In addition, the polymerization efficiency depended on the degree of alkylation ("Ln(AlMe$_4$)$_3$" > "Ln(AlMe$_4$)$_2$" > "Ln(AlMe$_4$)"). Finally, broad molecular weight distributions, which usually result from solution polymerizations with such *Ziegler*-type catalysts, are consistent with pronounced Ln-to-Al polymer chain transfer.

It is well-known that *ternary Ziegler Mischkatalysatoren* as so-to-speak the "1st generation" of transition-metal-based polymerization catalysts are often superior to more sophisticated metallocene and postmetallocene con-

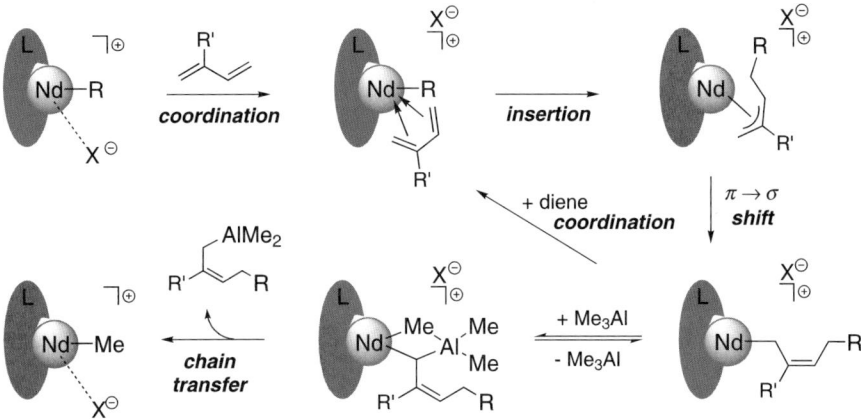

Scheme 76 Modified allyl insertion mechanism of the neodymium-based 1,3-diene polymerization taking into account organoaluminum-mediated chain transfer [148]

geners. However, their improvement and optimization have been hampered by their intricate nature and have been achieved only empirically. Given the latest structural knowledge of Ln/Al organometallics and the superb performance of the corresponding discrete binary Ln(III)-based *Ziegler* catalysts in diene polymerization, more diverse applications will be stimulated. Novel Al-to-Ln ligand transfer reactions can be envisaged providing catalyst mixtures, which can exhibit also good initiation activity in diene copolymerizations [341] as well as toward equally delicate monomers such as alkylene oxides (oxiranes), styrene, and acetylene.

Acknowledgements R.A. thanks his former and current coworkers Dr. Jörg Eppinger, Dr. Michael G. Klimpel, Dr. Frank Estler, Dr. Andreas Fischbach, Dr. H. Martin Dietrich, Melanie Zimmermann, and Christian Meermann for their shared enthusiasm for Ln/Al organometallics and much fun. Our research in this field has been generously supported by the Deutsche Forschungsgemeinschaft, the Fonds der Chemischen Industrie, DEGUSSA AG, and BAYER AG. A.F. thanks the Deutsche Forschungsgemeinschaft for a research fellowship and R.A. the Norwegian Research Council (project# 171245/V30) for financial support while writing this article.

References

1. Ziegler K, Holzkamp E, Breil H, Martin H (1955) Angew Chem 67:541
2. Fink G, Mühlhaupt R, Brintzinger H-H (eds) (1995) Ziegler Catalysts. Springer, Berlin Heidelberg New York
3. Natta G, Pino P, Mazzanti G, Giannini K (1957) J Am Chem Soc 79:2975
4. Breslow DS, Newburg NR (1957) J Am Chem Soc 79:5072
5. Ballard DGH, Pearce R (1975) J Chem Soc Chem Commun 621
6. Holton J, Lappert MF, Scollary GR, Ballard DGH, Pearce R, Atwood JL, Hunter WE (1976) J Chem Soc Chem Commun 425
7. Holton J, Lappert MF, Ballard DGH, Pearce R, Atwood JL, Hunter WE (1976) J Chem Soc Chem Commun 480
8. Ballard DGH, Courtis A, Holton J, McMeeking J, Pearce R (1978) J Chem Soc Chem Commun 994
9. Holton J, Lappert MF, Ballard DGH, Pearce R, Atwood JL, Hunter WE (1979) J Chem Soc Dalton Trans 45
10. Holton J, Lappert MF, Ballard DGH, Pearce R, Atwood JL, Hunter WE (1979) J Chem Soc Dalton Trans 54
11. Holton J, Lappert MF, Ballard DGH, Pearce R, Atwood JL, Hunter WE (1979) In: Marks TJ, Fischer RD (eds) Organometallics of the f-Elements. D. Reidel Publishing Company, Dordrecht
12. Watson PL, Herskovitz T (1983) ACS Symp Series 212:459
13. Brintzinger H-H, Fischer D, Mülhaupt R, Rieger B, Waymouth RW (1995) Angew Chem Int Ed Engl 34:1143
14. Bochmann M (1996) J Chem Soc Dalton Trans 255
15. Kaminsky WJ (1998) Chem Soc Dalton Trans 1413
16. Watson PL (1982) J Am Chem Soc 104:337
17. Watson PL, Parshall GW (1985) Acc Chem Res 18:51

18. Klimpel MG, Anwander R, Tafipolsky M, Scherer W (2001) Organometallics 20:3983
19. Casey CP, Hallenbeck SL, Wright JM, Landis CR (1997) J Am Chem Soc 119:9680
20. Abrams MB, Yoder JC, Loeber C, Day MW, Bercaw JE (1999) Organometallics 18:1389
21. Burger BJ, Thompson ME, Cotter WD, Bercaw JE (1990) J Am Chem Soc 112:1566
22. Piers WE, Bercaw JE (1990) J Am Chem Soc 112:9406
23. Watson PL, Roe DC (1982) J Am Chem Soc 104:6471
24. Klimpel MG, Eppinger J, Sirsch P, Scherer W, Anwander R (2002) Organometallics 21:4021
25. Bochmann M, Lancaster SJ (1994) Angew Chem Int Ed Engl 33:1634
26. Bolton PD, Clot E, Cowley AR, Mountford P (2005) Chem Commun 3313
27. Kaminsky W, Kopf J, Thirase G (1974) Liebigs Ann Chem 1531
28. Kaminsky W, Kopf J, Sinn H, Vollmer HJ (1976) Angew Chem Int Ed Engl 15:629
29. Kopf J, Vollmer HJ, Kaminsky W (1980) Cryst Struct Commun 9:271
30. Siedle AR, Neumark RA, Schroepfer JN, Lyon PA (1991) Organometallics 10:400
31. Tebbe FN, Parshall GW, Reddy GS (1978) J Am Chem Soc 100:3611
32. Herzog A, Roesky HW, Zak Z, Noltemeyer M (1994) Angew Chem Int Ed Engl 33:967
33. Herzog A, Roesky HW, Jäger F, Steiner A, Noltemeyer M (1996) Organometallics 15:909
34. Guérin F, Stephan DW (1999) Angew Chem Int Ed 38:3698
35. Kickham JE, Guérin F, Stewart JC, Stephan DW (2000) Angew Chem Int Ed 39:3263
36. Kickham JE, Guérin F, Stewart JC, Urbanska E, Stephan DW (2000) Organometallics 20:1175
37. Yue N, Hollink E, Guérin F, Stephan DW (2001) Organometallics 20:4424
38. Kickham JE, Guérin F, Stephan DW (2002) J Am Chem Soc 124:11486
39. Wei P, Stephan PW (2003) Organometallics 22:1992
40. Shen Z, Ouyang J (1987) In: Gschneidner KA Jr, Fleming L (eds) Handbook on the Physics and Chemistry of Rare Earths. Elsevier Science Publishers, Netherlands, chapter 61
41. Taube R, Sylvester G (2002) In: Cornils B, Herrmann WA (eds) Applied Homogeneous Catalysis with Organometallic Compounds. VCH, Weinheim, p 285
42. Thiele SK-H, Wilson DR (2003) J Macromol Sci Polym Rev C43:581
43. Porri L, Giarusso A (1989) In: Eastmond GC, Ledwith A, Russo S, Sigwalt P (eds) Comprehensive Polymer Science. Pergamon Press, Oxford, vol 4, part II, p 53
44. Boor J (1979) Ziegler–Natta Catalysts and Polymerizations, vol 17. Academic Press, New York, p 130
45. Anwander R (2002) In: Cornils B, Herrmann WA (eds) Applied Homogeneous Catalysis with Organometallic Compounds. VCH, Weinheim, p 974
46. Yasuda H (1999) Top Organomet Chem 2:255
47. Osakada K, Takeuchi D (2004) Adv Polym Sci 171:137
48. Wilson DJ (1993) Makromol Chem Macromol Symp 66:273
49. Lauretti E, Miani B, Mistrali F (1994) Rubber World 210:34
50. Porri L, Ricci G, Shubin N (1998) Macromol Symp 128:53
51. Oehme A, Gebauer U, Gehrke K, Beyer P, Hartmann B, Lechner MD (1994) Macromol Chem Phys 195:3773
52. Yunlu K, He M, Cuif J-P, Alas M (2000) US Patent 6 111 082
53. Shen Z, Gong C, Chung C, Ouyang J (1964) Sci Sin (Engl Trans) 13:1339
54. Yang JH, Hu J-Y, Feng S-F, Pan E-L, Xie D-I, Zhong C-O, Ouyang J (1980) Sci Sin (Engl Trans) 23:734
55. Yang J, Tsutsui M, Shen Z, Bergbreiter DE (1982) Macromolecules 15:230

56. Chen W, Jin Z, Xing Y, Fan Y, Yang G (1987) Inorg Chim Acta 130:125
57. Gallazzi M, Bianchi F, Depero L, Zocchi M (1988) Polymer 29:1516
58. Shen Z, Ouyang J, Wang F, Hu Z, Yu F, Qian B (1980) J Polym Sci Polym Chem Ed 18:3345
59. Hsieh HL, Yeh GHC (1986) Ind Eng Chem Prod Res Dev 25:456
60. Rao GSS, Upadhyay VK, Jain RC (1999) J Appl Polym Chem 71:595
61. Rao GSS, Upadhyay VK, Jain RC (1997) Angew Makromol Chem 251:193
62. Pang S, Li Y, Ding W, Xue J, Ouyang J (1984) Chin J Appl Chem 1:50
63. Yang J, Pang S, Sun T, Li Y, Ouyang J (1984) J Appl Chem 11
64. Yang J, Pang S, Sun T, Li Y, Ouyang J (1984) Polym Commun 73
65. Yang J, Pang S, Sun T, Li Y, Ouyang J (1984) Chin J Catal 291
66. Iovu H, Hubca C, Dimonie M, Simionescu E, Badea EG (1997) Mater Plast (Bucharest) 34:5
67. Ouyang J, Wang F-S, Shen Z (1981) Bilateral Symp Polym Chem Phys 382
68. Monakov YB, Marina NG, Khairullina RM, Kozlova OI, Tolstikov GA (1988) Inorg Chim Acta 142:161
69. Kozlov VG, Marina NG, Savel'eva IG, Monakov YB, Murinov YI, Tolstikov GA (1988) Inorg Chim Acta 154:239
70. Monakov YB, Sabirov ZM, Zaikov GE (2002) Aging of Polymers, Polymer Blends and Polymer Composites. Nova Science, Hauppauge, NY 1:237; and references therein
71. Monakov YB, Sabirov ZM, Zaikov GE (2003) In: Zaikov GE, Lobo VMM (eds) New Developments in Chemistry and Biochemistry. Nova Science, Hauppauge NY, p 53; and references therein
72. Urazbaev VN, Sabirov ZM, Efimov VP, Mullagaliev IR, Monakov YB (2003) In: Zaikov GE, Lobo VMM (eds) Studies in Chemistry and Biochemistry. Nova Science, Hauppauge NY, p 145
73. Sigaeva NN, Usmanov TS, Zaikov GE, Monakov YB (2003) In: Zaikov GE, Lobo VMM (eds) Reactions in Condensed Phases. Nova Science Publishers, Hauppauge NY, p 81
74. Monakov YB, Sabirov ZM, Urazbaev VN, Zaikov GE (2003) In: Pethrick RA, Zaikov GE (eds) Polymer Yearbook 18. Rapra Technology Limited, Shropshire, UK, p 75; and references therein
75. Dimonie M, Hubca G, Badea E, Simionescu E, Iovu H, Vasile I, Stan S (1995) Rev Roumaine Chim 40:33
76. Iovu H, Hubca G, Simionescu E, Badea EG, Dimonie M (1997) Angew Makromol Chem 249:59
77. Iovu H, Hubca G, Simionescu E, Badea EG, Hurst JS (1997) Eur Polym J 33:811
78. Iovu H, Hubca G, Racoti D, Hurst JS (1999) Eur Polym J 35:335
79. Iovu H, Hubca G, Badea E, Simionescu E, Dimonie M (2000) Rev Roumaine Chim 45:739
80. Bulgakov RG, Kuleshov SP, Zuzlov AN, Khalilov LM, Vafin RR (2002) Russ Chem Bull 51:904
81. Loo CC, Hsu CC (1974) Can J Chem Eng 52:381
82. Monakov YB, Marina NG, Sabirov ZM (1994) Polym Sci USSR (Engl Transl) 36:309
83. Rafikov SR, Monakov YB, Bieshev YK, Valitova IF, Muriov YI, Tolstikov GA, Nikitin YE (1976) Dokl Akad Nauk SSSR 229:1174
84. Sigaeva NN, Usmanov TS, Budtov VP, Spivak SI, Zaikov GE, Monakov YB (2001) Russ Polym News 6:11
85. Sigaeva NN, Usmanov TS, Budtov VP, Spivak SI, Monakov YB (2001) Int J Polym Mater 49:475
86. Sigaeva NN, Usmanov TS, Budtov VP, Monakov YB (2001) Russ J Appl Chem 74:1141

87. Monakov YB, Sabirov ZM, Urazbaev VN, Efimov VP (2001) Kinet Catal 42:310
88. Sigaeva NN, Usmanov TS, Budtov VP, Spivak SI, Zaikov GE, Monakov YB (2003) J Appl Polym Sci 87:358
89. Sigaeva NN, Usmanov TS, Zaikov GE, Monakov YB (2003) J Appl Polym Sci 89:674
90. Urazbaev VN, Efimov VP, Sabirov ZM, Monakov YB (1993) J Polym Sci A Polym Chem 89:601
91. Urazbaev VN, Mullagaliev IR, Efimov VP, Sabirov ZM, Monakov YB (2004) Oxidat Commun 27:770
92. Tikhonov AN, Goncharsky AV, Stepanov VV, Yagola AG (1995) Numerical Methods for the Solution of Ill-Posed Problems. Kluwer, Dordrecht
93. Tikhonov AN, Leonov AS, Yagola AG (1998) Nonlinear Ill-Posed Problems. Chapman and Hall, London
94. Urazbaev VN, Efimov VP, Savirov ZM, Monakov YB (2003) J Appl Polym Sci 89:601
95. Evans WJ, Giarikos DG, Allen NT (2003) Macromolecules 36:4256
96. Cotton FA, Schwotzer W (1986) J Am Chem Soc 108:4657
97. Cotton FA, Schwotzer W (1987) Organometallics 6:1275
98. Fan B, Shen T, Lin Y (1989) J Organomet Chem 376:61
99. Fan B, Shen T, Lin Y (1989) J Organomet Chem 377:51
100. Biagini P, Lugli G, Millini R (1994) Gazz Chim Ital 124:217
101. Liang H, Shen T, Guan J, Lin Y (1994) J Organomet Chem 474:113
102. Hu J, Tian H, Shen Q, Liang H (1992) Chin Sci Bull 37:566
103. Jin S, Guan J, Liang H, Shen Q (1993) Cuihua Xuebao 14:159
104. Jin S, Liang H, Shen Q (1993) Zhongguo Xitu Xuebao 11:274
105. Biagini P, Lugli G, Abis L, Millini R (1995) New J Chem 19:713
106. Garbassi F, Biagini P, Piero P, Lugli G (1995) US Patent 5 633 353
107. Scollary GR (1978) Austr J Chem 31:411
108. Evans WJ, Chamberlain LR, Ziller JW (1987) J Am Chem Soc 109:7209
109. Evans WJ, Chamberlain LR, Ulibarri TA, Ziller JW (1988) J Am Chem Soc 110:6423
110. Busch MA, Harlow R, Watson PL (1987) Inorg Chim Acta 140:15
111. Evans WJ, Leman JT, Clark RD, Ziller JW (2000) Main Group Met Chem 23:163
112. Evans WJ, Forrestal KJ, Ziller JW (1997) Angew Chem Int Ed Engl 36:774
113. Watson PL, Whitney JF, Harlow RL (1981) Inorg Chem 20:3271
114. Deacon GB, Shen Q (1996) J Organomet Chem 506:1
115. Evans WJ, Champagne TM, Giarikos DG, Ziller JW (2005) Organometallics 24:570
116. Evans WJ, Champagne TM, Ziller JW (2005) Organometallics 24:4882
117. den Haan KH, Teuben JH (1987) J Organomet Chem 322:321
118. Taube R, Maiwald S, Sieler J (2001) J Organomet Chem 621:327; and references therein
119. Kaita S, Hou Z, Wakatsuki Y (1999) Macromolecules 32:9078
120. Kaita S, Hou Z, Wakatsuki Y (2001) Macromolecules 34:1539
121. Kwag G, Lee H, Kim S (2001) Macromolecules 34:5367
122. Sabirov ZM, Minchenkova NK, Monakov YB (1990) Vysokomol Soedin Ser B 11:803
123. Porri L, Ricci G, Giarusso A, Shubin N, Lu Z (2000) ACS Symp Ser 749:15
124. Kwag G (2002) Macromolecules 35:4875
125. Friebe L, Nuyken O, Windisch H, Obrecht W (2002) Macromol Chem Phys 203:1055
126. Jin Y-T, Sun Y-F, Ouyang J (1979) Gaofenzi Tongxun 367
127. Li X, Jin Y, Li G, Shun Y, Ouyang J (1986) Yingyong Huaxue 3:77
128. Kobayashi E, Kaita S, Aoshima S, Furukawa J (1994) J Polym Sci A Polym Chem 32:1195

129. Kobayashi E, Kaita S, Aoshima S, Furukawa J (1995) J Polym Sci A Polym Chem 33:2175
130. Kobayashi E, Hayashi N, Aoshima S, Furukawa J (1998) J Polym Sci A Polym Chem 36:241
131. Kobayashi E, Hayashi N, Aoshima S, Furukawa J (1998) J Polym Sci A Polym Chem 36:1707
132. Li F, Jin Y, Song C, Lin Y, Pei F, Wang F, Hu N (1996) Appl Organomet Chem 10:761
133. Bradley DC, Mehrotra RC, Gaur DP (1978) Metal Alkoxides. Academic Press, New York
134. Evans WJ, Giarikos DG, Ziller JW (2001) Organometallics 20:5751
135. Evans WJ, Giarikos DG (2004) Macromolecules 37:5130
136. Porri L, Ricci G, Giarrusso A, Shubin N, Lu Z (2000) ACS Symp Ser 749:15
137. Cotton FA, Wilkinson G, Murillo CA, Bochmann M (1999) Advanced Inorganic Chemistry, chap 9. Wiley-Interscience, New York
138. Meitzner G, Via GH, Lytle FW, Sinfelt JH (1992) J Phys Chem 96:4960
139. Rao KJ, Wong J, Weber MJ (1983) J Chem Phys 78:6228
140. Jolly PW, Wilke G (1974) The Organic Chemistry of Nickel, vol 1. Academic Press, New York, p 157
141. Kwag G, Kim A, Lee S, Jang Y, Kim P, Baik H, Yoon D-I, Jeong H, Lee JG, Lee H (2002) Rubber Chemistry and Technology 75:907
142. Fischbach A (2003) PhD Thesis, Technische Universität München, Germany
143. Evans WJ, Champagne TM, Ziller JW (2005) Chem Commun 5925
144. Yamamoto H, Yasuda H, Yokota K, Nakamura A, Kai Y, Kasai N (1988) Chem Lett 1963
145. Porri L, Ricci G, Italia S, Cabassi F (1987) Polym Commun 28:223
146. Cabassi F, Italia S, Ricci G, Porri L (1988) In: Quirk RP (ed) Transition Metal Catalyzed Polymerization. Cambridge University Press, New York, p 655
147. Fischbach A, Perdih F, Sirsch P, Scherer W, Anwander R (2002) Organometallics 21:4569
148. Fischbach A, Perdih F, Herdtweck E, Anwander R (2006) Organometallics 25:1626
149. Evans WJ, Anwander R, Ziller JW (1995) Organometallics 14:1107
150. Fischbach A, Klimpel MG, Widenmeyer M, Herdtweck E, Scherer W, Anwander R (2004) Angew Chem Int Ed 43:2234
151. Dietrich HM, Raudaschl-Sieber G, Anwander R (2005) Angew Chem Int Ed 44:5303
152. Maiwald S, Weißenborn H, Windisch H, Sommer C, Müller G, Taube R (1997) Macromol Chem Phys 198:3305
153. Maiwald S, Taube R, Hemling H, Schumann H (1998) J Organomet Chem 552:195
154. Ward BD, Bellemin-Laponnaz S, Gade LH (2005) Angew Chem Int Ed 44:1668
155. Arndt S, Beckerle K, Zeimentz PM, Spaniol TP, Okuda J (2005) Angew Chem Int Ed 44:7473
156. Witte J (1981) Angew Makromol Chem 94:119
157. Friebe L, Nuyken O, Obrecht W (2005) J Macromol Sci A Pure Appl Chem 42:839
158. Boisson C, Barbotin F, Spitz R (1999) Macromol Chem Phys 200:1163
159. Bradley DC, Mehrotra RC, Rothwell IP, Singh A (2001) Alkoxo and Aryloxo Derivatives of Metals. Academic Press, London
160. Anwander R (1996) Top Curr Chem 179:149
161. Garbassi F, Biagini P, Andreussi P, Lugli G (1994) European Patent 0 638 598 A1
162. Pedretti U, Lugli G, Poggio S, Mazzei A (1979) German Patent 2 833 721 A1
163. Shan C, Li Y, Pang S, Ouyang J (1983) Huaxue Xuebao 41:490
164. Shan C, Li Y, Pang S, Ouyang J (1983) Huaxue Xuebao 41:498

165. Shan C, Ouyang J (1983) Gaofenzi Tongxun 238
166. Dong W, Masuda T (2002) J Polym Sci A Polym Chem 40:1838
167. Dong W, Endo K, Masuda T (2003) Macromol Chem Phys 204:104
168. Taniguchi Y, Dong W, Katsumata T, Shiotsuki M, Masuda T (2005) Polym Bull 54:173
169. Shan C, Lin Y, Ouyang J, Fan Y, Yang G (1987) Makromol Chem 188:629
170. Huang B, Shi Y, Fang W, Zhong C, Tang X (1992) Hecheng Xiangijiao Gongye 15:23
171. Li X-M, Sun Y-F, Jin Y-T (1986) Acta Chim Sin 44:1163
172. Shan C, Sun T, Pang S, Ji X, Yang J (1989) Cuihua Xuebao 10:294
173. Shan C, Sun T, Pang S, Ji X, Yang J (1989) Gaofenzi Xuebao 709
174. Dong W, Yang J, Shan C, Pang S, Huang B (1997) Cuihua Xuebao 18:234
175. Dong W, Yang J, Shan C, Pang S, Huang B (1998) Yingyong Huaxue 15:1
176. Greco A, Bertolini G, Cesca S (1977) Inorg Chim Acta 21:245
177. Kobayashi E, Shouzaki H, Aoshima S, Furukawa J (1992) Kobunshi Ronbunshu 49:535
178. Evans WJ, Boyle TJ, Ziller JW (1993) J Am Chem Soc 115:5084
179. Biagini P, Abis GL, Millini R (1994) J Organomet Chem 474:C16
180. Evans WJ, Boyle TJ, Ziller JW (1993) J Organomet Chem 462:141
181. Gromada J, Le Pichon L, Mortreux A, Leising F, Carpentier J-F (2003) J Organomet Chem 683:44
182. Carbonaro A, Ferraro D (1983) European Patent 92 270
183. Berg DJ, Andersen RA (2003) Organometallics 22:627
184. Evans WJ, Ansari MA, Ziller JW (1995) Inorg Chem 34:3079
185. Evans WJ, Ansari MA, Ziller JW (1997) Polyhedron 16:3429
186. Gordon JC, Giesbrecht GR, Brady JT, Clark DL, Keogh DW, Scott BL, Watkin JG (2002) Organometallics 21:127
187. Giesbrecht GR, Gordon JC, Brady JT, Clark DL, Keogh DW, Michalczyk R, Scott BL, Watkin JG (2002) Eur J Inorg Chem 723
188. Fischbach A, Herdtweck E, Anwander R, Eickerling G, Scherer W (2003) Organometallics 22:499
189. Murad E, Hildenbrand DL (1980) J Chem Phys 73:4005
190. Boncella JM, Andersen RA (1985) Organometallics 4:205
191. Evans WJ, Greci MA, Ziller JW (1998) Inorg Chem 37:5221
192. Gromada J, Mortreux A, Nowogrocki G, Leising F, Mathivet T, Carpentier J-F (2004) Eur J Inorg Chem 3247
193. Gromada J, Chenal T, Mortreux A, Carpentier J-F, Ziller JW, Leising F (2000) Chem Commun 2183
194. Gromada J, Mortreux A, Chenal T, Ziller JW, Leising F, Carpentier J-F (2002) Chem Eur J 8:3773
195. Gromada J, Chenal T, Mortreux A, Leising F, Carpentier J-F (2002) J Mol Catal A Chem 182–183:525
196. Abbenhuis HCL (2000) Chem Eur J 6:25
197. Duchateau R (2002) Chem Rev 102:3525
198. Scott S, Crudden CM, Jones CW (eds) (2002) Nanostructured Catalysts. Kluwer Academic/Plenum Publishers, New York
199. Copéret C, Chabanas M, Saint-Arroman RP, Basset J-M (2003) Angew Chem Int Ed 42:156
200. Fischbach A, Eickerling G, Scherer W, Herdtweck E, Anwander R (2004) Z Naturf 59b:1353
201. Thiele SK-H, Monroy VM, Wilson DR, Stoye H (1999) Patent WO 2003 033 545
202. Monteil V, Spitz R, Boisson C (2004) Polym Int 53:576

203. Anwander R (1996) Top Curr Chem 179:33
204. Eppinger J (1999) PhD Thesis, Technische Universität München, Germany
205. Nolan SP, Stern D, Marks TJ (1989) J Am Chem Soc 111:7844
206. Evans WJ, Anwander R, Doedens RJ, Ziller JW (1994) Angew Chem Int Ed Engl 33:1641
207. Evans WJ, Anwander R, Ziller JW (1995) Inorg Chem 34:5927
208. Klooster WT, Lu RS, Anwander R, Evans WJ, Koetzle TF, Bau R (1998) Angew Chem Int Ed Engl 37:1268
209. Anwander R (1999) Top Organomet Chem 2:1
210. Anwander R, Runte O, Eppinger J, Gerstberger G, Herdtweck E, Spiegler M (1998) J Chem Soc Dalton Trans 847
211. Anwander R, Klimpel MG, Dietrich HM, Shorokhov DJ, Scherer W (2003) Chem Commun 1008
212. Evans WJ, Meadows JH, Wayda AL, Hunter WE, Atwood JL (1982) J Am Chem Soc 104:2008
213. Klimpel MG, Sirsch P, Scherer W, Anwander R (2003) Angew Chem Int Ed 42:574
214. Chen EY-X, Marks TJ (2000) Chem Rev 100:1391
215. Evans WJ, Johnston MA, Greci MA, Gummersheimer TS, Ziller JW (2003) Polyhedron 22:119
216. Arndt S, Okuda J (2005) Adv Synth Catal 347:339
217. Evans WJ, Ansari MA, Ziller JW, Khan SI (1996) Inorg Chem 35:5435
218. Gordon JC, Giesbrecht GR, Clark DL, Hay PJ, Keogh DW, Poli R, Scott BL, Watkin JG (2002) Organometallics 21:4726
219. Arnold J, Hoffman CG, Dawson DY, Hollander FJ (1993) Organometallics 12:3645
220. Sugiyama H, Gambarotta S, Yap GPA, Wilson DR, Thiele SK-H (2004) Organometallics 23:5054
221. Lorenz V, Görls H, Thiele SK-H, Scholz J (2005) Organometallics 24:797
222. Estler F (2003) PhD Thesis, Technische Universität München, Germany
223. Zimmermann M, Törnroos KW, Anwander R (2006) Organometallics 25:3644
224. Cui C, Shafir A, Schmidt JAR, Oliver AG, Arnold J (2005) Dalton Trans 1387
225. Schaverien CJ (1991) J Chem Soc Chem Commun 459
226. Schaverien CJ, Orpen AG (1991) Inorg Chem 30:4968
227. Mazzei A (1981) Makromol Chem Suppl 4:61
228. Taube R, Windisch H, Maiwald S (1995) Macromol Symp 89:393
229. Taube R, Windisch H, Görlitz FH, Schumann H (1993) J Organomet Chem 445:85
230. Taube R, Maiwald S, Sieler J (1996) J Organomet Chem 513:37
231. Barbier-Baudry D, Dormond A, Desmurs P (1999) CR Acad Sci Ser IIc 375
232. Taube R, Windisch H (1994) J Organomet Chem 472:71
233. Barbier-Baudry D, Bonnet F, Dormond A, Visseaux M (2001) Entropie 37:96
234. Visseaux M, Barbier-Baudry D, Bonnet F, Dormond A (2001) Macromol Chem Phys 202:2485
235. Bonnet F, Barbier-Baudry D, Dormond A, Visseaux M (2002) Polym Int 51:986
236. Barbier-Baudry D, Bonnet F, Dormond A, Hafid A, Nyassi A, Visseaux M (2001) J Alloy Comp 323-324:592
237. Barbier-Baudry D, Bonnet F, Domenichini B, Dormond A, Visseaux M (2002) J Organomet Chem 647:167
238. Bonnet F, Visseaux M, Barbier-Baudry D, Dormond A (2002) Macromolecules 35:1143
239. Taube R, Windisch H, Maiwald S, Hemling H, Schumann H (1996) J Organomet Chem 513:49

240. Taube R, Windisch H, Weißenborn H, Hemling H, Schumann H (1997) J Organomet Chem 548:229
241. Taube R, Maiwald S, Sieler J (2001) J Organomet Chem 621:327
242. Maiwald S, Weißenborn H, Sommer C, Müller G, Taube R (2001) J Organomet Chem 640:1
243. Maiwald S, Sommer C, Müller G, Taube R (2002) Macromol Chem Phys 203:1029
244. Taube R, Windisch H, Hemling H, Schumann H (1998) J Organomet Chem 555:201
245. Woodman TJ, Schormann M, Bochmann M (2002) Isr J Chem 42:283
246. Evans WJ, Kozimor SA, Brady JC, Davis BL, Nyce GW, Seibel CA, Ziller JW, Doedens RJ (2005) Organometallics 24:2269
247. Dietrich HM, Zapliko C, Herdtweck E, Anwander R (2005) Organometallics 24:5767
248. Dietrich HM, Schuster O, Törnroos KW, Anwander R (2006) Angew Chem Int Ed 45:4858
249. Nakamura H, Nakayama Y, Yasuda H, Maruo T, Kanehisa N, Kai Y (2000) Organometallics 19:5392
250. Schumann H (1984) Angew Chem Int Ed Engl 23:474
251. Schumann H, Meese-Marktscheffel JA, Esser L (1995) Chem Rev 95:865
252. Schaverien CJ (1994) Adv Organomet Chem 36:283
253. Evans WJ (1987) Polyhedron 6:803
254. Qian C, Ye C, Lu H, Li Y, Zhou J, Ge Y, Tsutsui M (1983) J Organomet Chem 247:161
255. Schrems MG (2005) Diploma thesis, Technische Universität München, Germany
256. Schrems MG, Dietrich HM, Törnroos KW, Anwander R (2005) Chem Commun 5922
257. Gerteis RL, Dickerson RE, Brown TL (1964) Inorg Chem 3:872
258. Atwood JL, Stucky GD (1969) J Am Chem Soc 91:2538
259. Schumann H, Pickardt J, Bruncks N (1981) Angew Chem Int Ed Engl 20:120
260. Pauling L (1960) The Nature of the Chemical Bond, 3rd edn. Ithaca Press
261. Allred AL (1961) J Inorg Nucl Chem 17:215
262. Arnold J, Schumannn H, Meese-Marktscheffel JA, Dietrich A, Pickardt J (1992) J Organomet Chem 433:241
263. Hou Z, Kaita S, Wakatsuki Y (2001) Pure Appl Chem 73:291
264. Kaita S, Hou Z, Nishiura M, Doi Y, Kurazumi J, Horiuchi AC, Wakatsuki Y (2003) Macromol Rapid Commun 24:179
265. Kaita S, Yamanaka M, Horiuchi AC, Wakatsuki Y (2006) Macromolecules 39:1359
266. Kaita S, Takeguchi Y, Hou Z, Nishiura M, Doi Y, Wakatsuki Y (2003) Macromolecules 36:7923
267. Kaita S, Doi Y, Kaneko K, Horiuchi AC, Wakatsuki Y (2004) Macromolecules 37:5860
268. Bonnet F, Visseaux M, Barbier-Baudry D (2004) J Organomet Chem 689:264
269. Arndt S, Spaniol TP, Okuda J (2003) Angew Chem Int Ed 42:5075
270. Zhang L, Luo Y, Hou Z (2005) J Am Chem Soc 127:14562
271. Jeske G, Lauke H, Mauermann H, Swepston PN, Schumann H, Marks TJ (1985) J Am Chem Soc 107:8091
272. Evans WJ, Ulibarri TA, Ziller JW (1990) J Am Chem Soc 112:2314
273. Bulychev BM (1990) Polyhedron 9:387
274. Soloveichik GL, Knyazhanskii SY, Bulychev BM, Belsky VK (1992) Organomet Chem USSR 5:73
275. Soloveichik GL (1995) New J Chem 19:597
276. Ephritikhine M (1997) Chem Rev 97:2193
277. Kvostov AV, Bulychev BM, Belsky VK, Sizov AI (1998) J Organomet Chem 568:113
278. Knjazhanski SY, Kalyuzhnaya ES, Elizalde Herrera LE, Bulychev BM, Khvostov AV, Sizov AI (1997) J Organomet Chem 531:19

279. Knjazhanski SY, Elizalde L, Cadenas G, Bulychev BM (1998) J Organomet Chem 568:33
280. Shah SAA, Murugavel R, Roesky HW, Schmidt H-G (1998) Bull Pol Acad Sciences 46:157
281. Knjazhansky SY, Nomerotsky IY, Bulychev BM, Belsky VK, Soloveichik GL (1994) Organometallics 13:2075
282. Barbier-Baudry D, Blacque O, Hafid A, Nyassi A. Sitzmann H, Visseaux M (2000) Eur J Inorg Chem 2333
283. Bonnet F, Visseaux M, Barbier-Baudry D, Hafid A, Vigier E, Kubicki MM (2004) Inorg Chem 43:3682
284. Bonnet F, Visseaux M, Barbier-Baudry D, Vigier E, Kubicki MM (2004) Chem Eur J 10:2428
285. Bonnet F, Visseaux M, Pereira A, Bouyer F, Barbier-Baudry D (2004) Macromol Rapid Commun 25:873
286. Bonnet F, Visseaux M, Pereira A, Barbier-Baudry D (2005) Macromolecules 38:3162
287. Visseaux M, Chenal T, Roussel P, Mortreux A (2006) J Organomet Chem 691:86
288. Anwander R (2001) Chem Mater 13:4419
289. Robert P, Spitz R (1994) European Patent 599 096
290. Cann KJ, Apecetche MA, Moorhouse JH, Muruganadam N, Smith GG, Williams GH (1998) European Patent 856 530
291. Li Y, Ouyang J (1987) J Macromol Sci Chem A24:227
292. Yu G, Li Y, Qu Y, Li K (1993) Macromolecules 26:6702
293. Bergbreiter DE, Chen L-B, Chandran R (1985) Macromolecules 18:1055
294. Yu G, Li Y, Liu C (1992) Chin J Polym Sci 10:49
295. Zheng W, Li Y, Yu G, Li X (1996) Cuihua Xuebao 17:245
296. Zheng W, Li Y, Yu G, Li X (1996) Fenzi Cuihua 10:178
297. Li Y, Zheng W, Yu G, Li X (1994) Hecheng Xiangjiao Gongye 17:234
298. Liu G, Li Y (1990) Gaofenzi Xuebao 136
299. Yu G, Li Y (1988) Cuihua Xuebao 9:190
300. Zoellner K, Reichert K-H (2001) Chem Eng Sci 56:4099
301. Ling J, Ni XF, Zhang YF, Shen Z (2000) Polymer 41:8703
302. Ling J, Ni XF, Zhang YF, Shen Z (2003) Polym Int 52:213
303. Shen Z, Li WS, Zhang YF (2002) Sci China Ser B 30:227
304. Sun JZ, Eberstein C, Reichert K-H (1997) J Appl Polym Sci 64:203
305. Zhao JZ, Sun JZ, Zhou QY (2001) J Appl Polym Sci 81:719
306. Barbotin F, Spitz R, Boisson C (2001) Macromol Rapid Commun 22:1411
307. Rühmer T, Giesemann J, Schwieger W, Schmutzler K (1999) Kautsch Gummi Kunstst 52:420
308. Landmesser H, Berndt H, Kunath D, Lücke B (2000) J Mol Catal A Chem 16:257
309. Berndt H, Landmesser H (2003) J Mol Catal A Chem 197:245
310. Woodman TJ, Sarazin Y, Fink G, Hauschild K, Bochmann M (2005) Macromolecules 38:3060
311. Tajima K, Aida T (2000) Chem Commun 2399
312. O'Neill ME, Wade K (1982) In: Wilkinson G, Stone FGA, Abel EW (eds) Comprehensive Organometallic Chemistry. Pergamon Press, New York, p 593
313. Shan C-J, Lin Y-H, Jin S-C, Ouyang J, Fan Y-G, Yang G-D, Yu J-S (1987) Acta Chim Sin 45:949
314. Scherer W, McGrady GS (2004) Angew Chem Int Ed 44:1782
315. Klimpel MG (2001) PhD Thesis, Technische Universität München, Germany

316. Glanz M, Dechert S, Schumann H, Wolff D, Springer J (2000) Z Anorg Allg Chem 626:2467
317. Evans WJ, Forrestal KJ, Ansari MA, Ziller JW (1998) 120:2180
318. Dietrich HM, Grove H, Törnroos KW, Anwander R (2006) J Am Chem Soc 128:1458
319. Tong Y-J, Cheng Y-X, Guan H-M, Lin Y-H, Jin Y-T (1999) Chin J Struct Chem 18:47
320. Giesbrecht GR, Gordon JC, Clark DL, Scott BL, Watkin JG, Young KJ (2002) Inorg Chem 41:6372
321. Chai J, Jancik V, Singh S, Zhu H, He C, Roesky HW, Schmidt H-G, Noltemeyer M, Hosmane NS (2005) J Am Chem Soc 127:7521
322. Patzke GR, Wartchow R, Urland W (2000) Z Anorg Allg Chem 626:789
323. Hake D, Urland W (1990) Z Anorg Allg Chem 586:99
324. Hake D, Urland W (1989) Angew Chem Int Ed Engl 28:1364
325. Liu Q, Shen Q, Lin Y-H, Zhang Y (1998) Chin J Inorg Chem 14:194
326. Liu Q, Lin Y-H, Shen Q (1997) Acta Crystallogr Sect C Cryst Struct Commun 53:1579
327. Yao Y-M, Zhang Y, Shen Q, Liu Q-C, Meng Q-M, Lin Y-H (2001) Chin J Chem 19:588
328. Fan B, Shen Q, Lin Y (1989) Youji Huaxue 9:414
329. Troyanov SI (1998) Russ J Coord Chem 24:632
330. Troyanov SI (1998) Russ J Coord Chem 24:381
331. Liang H, Shen Q, Jin S, Lin Y-H (1992) J Chem Soc Chem Commun 480
332. Troyanov SI (1998) Russ J Coord Chem 24:373
333. Westin G, Moustiakimov M, Kritikos M (2002) Inorg Chem 41:3249
334. Wijk M, Norrestam R, Nygren M, Westin G (1996) Inorg Chem 35:1077
335. Mathur S, Shen H, Rapalaviciute R, Kareiva A, Donia N (2004) J Mater Chem 14:3259
336. Kritikos M, Wuk M, Westin G (2001) Acta Crystallogr Sect C Cryst Struct Commun 54:576
337. Moustiakimov M, Kritikos M, Westin G (2001) Acta Crystallogr Sect C Cryst Struct Commun 57:515
338. Veith M, Mathur S, Lecerf N, Bartz K, Heintz M, Huch V (2000) Chem Mater 12:271
339. Mathur S, Veith M, Shen H, Hüfner S, Jilavi MH (2002) Chem Mater 14:568
340. Tripathi UM, Singh A, Mehrotra RC, Goel SC, Chiang MY, Buhro WE (1992) J Chem Soc Chem Commun 152
341. Thuilliez J, Monteil V, Spitz R, Boisson C (2005) Angew Chem Int Ed 44:2593, and references therein

Author Index Volumes 201–204

Author Index Volumes 1–100 see Volume 100
Author Index Volumes 101–200 see Volume 200

Anwander, R. see Fischbach, A.: Vol. 204, pp. 155–290.
Ayres, L. see Löwik D. W. P. M.: Vol. 202, pp. 19–52.

Deming T. J.: Polypeptide and Polypeptide Hybrid Copolymer Synthesis via NCA Polymerization. Vol. 202, pp. 1–18.
Donnio, B. and *Guillon, D.*: Liquid Crystalline Dendrimers and Polypedes. Vol. 201, pp. 45–156.

Elisseeff, J. H. see Varghese, S.: Vol. 203, pp. 95–144.

Fischbach, A. and *Anwander, R.*: Rare-Earth Metals and Aluminum Getting Close in Ziegler-type Organometallics. Vol. 204, pp. 155–290.
Fischbach, C. and *Mooney, D. J.*: Polymeric Systems for Bioinspired Delivery of Angiogenic Molecules. Vol. 203, pp. 191–222.
Freier T.: Biopolyesters in Tissue Engineering Applications. Vol. 203, pp. 1–62.
Friebe, L., Nuyken, O. and *Obrecht, W.*: Neodymium Based Ziegler/Natta Catalysts and their Application in Diene Polymerization. Vol. 204, pp. 1–154.

García A. J.: Interfaces to Control Cell-Biomaterial Adhesive Interactions. Vol. 203, pp. 171–190.
Guillon, D. see Donnio, B.: Vol. 201, pp. 45–156.

Harada, A., Hashidzume, A. and *Takashima, Y.*: Cyclodextrin-Based Supramolecular Polymers. Vol. 201, pp. 1–44.
Hashidzume, A. see Harada, A.: Vol. 201, pp. 1–44.
Van Hest J. C. M. see Löwik D. W. P. M.: Vol. 202, pp. 19–52.

Jaeger, W. see Kudaibergenov, S.: Vol. 201, pp. 157–224.
Janowski, B. see Pielichowski, K.: Vol. 201, pp. 225–296.

Kataoka, K. see Osada, K.: Vol. 202, pp. 113–154.
Klok H.-A. and *Lecommandoux, S.*: Solid-State Structure, Organization and Properties of Peptide—Synthetic Hybrid Block Copolymers. Vol. 202, pp. 75–112.
Kudaibergenov, S., Jaeger, W. and *Laschewsky, A.*: Polymeric Betaines: Synthesis, Characterization, and Application. Vol. 201, pp. 157–224.

Laschewsky, A. see Kudaibergenov, S.: Vol. 201, pp. 157–224.
Lecommandoux, S. see Klok H.-A.: Vol. 202, pp. 75–112.
Löwik, D. W. P. M., Ayres, L., Smeenk, J. M., Van Hest J. C. M.: Synthesis of Bio-Inspired Hybrid Polymers Using Peptide Synthesis and Protein Engineering. Vol. 202, pp. 19–52.

Mooney, D. J. see Fischbach, C.: Vol. 203, pp. 191–222.

Njuguna, J. see Pielichowski, K.: Vol. 201, pp. 225–296.
Nuyken, O. see Friebe, L.: Vol. 204, pp. 1–154.

Obrecht, W. see Friebe, L.: Vol. 204, pp. 1–154.
Osada, K. and *Kataoka, K.*: Drug and Gene Delivery Based on Supramolecular Assembly of PEG-Polypeptide Hybrid Block Copolymers. Vol. 202, pp. 113–154.

Pielichowski, J. see Pielichowski, K.: Vol. 201, pp. 225–296.
Pielichowski, K., Njuguna, J., Janowski, B. and *Pielichowski, J.*: Polyhedral Oligomeric Silsesquioxanes (POSS)-Containing Nanohybrid Polymers. Vol. 201, pp. 225–296.
Pompe, T. see Werner, C.: Vol. 203, pp. 63–94.

Salchert, K. see Werner, C.: Vol. 203, pp. 63–94.
Schlaad H.: Solution Properties of Polypeptide-based Copolymers. Vol. 202, pp. 53–74.
Smeenk, J. M. see Löwik D. W. P. M.: Vol. 202, pp. 19–52.

Takashima, Y. see Harada, A.: Vol. 201, pp. 1–44.

Varghese, S. and *Elisseeff, J. H.*: Hydrogels for Musculoskeletal Tissue Engineering. Vol. 203, pp. 95–144.

Werner, C., Pompe, T. and *Salchert, K.*: Modulating Extracellular Matrix at Interfaces of Polymeric Materials. Vol. 203, pp. 63–94.

Zhang, S. see Zhao, X.: Vol. 203, pp. 145–170.
Zhao, X. and *Zhang, S.*: Self-Assembling Nanopeptides Become a New Type of Biomaterial. Vol. 203, pp. 145–170.

Subject Index

Acetylacetonates, Nd(III)-salts 13
Acrylonitrile-butadiene-styrene terpolymer (ABS) 7
Alcoholates, Nd(III)-salts 13
Alkyl aluminum cocatalysts 15
Alkyl→chloride exchange 228
Alkylation/cationization activation pathways, organoaluminum-promoted 156
Allyl compounds, Nd(III)-salts 13
Aluminate cleavage, donor-induced 158
Aluminum alkyl cocatalysts 15
Amide→alkyl exchange 210
Amide→chloride exchange 212
Amide→hydride exchange 212
Amides, Nd(III)-salts 13
Aryloxide (phenolate) ligands 195
Azaallyl ligands 27

Black incorporation time (BIT) 66
Boranes, Nd(III)-salts 13
Bulk/mass 93
Butadiene/1-alkenes 91
Butadiene/ethylene 91
Butadiene/isoprene 84
Butadiene/styrene 88
Butadiene polymerization, living 115
– –, microstructure 111
– –, molar mass regulation 124
– –, neodymium-catalyzed 99
– –, polyinsertion reaction 111
Butadiene rubber (BR) 7
Butadienes, magnesium-based catalyst mixtures 204
–, neodymium-based Ziegler/Natta-catalysts 1
–, substituted 85

Calix[4]arene 21
Carbon black filler 66
Carboxylates, Nd(III)-salts 13
Catalyst aging 47

Catalyst systems, ternary 60
Cerium 6
Co-BR (cobalt-BR) 8
Cocatalysts 2, 32
Coordinative polymerization, discovery 5
Cyclopentadienyl derivatives, Nd(III)-salts 13

Dialkylaluminum complexes 259
Didymium 10
Diene polymerization 156
– –, lanthanides 5
– –, Nd-catalyzed 2
Dienes, dimerization 65
Dimerization, dienes 65
Dimers 65
Dimethyl-di-2,4-pentadienyl-(E,E)-silane 52
Diphenylmethanediisocyanate (MDI) 67
Donor-induced aluminate cleavage 158

E-BR (emulsion-BR) 8
Electron donors 14
Ethylene polymerization, zirconocene-promoted 161
Europium 10
Europium(II) alkoxide 203

Glass transition temperature (T_g) 8

Half-lanthanidocene TMA-adduct 194
Halide donors 2, 35
Halides, Nd(III)-salts 13
Halogenoaluminates 165
High impact poly(styrene) (HIPS) 7
High-cis-1,4-polydienes 161
Homopolymerization in solution 81

Impurities 64
Indenyl neodymium dichloride systems 27
Isoprene (IP) 5

–, heterobimetallic Ln/Al siloxides   206
–, homoleptic lanthanide alkoxides   195
–, homopolymerization   82

Lanthanide acetates   173
Lanthanide alkoxides   189
Lanthanide alkylamides   209
Lanthanide alkyls   218
Lanthanide allyl complexes   218
Lanthanide amides   207
Lanthanide anilides   214
Lanthanide aryloxides   189
Lanthanide benzoates   178
– –, homoleptic   182
Lanthanide borohydrides   234
Lanthanide butanolates   192
Lanthanide butyroates   175
Lanthanide carboxylates   172
Lanthanide complexes, heterogenized   237
Lanthanide halides   162
– –, divalent   165
– –, heteroleptic   165
Lanthanide hydrides   234
Lanthanide methanolates   190
Lanthanide neodecanoates   176
Lanthanide pentanolates   192
Lanthanide phenolates   195
Lanthanide propanolates   190
Lanthanide siloxides   205
Lanthanide silylamides   210
Lanthanide tetraalkylaluminates   221
Lanthanide tris(tetramethylaluminate)   216
Lanthanide(II) phenolates   202
Lanthanide(III) phenolates   196
Lanthanides, diene polymerization   5
Lanthanidocene aluminohydrides   234
Lanthanidocene benzoates   178
Lanthanidocene chlorides   169
Lanthanum   10
– / Al heterobimetallics   156, 241
Lappert's, donor-induced aluminate cleavage   223
Li-BR (lithium-BR)   8

Methyl alumoxane (MAO)   18
Mixed-alkylated complexes   266
Moisture   64
Molar mass, control   74
Molar mass distribution (MMD)   8

Molar mass jumping   10
Monoalkylaluminum
Monomer concentration   63
Monomer conversion   64
Monomer–dimer equilibrium   241
Mooney jumping   10

$NdCl_3$ nanoparticles   14
Neodymium alcoholates   6, 20
Neodymium alk(aryl)oxide/ dialkylmagnesium diene   203
Neodymium allyl compounds   24
Neodymium amides   29
Neodymium-based cyclopentadienyl (Cp) derivatives   27
Neodymium bis(2-ethylhexyl)phosphate (NdP)   24
Neodymium bis(trimethylsilyl)amide   207
Neodymium butadiene rubber (Nd-BR)   7
– –, technology, evaluation   131
Neodymium carboxylate   12
Neodymium-catalyzed diene polymerization   2
Neodymium components   13
Neodymium 2-ethylhexanoate   16
Neodymium-halide-based systems, cocatalysts   15
Neodymium halides   13
Neodymium phosphate-based catalysts   22
Neodymium precursors   2
Neodymium(III) isooctanoate   16
Neodymium(III) naphthenate   17
Neodymium(III) octanoate   16
Neodymium(III) versatate   16
Neodymium-based butadiene rubbers   162
Ni-BR (nickel-BR)   8
Non-alkylaluminum complexes   261

Organoborane additives   68

Phosphates, Nd(III)-salts   13
Phosphonates, Nd(III)-salts   13
Poly(butadiene), cyclized   68
Poly(ethylene)s   5
Poly(propylene)s   5
Polydienes, Ln/Al heterobimetallics   156
Polymerization in solution   12

Subject Index

Polymerization temperature 68
Postlanthanidocene tetramethylaluminate 216
Post-polymerization modifications 66
Postsamarocene, β-diketiminato-derived 217
Praseodymium 10
Samarium 10
Shortstop 64
Silica fillers 66
Silica-supported complexes 238
Siloxides, alkylated 205
–, lanthanide 205
Stabilization of polymers 64

Ternary catalyst systems 60
Tetraalkylaluminate complexes, site mobility 241
Tetraalkylaluminates 247

Tetraisobutyl dialumoxane 18
Tetramethylaluminates 217
–, alkyl exchange 242
Ti-BR (titanium-BR) 8
Titanium aluminate complexes 160
Trialkylaluminum complexes 253
Trimethylaluminum adduct complexes, alkyl exchange 245

Uranium-based catalysts 6

Vinylcyclohexene (VCH), Diels–Alder-dimerization of BD 65

Yttrium aluminates 160

Ziegler/Natta catalysts, diene polymerization 5
Zirconium aluminate complexes 160

RETURN TO →	CHEMISTRY LIBRARY 100 Hildebrand Hall	642-3753 **1519**
LOAN PERIOD 1 ~~7 DAYS~~	2	3
4	**2 HOUR**	

~~ALL BOOKS MAY BE RECALLED AFTER 7 DAYS~~

DUE AS STAMPED BELOW		

FORM NO. DD5

UNIVERSITY OF CALIFORNIA, BERKELEY
BERKELEY, CA 94720